FRACTIONAL DIFFUSION EQUATIONS AND ANOMALOUS DIFFUSION

Anomalous diffusion has been detected in a wide variety of scenarios, from fractal media, systems with memory, transport processes in porous media to fluctuations of financial markets, tumour growth, and complex fluids. Providing a contemporary treatment of this process, this book examines the recent literature on anomalous diffusion and covers a rich class of problems in which surface effects are important, offering detailed mathematical tools of usual and fractional calculus for a wide audience of scientists and graduate students in physics, mathematics, chemistry, and engineering. Including the basic mathematical tools needed to understand the rules for operating with the fractional derivatives and fractional differential equations, this self-contained text presents the possibility of using fractional diffusion equations with anomalous diffusion phenomena to propose powerful mathematical models for a large variety of fundamental and practical problems in a fast-growing field of research.

LUIZ ROBERTO EVANGELISTA is Professor of Theoretical Physics at the University of Maringá (UEM), Brazil. His research interests lie in complex fluids, complex systems, and history of physics, and include mathematical physics of liquid crystals, diffusion problems, and adsorption-desorption phenomena.

ERVIN KAMINSKI LENZI is Associate Professor at the University of Ponta Grossa (UEPG), Brazil. His research interests are in complex systems and stochastic processes, and include anomalous diffusion processes, usual and fractional diffusion equations, and modern boundary value problems with applications in liquid-crystalline systems and impedance spectroscopy.

FRACTIONAL DIFFUSION EQUATIONS AND ANOMALOUS DIFFUSION

LUIZ ROBERTO EVANGELISTA
University of Maringá (UEM)

ERVIN KAMINSKI LENZI
University of Ponta Grossa (UEPG)

CAMBRIDGE
UNIVERSITY PRESS

CAMBRIDGE
UNIVERSITY PRESS

University Printing House, Cambridge CB2 8BS, United Kingdom

One Liberty Plaza, 20th Floor, New York, NY 10006, USA

477 Williamstown Road, Port Melbourne, VIC 3207, Australia

314–321, 3rd Floor, Plot 3, Splendor Forum, Jasola District Centre, New Delhi – 110025, India

79 Anson Road, #06–04/06, Singapore 079906

Cambridge University Press is part of the University of Cambridge.

It furthers the University's mission by disseminating knowledge in the pursuit of education, learning, and research at the highest international levels of excellence.

www.cambridge.org
Information on this title: www.cambridge.org/9781107143555
DOI: 10.1017/9781316534649

First published 2018

Printed in the United Kingdom by Clays, St Ives plc

Library of Congress Cataloging-in-Publication Data

Names: Evangelista, L. R., author. | Lenzi, Ervin Kaminski, 1975- author.
Title: Fractional diffusion equations and anomalous diffusion / Luiz Roberto Evangelista, Universidade Estadual de Maringá, Brazil ; Ervin Kaminski Lenzi, Universidade Estadual de Ponta Grossa.
Description: Cambridge : Cambridge University Press, [2018]
Identifiers: LCCN 2017035420 | ISBN 9781107143555 (hardback : alk. paper)
Subjects: LCSH: Differential equations, Partial. | Fractional calculus. | Diffusion.
Classification: LCC QA377 .E9448 2017 | DDC 515/.83–dc23 LC record available at https://lccn.loc.gov/2017035420

ISBN 978-1-107-14355-5 Hardback

To Lucas Evangelista Gomes,
who has just arrived among us
L. R. E.

To Alice and Pedro, with love
E. K. L.

Se tu se' or, lettore, a creder lento
ciò ch'io dirò, non sarà maraviglia,
ché io che 'l vidi, a pena il mi consento.

[If thou art, Reader, slow now to believe
What I shall say, it will no marvel be,
For I who saw it hardly can admit it.]

(Dante, Inferno XXV, 46–48)

Contents

Preface

The very irregular state of motion observed by Robert Brown for small pollen grains suspended in water initiated one of the the most fascinating fields of science. The importance of such discovery – the so-called diffusion process – is immeasurable; it has been found in many contexts and is widespread in nature. A characteristic feature of this random motion is the linear growth with time exhibited by the mean square displacement, which is typical of a Markovian process. In contrast with this situation, a large class of systems and processes present a diffusion behaviour characterised by a nonlinear time dependence of the same quantity, thus constituting what is called anomalous diffusion behaviour.

The last decades have witnessed an increased interest in the anomalous diffusion processes that seem to be indeed present in a variety of experimental scenarios in physics, chemistry, biology, and several other branches of engineering; it is a rapidly growing field of research, attracting the attention of the scientific community. This happens from the theoretical side – due to the new mathematical problems evoked – but also from the point of view of experimental or practical applications. It is noteworthy that the number of studies reporting experimental problems dealing with anomalous diffusion has strongly increased – this attests to the ubiquity of a phenomenon initially considered a rare event.

The power of the mathematical tools based on fractional calculus, on the other hand, has also attracted the attention of the community working with pure and applied mathematics. The association of these techniques with the diffusional problem represents in practice a new field of research. It was shown in several ways that fractional calculus, if it is not unique, is nevertheless a suitable or even the natural mathematical framework to use to face the high complexity represented by anomalous diffusion phenomena. One powerful way of using these mathematical tools to analyse diffusion processes leads naturally to the necessity to search for solutions of fractional linear and nonlinear diffusion equations.

The purpose of this book is to provide an updated literature on anomalous diffusion, covering also a rich class of problems in which surface effects in diffusion problems are important. Our motivations to write it come from the necessity to communicate recent consolidated advances of our research field to a wide audience of scientists and students in physics, mathematics, chemistry, engineering, and other, interdisciplinary areas. This is done by showing the possibility of using fractional diffusion equations in connection with anomalous diffusion phenomena to propose useful mathematical models for a large variety of fundamental and practical problems.

As a result, this book offers detailed mathematical tools of usual and fractional calculus to explore the deep significance of anomalous diffusion phenomena. Likewise, it also presents a discussion about the significance and meaning of fractional calculus. Some of the basic mathematical tools needed to understand the rules for operating with fractional derivatives and fractional differential equations are discussed in detail, in order to convey to the reader the essential techniques employed. It is hoped that it may be used by the reader to navigate the complex literature represented by these challenging problems, offering a comprehensive approach that combines, in a systematic and well-organised way, the study of contemporary anomalous diffusion problems with advanced techniques of calculus, emphasising the recent developments of fractional calculus.

The first part of the book is dedicated to presenting and discussing essential physical and mathematical concepts, forming the background material for the reading of the rest of the work.

In Chapter 1, we review some basic mathematical results, focusing on the essential properties of integral transforms of Fourier, Laplace, Hankel, and Mellin and their applications to simple problems. This presentation is followed by an account of special functions, emphasising the ones that arise more naturally in the framework of fractional calculus, such as the Mittag-Leffler, Wright, and the H-function of Fox, as well as their integral transforms. In Chapter 2, we propose a survey of fractional calculus, starting with a brief historical account of the evolution of the concepts of differentiation and integration of arbitrary order, and finishing with a concise discussion of the main properties of the fractional operators to be used in subsequent chapters. To complete this part of the book dedicated to fundamentals, in Chapter 3, we also present a brief history of the approaches to the diffusion phenomena, emphasising the first investigations of Brownian motion, the random walk problem, and its connection with the diffusion process, until the appearance of the concept of anomalous diffusion.

The second part of the book is devoted to presenting, in successive steps – each one incorporating a further degree of generalisation with respect to the previous one – a large number of solutions of the fractional diffusion equations in different applications.

Our presentation starts in Chapter 4 with the one-dimensional diffusion equation, written in terms of fractional operators in time and space variables, and focusing on some simple problems and applications. We discuss the concept of Lévy flights and the use of the continuous-time random walk formalism to understand the physical implications of the presence of fractional derivatives in the theoretical description of these systems. In Chapter 5, we consider the influence of the surfaces or membranes on diffusive processes. The problems we treat are intended to explore how the surface may modify the diffusive process of a system governed by a fractional diffusion equation. In the cases we analyse, the system may exhibit an anomalous diffusive behaviour for which surface effects play a nonnegligible role. In Chapter 6, we investigate situations in which nonlinear terms intervene in the diffusion equation as well as d-dimensional problems. These linear and nonlinear fractional diffusion equations take into account a diffusion coefficient with spatial dependences and external forces. Of particular importance in this regard is analysis of intermittent motion – the transition between diffusion and rest – to understand how it can can be described in the framework of the random walk approach, because this permits us to explore the manifestation of different diffusive regimes in the system. In Chapter 7, the phenomenon of anisotropic diffusion is explored by means of both usual and fractional diffusion equations. The analytical solutions of the two-dimensional comb-model with integer and fractional derivatives, and also with a drift term, are obtained. This opens different perspectives on the phenomenon, because it is shown that anomalous diffusion may be well described even in the framework of usual diffusion equations for some constrained systems. We show also that it can be investigated by means of some simplified picture of highly disordered systems. In Chapter 8, we investigate the time-dependent solutions of the fractional Schrödinger equation in presence of nonlocal terms. Extensions of the Schrödinger equation encompassing fractional derivatives in the spatial and temporal variables are an elegant way to tackle nonlocal and non-Markovian effects. For this reason, the continuous-time random walk approach is used to obtain consistent formulations of the fractional Schrödinger equation. A series of formal solutions is obtained for problems involving the presence of memory kernels, distributed order memory kernels, and nonlocal terms.

The third part of the book is formed by two chapters dedicated to a deep exploration of the role of the anomalous diffusion phenomena in the impedance spectroscopy response of liquid samples.

In Chapter 9, we present some analytical results obtained by means of a pioneering application of the fractional diffusion equations to the electrochemical impedance technique. After reviewing the fundamental equations of the continuum Poisson–Nernst–Planck (PNP) model, these main equations are rewritten in terms of time-fractional and distributed order derivatives, and the predictions of the model are analysed, emphasising the low frequency behaviour of the impedance by means of analytical solutions. This is the first step towards an extension of the PNP model to encompass anomalous diffusion. Some experimental data are invoked to test the robustness of the model in treating interfacial effects on the impedance in the low frequency region. Finally, in Chapter 10, the pathways towards the construction of PNP anomalous (PNPA) models are presented. It is analytically demonstrated that the effect of a constant-phase element in an equivalent electrical circuit may be represented by an appropriated term added to the boundary conditions of PNP or PNPA models. Thus, it is shown that the impedance spectroscopy models based on the fractional diffusion equations may be used to build an entire framework of continuum models general enough to analyse impedance data of high complexity.

The authors are in debt to many colleagues and students who participated in the successive developments of the subject. We are grateful to G. Barbero and A. Sapora (Torino), J. R. Macdonald (Chapel Hill), and P. Pasini, C. Chiccoli, and F. Mainardi (Bologna), for extremely encouraging discussions in the last years. It is our pleasure to thank the following for their scientific collaboration: R. S. Mendes, H. V. Ribeiro, R. S. Zola, L. C. Malacarne, P. A. Santoro, H. Mukai, P. R. G. Fernandes, B. F. de Oliveira, N. G. C. Astrath, A. T. Silva, A. A. Tateishi, M. A. F. dos Santos, M. F. de Andrade, V. G. Guimarães, R. Menechini Neto, T. Petrucci, F. R. G. B. Silva, and J. L. de Paula (Maringá); M. K. Lenzi (Curitiba); C. Tsallis (Rio de Janeiro); L. S. Lucena and L. R. da Silva (Natal); E. Capelas de Oliveira (Campinas), G. Gonçalves, C. A. R. Yednak, R. T. Teixeira-Souza, and R. Rossato (UTFPR); F. C. Zola (UFGD-Dourados); T. Sandev and I. Petreska (Skopje); F. Ciuchi, A. Mazzulla, and N. Scaramuzza (Calabria); and F. Mantegazza (Milano). We thank D. S. Vieira (Maringá) and M. R. Ribeiro (Ponta Grossa) who read and commented on preliminary versions of the manuscript, offering valuable suggestions, and M. P. Rosseto and R. L. Biagio (Maringá) for the help with the figures. Finally, we express our gratitude to our sponsors: CNPq – National Institute of Science and Technology for Complex Systems (Rio de Janeiro), and Complex Fluids (INCT-FCx, São Paulo), and the Brazilian agencies Capes (Brasília) and Fundação Araucária (Curitiba); the National Institute of Nuclear Physics (INFN) – Section of Bologna and Department of Applied Science and Technology – DISAT – Polytechnic of Turin (Italy) for partial financial support during the redaction of the book. We are thankful to Maria Elena Francalancia (Toro) for capturing the essence of our work in the cover illustration.

It is time to invite the reader to explore with us the new class of problems and scenarios treated in this book. We are sure it can be used as a textbook in scientific and technological areas as well as an advanced monograph in frontier physics and applied mathematics, helping the reader to discover or to deepen their knowledge of fractional calculus and anomalous diffusion phenomena. The warning represented by Dante's quotation at the beginning of the enterprise may be slightly modified just to remember that, if the reader is slow now to believe what we shall tell, that should be no cause for wonder, for we, who saw it slightly before, remain enlightened – as we hope the reader becomes, after reading it!

1
Mathematical Preliminaries

This preliminary chapter is dedicated to reviewing some basic mathematical results to be used in this book. The first half of this chapter is concerned with the integral transforms of Fourier, Laplace, Hankel, and Mellin and their applications to simple problems of mathematical physics with illustrative purposes. The second half contains a comprehensive review of some properties of more commonly known special functions, such as gamma, beta, and Bessel functions, followed by the presentation of other less known special functions, which arise more naturally in the framework of fractional calculus, as the Mittag-Leffler function, Wright function, and the H-function of Fox. Some integral transforms of these special functions are presented at the end of the chapter.

1.1 Integral Transforms

In mathematics, an integral transform $\mathcal{T}$ of a given function $f(t)$ has the general form

$$\mathcal{T}\{f(t);s\} := F(s) = \int_{t_1}^{t_2} K(t,s)f(t)\,dt,$$

such that the input is some function $f(t)$ and the output is another function $F(s)$. The choice of the *kernel* $K(t,s)$ defines different types of transforms. In this section, we present and review some definitions and properties of the integral transforms of Fourier, Laplace, Hankel, and Mellin, indicating a few applications to classical problems in mathematical physics. More detailed information may be found in the References at the end of the book.

1.1.1 Fourier Transforms

In the one-dimensional case, the *Fourier transform* of a function $f(x)$ of a real variable $x \in \mathbb{R} = (-\infty,\infty)$ is defined as

$$\mathcal{F}\{f(x);k\} = F(k) = \int_{-\infty}^{\infty} e^{-ikx} f(x)\,dx, \qquad k \in \mathbb{R}, \tag{1.1}$$

whereas the *inverse Fourier transform* is given by

$$\mathcal{F}^{-1}\{F(k);x\} = f(x) = \frac{1}{2\pi}\int_{-\infty}^{\infty} e^{ikx} F(k)\,dk, \qquad k \in \mathbb{R}. \tag{1.2}$$

In Eqs. (1.1) and (1.2), $i^2 = -1$ and sometimes both are defined with a $1/\sqrt{2\pi}$ in front of the integration symbol and also with the reversed sign in the exponent. In this case, the only difference between the Fourier transform and its inverse is the signal of the exponential. The existence of the transform $F(k)$ is guaranteed if $f(x)$ is an integrable function and the integral converges. A sufficient (but not necessary) condition is to require that $f(x)$ be absolutely integrable, i.e., the integral

$$\int_{-\infty}^{\infty} |f(x)|\,dx$$

exists. In this case, $F(k)$ is absolutely convergent and, thus, it is convergent [1]. The Fourier convolution operator of two functions f and g is defined as

$$(f * g)(x) = f(x) * g(x) = \int_{-\infty}^{\infty} f(x-\tau)g(\tau)d\tau, \qquad x \in \mathbb{R}, \tag{1.3}$$

which has the commutative property, i.e.,

$$f * g = g * f.$$

The Fourier transform of the convolution (1.3) is given by the Fourier convolution theorem, which states that

$$\begin{aligned}
\mathcal{F}\{(f * g);k\} &= \int_{-\infty}^{\infty}\left[\int_{-\infty}^{\infty} f(x-\tau)\,g(\tau)\,d\tau\right] e^{-ikx}\,dx \qquad (1.4)\\
&= \int_{-\infty}^{\infty}\left[\int_{-\infty}^{\infty} f(z)\,g(\tau)\,d\tau\right] e^{-ik(z+\tau)}\,dz\\
&= \left[\int_{-\infty}^{\infty} f(z)\,e^{-ikz}\right]\left[\int_{-\infty}^{\infty} g(\tau)\,d\tau\,e^{-ik\tau}\right] dz\\
&= F(k)G(k),
\end{aligned}$$

in which we have introduced $x = z + \tau$. Symbolically, the above expression may be rewritten as

$$\mathcal{F}\{(f * g); k\} = F(k)G(k)$$

and is well defined when the transforms $F(k)$ and $G(k)$ exist. Conversely, a useful result may be obtained when a function $W(k)$ is known in the Fourier space. If one is able to rewrite it as a product of two functions, namely, $W(k) = F(k)G(k)$, then it is possible to apply the inverse Fourier transform to (1.4), in order to obtain the identity

$$\mathcal{F}^{-1}\{F(k)G(k); x\} = (f * g)(x) = \int_{-\infty}^{\infty} f(x-\tau)\, g(\tau)\, d\tau. \tag{1.5}$$

Another useful property of the Fourier transform is the transform of the derivatives of a function $f(x)$, when $\lim_{x\to\pm\infty} f^{(p)}(x) = 0$, for $p = 0, 1, \ldots, n-1$. For the first derivative, we have

$$\mathcal{F}\left\{\frac{df}{dx}; k\right\} = \int_{-\infty}^{\infty} \frac{df}{dx} e^{ikx}\, dx.$$

After an integration by parts, it can be cast in the form

$$\begin{aligned}\mathcal{F}\left\{\frac{df}{dx}; k\right\} &= \left[ik\, e^{ikx} f(x)\Big|_{-\infty}^{\infty} - \int_{-\infty}^{\infty} f(x)\, ik\, e^{ikx}\, dx\right] \\ &= -ik \int_{-\infty}^{\infty} f(x)\, e^{ikx}\, dx = -ikF(k).\end{aligned} \tag{1.6}$$

Following the same procedure, the Fourier transform of the n-derivative of $f(x)$ is shown to be

$$\mathcal{F}\left\{\frac{d^n f}{dx^n}; k\right\} = (-ik)^n F(k). \tag{1.7}$$

The transform may be considered as a linear integral operator acting on a given function, such that if $f(x) = a\, g(x) + b\, h(x)$, with $(a, b) \in \mathbb{C}$, we can write

$$\begin{aligned}\mathcal{F}\{f(x); k\} &= \mathcal{F}\{a\, g(x) + b\, h(x); k\} \\ &= \int_{-\infty}^{\infty} \left[a\, g(x) + b\, h(x)\right] e^{ikx}\, dx \\ &= a \int_{-\infty}^{\infty} g(x)\, e^{ikx}\, dx + b \int_{-\infty}^{\infty} h(x)\, e^{ikx}\, dx \\ &= a\, G(k) + b\, H(k).\end{aligned} \tag{1.8}$$

Finally, we may introduce the cosine- and sine-Fourier transforms of a function $f(t)$, $t \in \mathbb{R}^+ = (0, \infty)$, defined, respectively, as

$$\mathcal{F}_c\{f(t); k\} = F_c(k) = \int_0^\infty f(t) \cos(kt)\, dt, \qquad k \in \mathbb{R}^+ \tag{1.9}$$

and

$$\mathcal{F}_s\{f(t); k\} = F_s(k) = \int_0^\infty f(t) \sin(kt)\, dt, \qquad k \in \mathbb{R}^+. \tag{1.10}$$

The Fourier transform is an important tool in many branches of mathematical physics. In particular, in linear dynamical systems, it transforms the function from the time-domain to the frequency-domain, as we discuss in Chapters 9 and 10, dedicated to some applications of fractional calculus in problems of electrochemical impedance.

Let us illustrate some of the presented results involving Fourier transforms by applying them to the classical boundary-value problem of obtaining the temperature profile $u(x, t)$ of an infinitely long bar at a point x and time t, when the initial temperature $u(x, 0) = f(x)$, such that $u(x, t) \to 0$, as $t \to \infty$. The differential equation to be solved may be written as

$$\frac{\partial u}{\partial t} = \alpha^2 \frac{\partial^2 u}{\partial x^2}, \qquad t > 0, \qquad -\infty < x < \infty, \tag{1.11}$$

in which, for simplicity, we assume $\alpha = 1$. A trial solution of the form $u_{q,p}(x, t) = e^{px+qt}$, when substituted in (1.11), permits us to conclude that $q = p^2$ and $p = ik$, with $k \in \mathbb{R}$, such that

$$u_k(x, t) = e^{ikx} e^{-k^2 t}. \tag{1.12}$$

The initial condition $u(x, 0) = f(x)$ may be satisfied by the linear combination of solutions like (1.12), in the form

$$u(x, t) = \int_{-\infty}^{\infty} e^{ikx} e^{-k^2 t} g(k)\, dk. \tag{1.13}$$

Expression (1.13) is a solution of (1.11) provided we can differentiate under the integral sign and is valid for an arbitrary $g(k)$, which can be determined from the initial condition; i.e., since

$$u(x, 0) = f(x) = \int_{-\infty}^{\infty} e^{ikx} g(k)\, dk,$$

we have

$$g(k) = \frac{1}{2\pi} \int_{-\infty}^{\infty} e^{-ikx} f(x)\, dx, \tag{1.14}$$

which is the inverse Fourier transform of $f(x)$. By substituting (1.14) in (1.13), we easily obtain

$$\begin{aligned} u(x,t) &= \int_{-\infty}^{\infty} \left[\frac{1}{2\pi} \int_{-\infty}^{\infty} e^{-k^2 t + ik(x-x')}\, dk \right] f(x')\, dx' \\ &= \frac{1}{\sqrt{4\pi t}} \int_{-\infty}^{\infty} e^{-(x-x')^2/4t} f(x')\, dx' \\ &= \int_{-\infty}^{\infty} f(x-x') g(x') dx', \end{aligned} \tag{1.15}$$

which can also be recognised as the convolution, (1.3), in which

$$f(x-x') = \frac{1}{\sqrt{4\pi t}} e^{-(x-x')^2/4t} \quad \text{and} \quad g(x') = f(x'). \tag{1.16}$$

1.1.2 The Laplace Transform

The Laplace transform of a function $f(t)$ of a real variable $t \in \mathbb{R}^+ = (0, \infty)$ is defined as

$$\mathcal{L}\{f(t); s\} = F(s) = \int_0^{\infty} e^{-st} f(t)\, dt, \qquad s \in \mathbb{C}. \tag{1.17}$$

It is an integral transform as useful as the Fourier transform in solving physical problems. For its existence, the function $f(t)$ must be such that

$$e^{-\alpha t}|f(t)| \leq M, \qquad \text{for all} \quad t \in [0, \infty),$$

where M is a positive constant and $\mathrm{Re}\,(\alpha) > 0$, indicating that the function $f(t)$ must not grow faster than a certain exponential function when $t \to \infty$. This point can be made more clearly for $s \in \mathbb{R}$ and $\alpha \in \mathbb{R}$, because

$$\int_0^L |f(t)| e^{-st}\, dt \leq \int_0^L M e^{\alpha t} e^{-st}\, dt \leq \frac{M}{s-\alpha}.$$

The integrand remains bounded as $L \to \infty$, if $s > \alpha$. When $s \leq \alpha$, the integral diverges. Certain functions, like $f(t) = e^{t^2}$, do not have a Laplace transform.

Another feature influencing the existence of this integral transform is the presence of singularities in $f(t)$. Consider, for instance, a typical power-law behaviour $f(t) = t^n$, with $n \in \mathbb{R}$. Its Laplace transform is given by

$$\mathcal{L}\{t^n; s\} = F(s) = \int_0^\infty t^n \, e^{-st} \, dt, \tag{1.18}$$

which diverges for $t \to 0$, when $n \leq -1$.

The *inverse Laplace transform* is given for $t \in \mathbb{R}^+$ by the formula

$$\mathcal{L}^{-1}\{F(s); t\} = \frac{1}{2\pi i} \int_{\gamma - i\infty}^{\gamma + i\infty} e^{st} \, F(s) \, ds, \qquad \gamma = \mathrm{Re}\,(s) > \sigma, \tag{1.19}$$

where σ is the infimum of s values for which the Laplace integral (1.17) converges, and is called the abscissa of convergence [2]. This integral is known as the Bromwich integral, sometimes known as the Fourier–Mellin integral, as we discuss in Section 1.1.4.

The Laplace transform is linear since

$$\begin{aligned} \mathcal{L}\{af(t) + b\,g(t); s\} &= \int_0^\infty \big[af(t) + b\,g(t)\big] \, e^{-st} \, dt \\ &= a \int_0^\infty f(t) \, e^{-st} \, dt + b \int_0^\infty g(t) \, e^{-st} \, dt \\ &= a\,\mathcal{L}\{f(t); s\} + b\,\mathcal{L}\{g(t); s\} = a\,F(s) + b\,G(s). \end{aligned} \tag{1.20}$$

The Laplace convolution operator of two functions $f(t)$ and $g(t)$, given on $\mathbb{R}^+$, is defined by the integral

$$f * g = (f * g)(t) = \int_0^t f(t - x) \, g(x) \, dx = g * f, \tag{1.21}$$

which, as indicated before, has the commutative property. Now, let $F(s)$ and $G(s)$ be, respectively, the Laplace transforms of the functions $f(t)$ and $g(t)$. We can write

$$\begin{aligned} G(s)\,F(s) &= \int_0^\infty g(x) \, e^{-sx} \, dx \int_0^\infty f(y) \, e^{-sy} \, dy \\ &= \int_0^\infty dx \int_0^\infty g(x) f(y) \, e^{-s(x+y)} \, dy \end{aligned}$$

$$= \int_{y=0}^{\infty} dy \int_{t=y}^{\infty} g(t-y)f(y)\,e^{-st}\,dt$$

$$= \mathcal{L}\left\{\int_{0}^{t} g(t-y)f(y)\,dy; s\right\}, \tag{1.22}$$

where we have changed the variables by inserting $t = x + y$. In Eq. (1.22), we recognise the definition (1.21), such that

$$G(s)F(s) = \int_{0}^{\infty} \left[(g * f)(t)\right] e^{-st}\,dt = \mathcal{L}\{(g * f)(t); s\}$$
$$= \mathcal{L}\{g(t); s\}\,\mathcal{L}\{f(t); s\}.$$

Some formulas for the Laplace transform of elementary and generalised function are as follows. If $f(t) = 1$, the Laplace transform is

$$\mathcal{L}\{1; s\} = \int_{0}^{\infty} 1\,e^{-st}\,dt = \frac{1}{s}.$$

For $f(t) = t^{\alpha}$, with $\mathrm{Re}\,(\alpha) > -1$, we have

$$\mathcal{L}\{t^{\alpha}; s\} = \int_{0}^{\infty} t^{\alpha}\,e^{-st}\,dt = \frac{\Gamma(\alpha+1)}{s^{\alpha+1}}, \qquad \mathrm{Re}\,(s) > 0.$$

In particular, if $\alpha = k \in \mathbb{N}_0 = \{0, 1, 2, \ldots, \}$, we obtain

$$\mathcal{L}\{t^{k}; s\} = \frac{k!}{s^{k+1}}. \tag{1.23}$$

For $f(t) = t^{-1-k}$, we can show that

$$\mathcal{L}\{t^{-1-k}; s\} = \int_{0}^{\infty} t^{-1-k}\,e^{-st}\,dt$$
$$= (-1)^{k+1}\frac{s^{k}}{k!}\left[\ln(s) - \psi(k+1)\right],$$

with $\psi(k)$ being the Euler's psi function, i.e., the logarithm derivative of the gamma function, to be defined in Eq. (1.85).

The Laplace transform of the Dirac delta function is

$$\mathcal{L}\{\delta(t); s\} = \int_{0}^{\infty} \delta(t)\,e^{-st}\,dt = 1, \qquad s \in \mathbb{C}.$$

For the derivatives, we simply have

$$\mathcal{L}\{\delta^k(t); s\} = s^k, \qquad s \in \mathbb{C} \quad \text{and} \quad k \in \mathbb{N}.$$

Finally, the Laplace transform of a derivative of a function $f(t)$ is given by

$$\mathcal{L}\left\{\frac{df}{dt}; s\right\} = -e^{-st}f(t)\Big|_0^\infty + s\int_0^\infty f(t)\, e^{-st}\, dt = -f(0) + s\,F(s), \tag{1.24}$$

where an integration by parts has been done. This result may be generalised for the k-derivative, as

$$\mathcal{L}\left\{\frac{d^k f}{dt^k}; s\right\} = s^k F(s) - \sum_{p=0}^{k-1} s^p \left.\frac{d^{(k-p-1)}}{dt^{(k-p-1)}} f(t)\right|_{t=0}. \tag{1.25}$$

For those functions such that $f(0) = 0$, we obtain the very useful result:

$$\mathcal{L}\left\{\frac{d^k f}{dt^k}; s\right\} = s^k F(s).$$

The Laplace transform method is particularly suitable for studying wave propagation along transmission lines and physical problems with boundary conditions involving time derivatives. One illustration of the use of the Laplace transform method may be worked out from the same example treated in the previous section, i.e., to solve the problem:

$$\begin{aligned} \frac{\partial u(x,t)}{\partial t} &= \frac{\partial^2 u(x,t)}{\partial x^2}, \qquad -\infty < x < \infty, \quad t > 0, \\ u(x,0) &= f(x), \quad u(x,t) \to 0, \quad t \to \infty. \end{aligned} \tag{1.26}$$

Since $u(x,t)$ is assumed as bounded, $|u(x,t)| \leq M$, the Laplace transform exists in the form

$$U(x,s) = \int_0^\infty e^{-st} u(x,t) dt, \tag{1.27}$$

such that

$$|U(x,s)| \leq \int_0^\infty e^{-st}\, |u(x,t)| dt \leq \frac{M}{s}.$$

Applying the Laplace transform to the first of Eqs. (1.26), we obtain:

$$\frac{d^2 U}{dx^2} = sU(x,s) - u(x,0) = sU(x,s) - f(x), \tag{1.28}$$

where we have used (1.24) and the initial condition in (1.26). Equation (1.28) may be rewritten as

$$\frac{d^2U}{dx^2} - sU(x,s) = -f(x), \tag{1.29}$$

and its general solution has the form

$$U(x,s) = U_h(x,s) + U_p(x,s), \tag{1.30}$$

where

$$U_h(x,s) = c_1U_1(x,s) + c_2U_2(x,s) = c_1e^{\sqrt{s}x} + c_2e^{-\sqrt{s}x} \tag{1.31}$$

is the general solution of the homogeneous problem, whereas $U_p(x,s)$ is any particular solution of the nonhomogenous problem. The Wronskian may be calculated as

$$W(U_1,U_2) = \begin{vmatrix} U_1(x,s) & U_2(x,s) \\ U_1'(x,s) & U_2'(x,s) \end{vmatrix} = -2\sqrt{s}.$$

The particular solution is thus

$$\begin{aligned} U_p(x,s) &= \int_0^x \frac{-U_1(x,s)U_2(y,s) + U_2(x,s)U_1(y,s)}{W(y,s)} \left[-f(y)\right] dy \\ &= \frac{e^{\sqrt{s}x}}{2\sqrt{s}} \int_0^x e^{-\sqrt{s}y} f(y)dy + \frac{e^{-\sqrt{s}x}}{2\sqrt{s}} \int_0^x e^{\sqrt{s}y} f(y)dy. \end{aligned} \tag{1.32}$$

The general solution (1.30) may be written as

$$U(x,s) = e^{\sqrt{s}x}\left[c_1 - \frac{1}{2\sqrt{s}}u_-(s)\right] + e^{-\sqrt{s}x}\left[c_2 + \frac{1}{2\sqrt{s}}u_+(s)\right], \tag{1.33}$$

where

$$u_\pm(s) = \int_0^x e^{\pm\sqrt{s}y} f(y)dy.$$

Since the solution (1.33) has to stay bounded when $|x| \to \infty$, we require that

$$\lim_{x\to\infty}\left[c_1 - \frac{1}{2\sqrt{s}}\int_0^x e^{-\sqrt{s}y} f(y)dy\right] = 0$$

and

$$\lim_{x\to-\infty}\left[c_2+\frac{1}{2\sqrt{s}}\int_0^x e^{+\sqrt{s}y}f(y)dy\right]=0,$$

which implies

$$c_1=\frac{1}{2\sqrt{s}}\int_0^\infty e^{-\sqrt{s}y}f(y)dy \quad \text{and} \quad c_2=\frac{1}{2\sqrt{s}}\int_{-\infty}^0 e^{\sqrt{s}y}f(y)dy. \tag{1.34}$$

The solution (1.33) becomes:

$$\begin{aligned}U(x,s)&=\frac{e^{\sqrt{s}x}}{2\sqrt{s}}\int_x^\infty e^{-\sqrt{s}y}f(y)dy+\frac{e^{-\sqrt{s}x}}{2\sqrt{s}}\int_{-\infty}^x e^{\sqrt{s}y}f(y)dy\\&=\frac{1}{2\sqrt{s}}\int_{-\infty}^\infty e^{-\sqrt{s}|x-y|}f(y)dy.\end{aligned} \tag{1.35}$$

The problem is formally solved in the Laplace space (s-domain). To find the solution in the t-domain, we have to obtain the inverse Laplace transform of (1.35), i.e.,

$$u(x,t)=\mathcal{L}^{-1}\{U(x,s);t\}=\int_{-\infty}^\infty \mathcal{L}^{-1}\left\{\frac{e^{-\sqrt{s}|x-y|}}{2\sqrt{s}};t\right\}f(y)dy. \tag{1.36}$$

If we put $a=|x-y|$, from the tables of Laplace transforms [3] we obtain

$$\mathcal{L}^{-1}\left\{\frac{e^{-a\sqrt{s}}}{2\sqrt{s}};t\right\}=\frac{e^{-a^2/4t}}{\sqrt{\pi t}}.$$

Finally, the solution may be written as

$$u(x,t)=\frac{1}{\sqrt{4\pi t}}\int_{-\infty}^\infty e^{-(x-y)^2/4t}f(y)dy, \tag{1.37}$$

which coincides, obviously, with the result stated in (1.15).

1.1.3 The Hankel Transform

The Hankel transform (or Fourier–Bessel integral) is appropriate for those problems in which there is axial symmetry and the radial variable ranges from 0 to ∞. It is defined as

$$\mathcal{H}_\nu\{f(r);s\} = F_\nu(s) = \int_0^\infty f(r)\,\mathrm{J}_\nu(s\,r)\,r\,dr, \quad 0 \le s < \infty, \quad \mathrm{Re}\,(\nu) > -\frac{1}{2}, \tag{1.38}$$

where $\mathrm{J}_\nu(z)$ is the Bessel function of order ν, to be defined in (1.91). The *inverse Hankel transform* is written as

$$\mathcal{H}_\nu^{-1}\{F_\nu(s);r\} = f(r) = \int_0^\infty F_\nu(s)\,\mathrm{J}_\nu(s\,r)\,s\,ds, \quad 0 < r < \infty. \tag{1.39}$$

The *Hankel convolution* of the function $f(r)$ with the function $g(r)$ is

$$h(r) = \frac{2^{1-3\nu}r^{-\nu}}{\sqrt{\pi}\Gamma(\nu+1/2)} \int\limits_{u+v>r}\int\limits_{|u-v|<r} \left[r^2 - (u-v)^2\right]^{\nu-1/2}$$
$$\times \left[(u+v)^2 - r^2\right]^{\nu-1/2} (uv)^{1-\nu} f(u)g(v)\,du\,dv, \quad r \in (0,\infty). \tag{1.40}$$

It is possible to demonstrate that the function $h(r)$ in (1.40) exists and there holds the convolution property:

$$\mathcal{H}_\nu\{h(r);s\} = r^{-\nu}\mathcal{H}_\nu\{f(r);s\}\,\mathcal{H}_\nu\{g(r);s\},$$

such that [4]

$$h(r) = \int_0^\infty t\mathrm{J}_\nu(rt)t^{-\nu}r^{-\nu}\mathcal{H}_\nu\{f(t);s\}\,\mathcal{H}_\nu\{g(t);s\}\,dt.$$

If we define Δ_ν as

$$\Delta_\nu = \frac{d^2}{dr^2} + \frac{1}{r}\frac{d}{dr} - \frac{\nu^2}{r^2},$$

then, for those functions such that $\lim_{r\to\infty} f(r) = 0$, we have

$$\mathcal{H}_\nu\{\Delta_\nu f(r);s\} = -s^2 F_\nu(s).$$

Some other useful properties of the Hankel transform are

$$\mathcal{H}_\nu\{f'(r);s\} = \frac{s}{2\nu}\left[(\nu+1)F_{\nu-1}(s) - (\nu-1)F_{\nu+1}(s)\right],$$
$$\mathcal{H}_\nu\left\{\frac{1}{r}f(r);s\right\} = \frac{s}{2\nu}\left[F_{\nu-1}(s) + F_{\nu+1}(s)\right],$$
$$\mathcal{H}_\nu\left\{\frac{1}{r^2}f(r);s\right\} = \frac{s^2}{4\nu(\nu-1)(\nu+1)}$$
$$\times\left[(\nu+1)F_{\nu-2}(s) + 2\nu F_\nu(s) + (\nu-1)F_{\nu+2}(s)\right],$$

$$\mathcal{H}_\nu\left\{r^{\nu-1}\frac{d}{dr}r^{1-\nu}f(r);s\right\} = -sF_{\nu-1}(s),$$

$$\mathcal{H}_\nu\left\{r^{-\nu-1}\frac{d}{dr}r^{\nu+1}f(r);s\right\} = sF_{\nu+1}(s).$$

Simple examples of Hankel transforms are

$$\mathcal{H}_0\{e^{-ar^2};s\} = \frac{1}{2a}e^{-s^2/4a}, \quad a > 0,$$
$$\mathcal{H}_0\{\delta(r-a);s\} = a\mathrm{J}_\nu(s\,a), \quad a > 0,$$
$$\mathcal{H}_0\{a\mathrm{J}_\nu(a\,r);s\} = \delta(s-a), \quad a > 0,$$
$$\mathcal{H}_\nu\{r^{\nu-1}e^{-ar};s\} = \int_0^\infty r^\nu\, e^{-ar}\,\mathrm{J}_\nu(s\,r)\,dr$$
$$= \frac{1}{s^{\nu+1}}\mathcal{L}\left[x^\nu \mathrm{J}_\nu(x), p = \frac{a}{s}\right], \quad a > 0,$$

where $\mathcal{L}$ is the Laplace transform:

$$\mathcal{L}\{x\mathrm{J}_\nu(x);p\} = \frac{2^\nu\Gamma(\nu+1/2)}{\sqrt{\pi}(p^2+1)^{\nu+1/2}}.$$

As an illustrative example of application of Hankel transform, consider the problem in which heat is transmitted solely by conduction along a cylindrical bar of radius a, infinitely long, in a steady state. It is required to find how the temperature $u(r,z)$ at different points of the bar varies with time, when the problem is subject to the boundary conditions:

$$-\kappa\frac{\partial u}{\partial z} = q\theta(a-r), \qquad z = 0, \tag{1.41}$$

where κ and q are phenomenological parameters and $\theta(x)$ is the Heaviside step function. The equation to be solved is

$$\left(\frac{\partial^2}{\partial r^2} + \frac{1}{r}\frac{\partial}{\partial r} + \frac{\partial^2}{\partial z^2}\right)u(r,z) = 0, \qquad z > 0. \tag{1.42}$$

Using the Hankel transform $U_0\{u(r,z);s\} = U_0(s,z)$, Eq. (1.42) becomes

$$\left(-s^2 + \frac{\partial^2}{\partial z^2}\right)U_0(s,z) = 0, \tag{1.43}$$

whereas Eq. (1.41) is

$$-\kappa\frac{\partial U_0}{\partial z} = q\,a\frac{\mathrm{J}_1(a\,s)}{s}, \qquad z = 0. \tag{1.44}$$

Equation (1.43) may be easily solved to give $U(s,z) = ce^{-sz}$, which, using (1.44), becomes

$$U_0(s,z) = \frac{qa}{\kappa}\frac{J_1(a\,s)}{s^2}e^{-sz}.$$

By performing the inverse transformation, we obtain the formal solution

$$u(r,z) = \frac{qa}{\kappa}\int_0^\infty e^{-s\,z}\frac{J_1(s\,a)}{s}J_0(s\,r)\,ds. \tag{1.45}$$

This result is a simple illustration that the Hankel transform may be a useful tool in solving problems with cylindrical symmetry and involving the Laplace operator [5].

1.1.4 The Mellin Transform

The *Mellin transform* of a function $f(t)$ of a real variable $t \in \mathbb{R}^+$ is defined by

$$\mathcal{M}\{f(t);s\} = F_{\mathrm{M}}(s) = \int_0^\infty t^{s-1}f(t)\,dt, \qquad s \in \mathbb{C}. \tag{1.46}$$

The complex variable s must be restricted to those values for which the integral converges. This means that we have convergence at $t = 0$ only if $\mathrm{Re}(s)$ is greater than a certain value, γ_1, and at $t = \infty$ only if $\mathrm{Re}(s)$ is smaller than a certain value, γ_2 [6]. Thus, we have the strip $\gamma_1 < \mathrm{Re}\,(s) < \gamma_2$, and the existence of the Mellin transform (1.46) requires a "sufficiently good" function $f(t)$; it has to be piecewise continuous in the closed interval $[a,b] \subset (0,\infty)$, with

$$\int_0^1 |f(t)|t^{\gamma_1-1}\,dt < \infty \quad \text{and} \quad \int_1^\infty |f(t)|t^{\gamma_2-1}\,dt < \infty.$$

Let us consider two simple examples. Consider first the function $f(t) = e^{-t}$. Its Mellin transform coincides with the Euler Gamma function, to be defined in Eq. (1.63):

$$\mathcal{M}\{e^{-t};s\} = \int_0^\infty e^{-t}\,t^{s-1}\,dt, \qquad \mathrm{Re}\,(s) > 0.$$

Consider now the function $f(t) = (1+t)^{-1}$. We have

$$\mathcal{M}\{(1+t)^{-1};s\} = \int_0^\infty t^{s-1}\,(1+t)^{-1}\,dt.$$

An alternative method to the calculus of residues to compute the transform consists in changing the variable $1 + t = 1/(1 - x)$, in such a way that

$$x = \frac{t}{t+1} \quad \text{and} \quad dx = \frac{dt}{(1+t)^2}.$$

The Mellin transform can be rewritten as [7]

$$\mathcal{M}\{(1+t)^{-1}; s\} = \int_0^1 x^{s-1} (1-x)^{-s} \, dx, \quad 0 < \text{Re}(s) < 1,$$

which is nothing but the definition of the beta function $B(s, 1 - s)$ given by (1.77). Thus,

$$\mathcal{M}\{(1+t)^{-1}; s\} = B(s, 1-s). \tag{1.47}$$

For an arbitrary power $f(t) = (1 + t)^{-\alpha}$, it is possible to show that [2]

$$\mathcal{M}\{(1+t)^{-\alpha}; s\} = \frac{\Gamma(s)\Gamma(\alpha - s)}{\Gamma(\alpha)}, \qquad 0 < \text{Re}(s) < \text{Re}(\alpha) < 1. \tag{1.48}$$

The *inverse Mellin transform* is given for $t \in \mathbb{R}^+$ by the formula

$$\mathcal{M}^{-1}\{F_{\text{M}}(s); t\} = f(t) = \frac{1}{2\pi i} \int_{\gamma - i\infty}^{\gamma + i\infty} F_{\text{M}}(s) t^{-s} ds, \quad \gamma_1 < \gamma < \gamma_2. \tag{1.49}$$

In Eq. (1.49), γ is a vertical contour in the complex plane, chosen so that all singularities of $F_{\text{M}}(s)$ are to the left of it. It is known as the Bromwich contour. Notice that if all the singularities are in the left half-plane, or $F_{\text{M}}(s)$ is a function with no singularities in $-\infty < \text{Re}(s) < \infty$, then γ can be set to zero and the integration in Eq. (1.49) becomes identical to the inverse Fourier transform. In addition, if we replace in Eqs. (1.1) and (1.2) $f(t) \to f(e^t)$ and $it \to s$, we obtain, respectively, Eqs. (1.46) and (1.49). Thus, the conditions for the existence of Eqs. (1.46) and (1.49) can be derived from the direct and inverse Fourier transform [2]. A simple but very useful property of (1.46) is

$$\mathcal{M}\{t^\alpha f(t); s\} = \int_0^\infty f(t) t^{s+\alpha-1} \, dt = \mathcal{M}\{f(t); s+\alpha\} = F_{\text{M}}(s+\alpha). \tag{1.50}$$

The *Mellin convolution operator* of two functions $f(t)$ and $g(t)$, defined for $x \in \mathbb{R}^+$, is given by

$$\begin{aligned} (f * g)(x) = f * g &= \int_0^\infty f\left(\frac{x}{t}\right) g(t) \, \frac{dt}{t} \\ &= g * f; \end{aligned} \tag{1.51}$$

i.e., it has the commutative property. Similar to what we have done before with the Fourier transform, we apply the Mellin transform to the Mellin convolution operator, to obtain

$$
\begin{aligned}
\mathcal{M}\left\{\int_0^\infty f\left(\frac{x}{t}\right) g(t)\,\frac{dt}{t}; s\right\} &= \int_0^\infty x^{s-1}\left[\int_0^\infty f\left(\frac{x}{t}\right) g(t)\,\frac{dt}{t}\right] dx \\
&= \int_0^\infty g(t)\left[\int_0^\infty x^{s-1} f\left(\frac{x}{t}\right)\right]\frac{dt}{t} \\
&= \int_0^\infty u^{s-1} f(u)\,du \int_0^\infty g(t)\,t^{s-1}\,dt \\
&= F_{\mathrm{M}}(s) G_{\mathrm{M}}(s),
\end{aligned} \tag{1.52}
$$

after promoting the substitution $u = x/y$ and $dx/x = du/u$. The Mellin transform of the derivative of a function $f(t), f^{(1)}(t) = df/dt$, may be established as follows:

$$
\mathcal{M}\{f^{(1)}(t); s\} = \int_0^\infty \frac{df}{dt}\,t^{s-1}\,dt.
$$

Integration by parts of the preceding expression yields

$$
\begin{aligned}
\int_0^\infty \frac{df}{dt}\,t^{s-1}\,dt &= t^{s-1} f(t)\Big|_0^\infty - (s-1)\int_0^\infty f(t)\,t^{s-2}\,dt \\
&= -(s-1)\mathcal{M}\{f(t); s-1\},
\end{aligned}
$$

if we assume that $f(t)$ and $\mathrm{Re}\,(s)$ are such that the first term, when evaluated in the indicated extremes, vanishes. By using the property (1.50), we may write

$$
\mathcal{M}\{f^{(1)}(t); s\} = -(s-1) F_{\mathrm{M}}(s-1).
$$

For the second derivative, we obtain

$$
\mathcal{M}\{f^{(2)}(t); s\} = +(s-1)(s-2) F_{\mathrm{M}}(s-2).
$$

By operating in the same way, for the n-derivative, we can show that

$$
\mathcal{M}\{f^{(n)}(t); s\} = (-1)^n \frac{\Gamma(s)}{\Gamma(s-n)} F_{\mathrm{M}}(s-n). \tag{1.53}
$$

As an illustration of the use of Mellin transform, consider the solution of a potential problem in an infinite two-dimensional wedge with Dirichlet boundary conditions [7]. We use polar coordinates with origin at the apex of the wedge and the sides are located at $\theta = \pm\Theta$. The Laplace equation in polar coordinates for $u(r,\theta)$ is

$$\frac{\partial^2 u}{\partial r^2} + \frac{1}{r}\frac{\partial u}{\partial r} + \frac{1}{r^2}\frac{\partial^2 u}{\partial \theta^2} = 0, \quad 0 < r < \infty, \quad -\Theta < \theta < \Theta. \tag{1.54}$$

The boundary conditions are such that

$$u(r, \pm\Theta) = \begin{cases} 1, & 0 < r < R \\ 0, & r > R \end{cases} \tag{1.55}$$

and $u(r,\theta)$ is bounded for r finite, whereas

$$\lim_{r\to\infty} u(r,\theta) \sim r^{-\beta}, \quad \beta > 0.$$

The Mellin transform $U(s,\theta)$ exists with respect to r in the region $0 < \mathrm{Re}\,(s) < \beta$. When applied to Eq. (1.54), it yields

$$\frac{d^2 U(s,\theta)}{d\theta^2} + s^2 U(s,\theta) = 0; \tag{1.56}$$

the transformed boundary conditions (1.55) read

$$U(s, \pm\Theta) = R^s s^{-1}, \qquad \mathrm{Re}\,(s) > 0. \tag{1.57}$$

Equation (1.56) has the general solution

$$U(s,\theta) = A(s)e^{is\theta} + B(s)e^{-is\theta}, \tag{1.58}$$

in which $A(s)$ and $B(s)$ are to be determined by the boundary condition (1.57), i.e.,

$$\begin{aligned} U(s,\Theta) &= A(s)e^{is\Theta} + B(s)e^{-is\Theta} = R^s s^{-1}, \\ U(s,-\Theta) &= A(s)e^{-is\Theta} + B(s)e^{+is\Theta} = R^s s^{-1}, \end{aligned} \tag{1.59}$$

from which we obtain

$$A(s) = B(s) = \frac{R^s}{2s\cos(s\Theta)}. \tag{1.60}$$

The solution in the Mellin space (s-domain) is given by

$$U(s,\theta) = \frac{R^s\,\cos(s\theta)}{s\cos(s\Theta)}. \tag{1.61}$$

To obtain the solution $u(r,\theta)$ we have to invert the expression

$$F(s) = \frac{\cos(s\theta)}{s\cos(s\Theta)},$$

for $\Theta \in \mathbb{R}$ in the strip $0 < \mathrm{Re}(s) < \pi/2\Theta$. It is possible to show that [7]:

$$\mathcal{M}^{-1}\left\{\frac{\cos(s\theta)}{s\cos(s\Theta)}; s\right\} = \begin{cases} 1 - \frac{1}{\pi}\arctan\left[\frac{2r^\nu\cos(\nu\theta)}{1-r^{2\nu}}\right], & 0 \le r < 1 \\ \frac{1}{\pi}\arctan\left[\frac{2r^\nu\cos(\nu\theta)}{r^{2\nu}-1}\right], & r > 1, \end{cases} \tag{1.62}$$

where $\nu = \pi/2\Theta$ and $\theta < \pi/2$.

1.2 Special Functions of Fractional Calculus

In this section we present the definitions and some properties of special functions of mathematical physics like gamma, beta, and Bessel functions. The final part of the section is dedicated to the special functions more strictly related to the fractional calculus, like the classical and generalised Mittag-Leffler functions, the Wright function, and the H-function of Fox.

1.2.1 Gamma Function and Related Special Functions

When $\alpha = m$, with $m \in \mathbb{N}$, the definition of $m! = 1 \cdot 2 \cdot 3 \ldots$ is well known. In the problems in which we will be interested, a generalisation of the results to non-integer indices is desired. To accomplish this task, we introduce the gamma function, which is the appropriate generalisation of the factorial function.

The *Euler gamma function* $\Gamma(z)$ is defined by the so-called Euler integral of second kind [2]:

$$\Gamma(z) = \int_0^\infty t^{z-1}\, e^{-t}\, dt, \qquad \mathrm{Re}\,(z) > 0, \tag{1.63}$$

where $t^{z-1} = e^{(z-1)\ln t}$. This integral converges for all $z \in \mathbb{C}$, with $\mathrm{Re}\,(z) > 0$.

The gamma function may be also defined according to Gauss as:

$$\Gamma(z) = \lim_{n\to\infty} \frac{n!\, n^z}{z(z+1)\ldots(z+n)}, \qquad z \notin \mathbb{Z}_0^- = (0, -1, -2, \ldots). \tag{1.64}$$

If this limit exists, $\Gamma(z)$ is defined for all z, except possibly for nonpositive integers [8].

Another definition, due to Weierstrass, is such that

$$\frac{1}{\Gamma(z)} = z e^{\gamma z} \prod_{n=1}^{\infty} \left(1 + \frac{z}{n}\right) e^{-z/n}, \tag{1.65}$$

where γ is the Euler–Mascheroni constant, defined by

$$\gamma = \lim_{n\to\infty} \sum_{k=1}^{n} \frac{1}{k} - \ln(n+1) \approx 0.5772157. \tag{1.66}$$

Using (1.63) we can derive a recurrence formula for $\Gamma(z)$. Integration by parts yields

$$\Gamma(z+1) = \int_0^\infty t^z\, e^{-t}\, dt = z \int_0^\infty e^{-t}\, t^{z-1}\, dt = z\Gamma(z). \tag{1.67}$$

Since $\Gamma(z+1)$ is defined for all z, such that $\mathrm{Re}\,(z) > 1$, we see that the expression

$$\Gamma(z) = \frac{\Gamma(z+1)}{z} \tag{1.68}$$

permits us to obtain the analytic continuation of $\Gamma(z)$ into the strip $-1 < \mathrm{Re}(z) < 0$. The definition (1.64) indicates that the gamma function is analytic everywhere in the complex plane $\mathbb{C}$ except at $z = 0, -1, -2, \ldots$, where $\Gamma(z)$ has simple poles.

Other useful properties of the gamma function are discussed below. The functional equation is defined as

$$\Gamma(z)\Gamma(1-z) = \frac{\pi}{\sin(\pi z)}, \qquad z \notin \mathbb{Z}_0 = (0, \pm 1, \pm 2, \ldots). \tag{1.69}$$

A particular result for $z = 1/2$ may be obtained from (1.69), in the form

$$\left[\Gamma\left(\frac{1}{2}\right)\right]^2 = \frac{\pi}{\sin(\pi/2)} \quad \Rightarrow \quad \Gamma\left(\frac{1}{2}\right) = \sqrt{\pi}. \tag{1.70}$$

Evidently, the same result may be obtained from (1.63):

$$\Gamma\left(\frac{1}{2}\right) = \int_0^\infty e^{-t}\, t^{-1/2}\, dt = 2\int_0^\infty e^{-t^2}\, dt = \sqrt{\pi}.$$

We have also the *Legendre duplication formula*:

$$\Gamma(z)\Gamma\left(z + \frac{1}{2}\right) = 2^{1-2z}\sqrt{\pi}\,\Gamma(2z), \qquad z \in \mathbb{C}, \tag{1.71}$$

which is a special case of the *Gauss–Legendre multiplication theorem*:

$$\begin{aligned}\prod_{k=0}^{m-1} \Gamma\left(z + \frac{k}{m}\right) &= \Gamma(z)\Gamma\left(z + \frac{1}{m}\right)\Gamma\left(z + \frac{2}{m}\right)\ldots\Gamma\left(z + \frac{m-1}{m}\right) \\ &= (2\pi)^{1/2(m-1)} m^{1/2-mz}\Gamma(mz), \quad z \in \mathbb{C}, m \in \mathbb{N},\end{aligned} \tag{1.72}$$

for $m \neq 1$. Finally, we have also the *Stirling's formula*:

$$\Gamma(z) \approx \sqrt{\frac{2\pi}{z}}\left(\frac{z}{e}\right)^z \sum_{k=0}^{\infty} \frac{a_k}{z^k}, \qquad \mathrm{Re}\,(z) > 0, \tag{1.73}$$

where $a_0 = 1, a_1 = 1/12, \ldots$. Particular cases are

$$n! = (2\pi n)^{1/2}\left(\frac{n}{e}\right)^n \left[1 + O\left(\frac{1}{n}\right)\right], \quad n \in \mathbb{N}, \quad n \to \infty, \tag{1.74}$$

and

$$|\Gamma(x + iy)| = (2\pi)^{1/2}|y|^{x-1/2}e^{-x-\pi|y|/2}\left[1 + O\left(\frac{1}{n}\right)\right], \quad y \to \infty.$$

We can write for the quotient expansion of two gamma functions at infinity the following expression:

$$\frac{\Gamma(z+a)}{\Gamma(z+b)} = z^{a-b}\left[1 + O\left(\frac{1}{z}\right)\right], \qquad |\arg(z+a)| < \pi; \quad |z| \to \infty,$$

and, finally,

$$\Gamma\left(n + \frac{1}{2}\right) = \frac{(2n-1)!!}{2^n}\sqrt{\pi},$$

where the double factorial is defined as

$$(2n-1)!! = 1 \cdot 3 \cdots (2n-1), \qquad n \in \mathbb{N}.$$

The *incomplete gamma function* is defined as

$$\gamma(\alpha, x) = \int_0^x e^{-t}\, t^{\alpha-1}\, dt, \qquad \mathrm{Re}\,(\alpha) > 0, \tag{1.75}$$

or

$$\Gamma(\alpha, x) = \int_x^\infty e^{-t}\, t^{\alpha-1}\, dt, \tag{1.76}$$

such that

$$\gamma(\alpha, \infty) = \Gamma(\alpha, 0) = \gamma(\alpha, x) + \Gamma(\alpha, x) = \Gamma(\alpha).$$

These expressions also have power-series expansions for noninteger α, when x is small,

$$\gamma(\alpha, x) = x^\alpha \sum_{n=0}^{\infty} (-1)^n \frac{x^n}{n!\,(\alpha+n)},$$

and for asymptotic values

$$\Gamma(\alpha, x) = x^{\alpha-1} e^{-x} \sum_{n=0}^{\infty} (-1)^n \frac{(n-\alpha)!}{(-\alpha)} \frac{1}{x^n}.$$

A function closely related to the gamma function is the *beta function*, defined by the Euler integral of first kind:

$$B(x, y) = \int_0^1 t^{x-1}\, (1-t)^{y-1}\, dt, \quad \mathrm{Re}\,(x) > 0, \quad \mathrm{Re}\,(y) > 0. \tag{1.77}$$

This function is connected with the gamma function by the relation

$$B(x,y) = \frac{\Gamma(x)\Gamma(y)}{\Gamma(x+y)}, \qquad (x,y) \in \mathbb{Z}_0^-. \tag{1.78}$$

To prove it, consider

$$\begin{aligned}\Gamma(x)\Gamma(y) &= \int_0^\infty e^{-t}\, t^{x-1}\, dt \int_0^\infty e^{-\tau}\, \tau^{x-1}\, d\tau \\ &= \int_0^\infty \int_0^\infty e^{-(t+\tau)}\, t^{x-1}\, \tau^{y-1} dt\, d\tau.\end{aligned}$$

Now, introducing the new coordinates $t = r\cos^2\theta$ and $\tau = r\sin^2\theta$, we have

$$\begin{aligned}\Gamma(x)\Gamma(y) &= 2\int_0^\infty \int_0^{\pi/2} e^{-r}\, r^{x+y-1}\, \cos^{2x-1}\theta\, \sin^{2y-1}\theta\, dr\, d\theta \\ &= 2\Gamma(x+y) \int_0^{\pi/2} \cos^{2x-1}\theta\, \sin^{2y-1}\theta\, d\theta.\end{aligned}$$

The last integral may be evaluated by changing the variable again, now letting $\cos\theta = t^{1/2}$. Thus,

$$\begin{aligned}\Gamma(x)\Gamma(y) &= \Gamma(x+y) \int_0^1 t^{x-1}\, (1-t)^{y-1}\, dt \\ &= \Gamma(x+y)B(x,y).\end{aligned}$$

The *incomplete beta function* is defined as

$$B_x(p,q) = \int_0^x t^{p-1}\, (1-t)^{q-1}\, dt, \quad 0 \le x \le 1, \quad p > 0, \quad q > 0. \tag{1.79}$$

When $x = 1$, $B_1(p,q)$ is the (complete) beta function defined in Eq. (1.77). In probability theory, the ratio $B_x(p,q)/B_1(p,q)$ is often identified as a distribution function [9], with μ and σ^2, respectively, the mean and the variance, given by

$$\mu = \frac{p}{p+q} \quad \text{and} \quad \sigma^2 = \frac{pq}{(p+q+1)(p+q)^2}.$$

Another useful definition to be employed in this book refers to the *binomial coefficients*, which usually are written as

$$\binom{\alpha}{n} = \frac{\alpha(\alpha-1)\cdots(\alpha-n+1)}{n!}, \qquad \binom{\alpha}{0} = 1, \qquad n \in \mathbb{N}. \tag{1.80}$$

When $\alpha = m$, with $m \in \mathbb{N}_0$, we have

$$\binom{m}{n} = \frac{m!}{n!\,(m-n)!}, \qquad m \geq n, \tag{1.81}$$

such that

$$\binom{m}{n} = 0, \qquad \text{if} \qquad 0 \leq m < n. \tag{1.82}$$

The general representation for $\alpha \notin \mathbb{Z}_0^-$ is

$$\binom{\alpha}{n} = \frac{\Gamma(\alpha+1)}{n!\,\Gamma(\alpha-n+1)}, \qquad \alpha \in \mathbb{C}, \quad n \in \mathbb{N}_0. \tag{1.83}$$

For arbitrary complexes $\alpha \in \mathbb{C}$ and $\beta \in \mathbb{C}$, we have

$$\binom{\alpha}{\beta} = \frac{\Gamma(\alpha+1)}{\Gamma(\alpha-\beta+1)\Gamma(\beta+1)}. \tag{1.84}$$

Finally, we define the *Euler's psi function* as the logarithmic derivative of the gamma function [2]:

$$\psi(z) = \frac{d}{dz}\log\Gamma(z) = \frac{\Gamma'(z)}{\Gamma(z)}, \qquad z \in \mathbb{C}, \tag{1.85}$$

such that

$$\psi(z+m) = \psi(z) + \sum_{k=0}^{m-1}\frac{1}{z+k}, \quad z \in \mathbb{C} \quad \text{and} \quad m \in \mathbb{N}.$$

For $m = 1$, we have

$$\psi(z+1) = \psi(z) + \frac{1}{z}, \qquad z \in \mathbb{C}.$$

1.2.2 Bessel Functions

Bessel functions are named after the German astronomer, mathematician, physicist, and geodesist Friedrich Wilhelm Bessel (1784–1846), even if the concept of Bessel function was first introduced in 1732 by the Swiss mathematician and physicist Daniel Bernoulli (1700–1782). Bernoulli treated the solution of the problem of an oscillating chain suspended at one end. In 1764, Leonhard Euler (1707–1783) also employed Bessel functions of both zero and integral orders to the analysis of vibrations of a stretched membrane.

To motivate the introduction of Bessel functions, consider the known problem of solving the diffusion equation

$$\frac{\partial u}{\partial t} = D\nabla^2 u \tag{1.86}$$

in a disk of radius a, in polar coordinates, subjected to a typical boundary condition $u(r = a, \theta, t) = 0$ and to the initial condition $u(r, \theta, 0) = \phi(r, \theta)$. The differential equation (1.86) may be rewritten as

$$\frac{\partial u}{\partial t} = D\left(\frac{1}{r}\frac{\partial}{\partial r}r\frac{\partial u}{\partial r} + \frac{1}{r^2}\frac{\partial^2 u}{\partial \theta^2}\right), \tag{1.87}$$

and admits a partial solution in the form

$$u(r, \theta, t) = e^{i\nu\theta}e^{-Dk^2 t}R(r), \qquad (\nu, k) \in \mathbb{R}, \tag{1.88}$$

to assure that $u(r, \theta, t)$ remains bounded, i.e., $u(r, \theta, t \to \infty) = 0$. By assuming that $z = kr$, the radial part $R(z)$ satisfies the equation

$$\frac{1}{z}\frac{\mathrm{d}}{\mathrm{d}z}\left[z\frac{\mathrm{d}R(z)}{\mathrm{d}z}\right] + \left(1 - \frac{\nu^2}{z^2}\right)R(z) = 0, \tag{1.89}$$

which is known as Bessel's equation. The solution to Bessel's equation yields Bessel functions of the first and second kind, respectively denoted by $\mathrm{J}_\nu(z)$ and $\mathrm{N}_\nu(z)$, as follows:

$$R(z) = A\mathrm{J}_\nu(z) + B\mathrm{N}_\nu(z),$$

where A and B are arbitrary constants and the solutions are defined for all real values $-\infty < \nu < \infty$. Let us analyse these solutions more closely.

The Bessel Function of First Kind

Bessel's equation (1.89) may be rewritten for $y(z) = R(z)$ as

$$y''(z) + \frac{1}{z}y'(z) + \left[1 - \frac{\nu^2}{z^2}\right]y(z) = 0 \tag{1.90}$$

and is called Bessel's equation for functions of order ν [10]. It has two singular points, a regular singularity at $z = 0$ and an irregular singularity at $z = \infty$. One solution that is regular near the origin may be obtained in the form

$$y(z) = z^{\pm\nu}\sum_{k=0}^{\infty} a_k z^k,$$

which leads to the expression

$$\mathrm{J}_\nu(z) = \left(\frac{z}{2}\right)^\nu \sum_{k=0}^{\infty} \frac{(iz/2)^{2k}}{k!\,\Gamma(\nu + k + 1)}, \qquad \nu \in \mathbb{C}, \qquad \mathrm{Re}\,(\nu) > 0. \tag{1.91}$$

This is the regular Bessel function, or the *Bessel function of the first kind*. If we replace ν by $-\nu$ in Eq. (1.90), it follows that $\mathrm{J}_{-\nu}(z)$ is also its solution. The series (1.91) converges for all $z \in \mathbb{C}$. For $\nu = n \in \mathbb{N}$, we have

$$\mathrm{J}_{-n}(z) = (-1)^n \mathrm{J}_n(z). \tag{1.92}$$

In this case, since the solutions are not independent, we must look for a second solution of (1.90) (see the Neumann function). However, if ν is not an integer, $\mathrm{J}_\nu(z)$ and $\mathrm{J}_{-\nu}(z)$ are linearly independent. In this case, the general solution of (1.90) has the form

$$y(z) = A\,\mathrm{J}_\nu(z) + B\,\mathrm{J}_{-\nu}(z),$$

where A and B are constants [11]. Particular illustrative cases for $\nu = 1/2$ and $\nu = -1/2$ are, respectively,

$$\mathrm{J}_{1/2}(z) = \left(\frac{2}{\pi z}\right)^{1/2} \sin z \quad \text{and} \quad \mathrm{J}_{-1/2}(z) = \left(\frac{2}{\pi z}\right)^{1/2} \cos z.$$

For $\nu = 3/2$, we obtain

$$\mathrm{J}_{3/2}(z) = -\left(\frac{2}{\pi z}\right)^{1/2} \left[\cos z - \frac{\sin z}{z}\right].$$

In general, when n is an integer, for $\nu = n + 1/2$, the Bessel function may be expressed as:

$$\mathrm{J}_{n+1/2}(z) = \left(\frac{2}{\pi z}\right)^{1/2} \left[P_n\left(\frac{1}{z}\right) \cos z + Q_n\left(\frac{1}{z}\right) \sin z\right],$$

where P_n and Q_n are suitable polynomials of degree n [8].

The Bessel function $\mathrm{J}_\nu(z)$ is also defined by means of the Poisson integral representation

$$\begin{aligned}\mathrm{J}_\nu(z) &= \frac{(z/2)^\nu}{\sqrt{\pi}\,\Gamma(\nu + 1/2)} \int_{-1}^{1} (1 - t^2)^{\nu - 1/2}\, e^{izt}\, dt \\ &= \frac{(z/2)^\nu}{\sqrt{\pi}\,\Gamma(\nu + 1/2)} \int_{-1}^{1} (1 - t^2)^{\nu - 1/2} \cos(tz)\, dt,\end{aligned} \tag{1.93}$$

for $\mathrm{Re}\,(z) > -\frac{1}{2}$. By manipulating the integral representation (1.93) or the series (1.91), we can obtain the following recurrence formulas for the Bessel function:

$$\begin{aligned}\mathrm{J}_{\nu-1}(z) + \mathrm{J}_{\nu+1}(z) &= \frac{2\nu}{z} \mathrm{J}_\nu(z) \\ \mathrm{J}_{\nu-1}(z) - \mathrm{J}_{\nu+1}(z) &= 2\frac{d}{dz} \mathrm{J}_\nu(z).\end{aligned} \tag{1.94}$$

Since these formulas can be established from the integral representation, they will hold for other solutions of the Bessel equation obtained by changing the limit of integration; i.e., they will be valid for solutions of the second and third kind [12]. The main terms of the asymptotic behaviour of $J_\nu(z)$ are [10, 13]

$$J_\nu(z) = \frac{1}{\Gamma(\nu+1)}\left(\frac{z}{2}\right)^\nu [1 + O(z)], \quad \nu \in \mathbb{C}, \quad \nu \neq -1, -2, -3, \ldots, \tag{1.95}$$

for $z \to 0$, and

$$J_\nu(z) = \left(\frac{2}{\pi z}\right)^{1/2} \left\{ \left[\cos\left(z - \frac{\nu\pi}{2} - \frac{\pi}{4}\right) + O\left(\frac{1}{z}\right) \right] + \left[-\sin\left(z - \frac{\nu\pi}{2} - \frac{\pi}{4}\right) + O\left(\frac{1}{z}\right) \right] \right\}, \tag{1.96}$$

when $|z| \to \infty$. For those problems involving the wave equation in spherical coordinates, it is convenient to use *spherical Bessel functions*, defined by

$$j_n(z) = \left(\frac{2}{\pi z}\right)^{1/2} J_{n+1/2}(z). \tag{1.97}$$

The Hankel Functions

If we consider again the integral representation (1.93), it is not difficult to show that

$$y(z) = z^\nu \int_\alpha^\beta e^{izt} (t^2 - 1)^{\nu - 1/2} \, dt$$

satisfies the Eq. (1.90) provided that α and β are such that

$$(t^2 - 1)^{\nu - 1/2} e^{izt} \Big|_\alpha^\beta = 0.$$

Indeed, for $\mathrm{Re}(\alpha) > -1/2$, $\alpha = -1$ and $\beta = 1$, we obtain (1.93). Now, for $\mathrm{Re}(z) > 0$, we may introduce two solutions for the Bessel equation:

$$H_\nu^{(1)} = \frac{\Gamma(1/2 - \nu)(z/2)^\nu}{i\pi\Gamma(1/2)} \int_{C_1} e^{izt} (t^2 - 1)^{\nu - 1/2} \, dt \quad \text{and}$$

$$H_\nu^{(2)} = \frac{\Gamma(1/2 - \nu)(z/2)^\nu}{i\pi\Gamma(1/2)} \int_{C_2} e^{izt} (t^2 - 1)^{\nu - 1/2} \, dt, \tag{1.98}$$

for $\mathrm{Re}(z) > 0$, where C_1 and C_2 are the contours shown in Fig. 1.1. Using the relation for the gamma function

$$\Gamma(1/2 - \nu)\Gamma(1/2 + \nu) = \frac{\pi}{\cos(\pi\nu)},$$

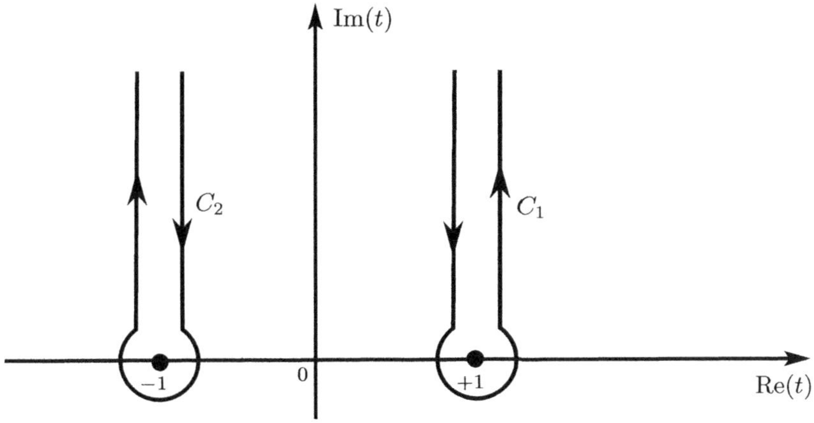

Figure 1.1 Contours in the t-plane for the integrations to be performed in (1.98) [8].

we may rewrite (1.98) as

$$\mathrm{H}_\nu^{(1,2)}(z) = \frac{(z/2)^\nu}{i\Gamma(1/2)\Gamma(1/2+\nu)\cos(\pi\nu)} \int_{C_1,C_2} e^{izt}(t^2-1)^{\nu-1/2}\,dt, \tag{1.99}$$

for Re $(z) > 0$. In (1.99), $\mathrm{H}_\nu^{(1)}(z)$ and $\mathrm{H}_\nu^{(2)}(z)$ are known as the *Hankel functions of the first and second kind*, respectively. Since they are linearly independent, it is possible to express the Bessel function of the first kind as the linear combination:

$$\mathrm{J}_\nu(z) = \frac{1}{2}\left[\mathrm{H}_\nu^{(1)}(z) + \mathrm{H}_\nu^{(2)}(z)\right]. \tag{1.100}$$

Some asymptotic expansions are [10]

$$\mathrm{H}_\nu^{(1)}(z) \approx \left(\frac{2}{\pi z}\right)^{1/2} e^{i(z-\nu\pi/2-\pi/4)} \sum_{k=0}^{\infty} \frac{\Gamma(\nu+k+1/2)(i/2z)^k}{k!\,\Gamma(\nu-k+1/2)}, \tag{1.101}$$

for $-\pi < \arg(z) < 2\pi$, and

$$\mathrm{H}_\nu^{(2)}(z) \approx \left(\frac{2}{\pi z}\right)^{1/2} e^{-i(z-\nu\pi/2-\pi/4)} \sum_{k=0}^{\infty} \frac{\Gamma(\nu+k+1/2)(-i/2z)^k}{k!\,\Gamma(\nu-k+1/2)}, \tag{1.102}$$

for $-2\pi < \arg(z) < \pi$. When $\nu = m + 1/2$, where m is not a negative integer, the series in (1.101) and (1.102) terminate, and the expressions become the exact representation for the Hankel functions [8]. Some illustrative values for $\nu = 1/2$ are

$$\mathrm{H}_{1/2}^{(1)}(z) = -i\left(\frac{2}{\pi z}\right)^{1/2} e^{iz} \quad \text{and} \quad \mathrm{H}_{1/2}^{(2)}(z) = \left(\frac{2}{\pi z}\right)^{1/2} e^{-iz}; \tag{1.103}$$

for $\nu = 3/2$, we have

$$\mathrm{H}^{(1)}_{3/2}(z) = -\left(\frac{2}{\pi z}\right)^{1/2} e^{iz}\left[1+\frac{i}{z}\right] \tag{1.104a}$$

and

$$\mathrm{H}^{(2)}_{3/2}(z) = -\left(\frac{2}{\pi z}\right)^{1/2} e^{-iz}\left[1-\frac{i}{z}\right]. \tag{1.104b}$$

The *spherical Hankel functions of the first and second kind* are defined, respectively, as

$$h^{(1)}_{\nu}(z) = \sqrt{\frac{\pi}{2z}}\left[\mathrm{J}_{l+1/2}(z) + i\,\mathrm{N}_{l+1/2}(z)\right] \tag{1.105a}$$

and

$$h^{(2)}_{\nu}(z) = \sqrt{\frac{\pi}{2z}}\left[\mathrm{J}_{l+1/2}(z) - i\,\mathrm{N}_{l+1/2}(z)\right], \tag{1.105b}$$

where $\mathrm{J}_{\nu}(z)$ is the Bessel function of first kind, introduced in Eq. (1.91), and $\mathrm{N}_{\nu}(z)$ is the Bessel function of the second kind, or the Neumann function, to be defined in the next section. They can be connected with the Green's function of the Helmholtz equation in the form [14]:

$$G_{\mathrm{out}} = -\frac{1}{4\pi}\frac{e^{ik|\mathbf{r}-\mathbf{r}'|}}{|\mathbf{r}-\mathbf{r}'|} \propto h^{(1)}_{0}(k|\mathbf{r}-\mathbf{r}'|) \tag{1.106}$$

and

$$G_{\mathrm{in}} = -\frac{1}{4\pi}\frac{e^{-ik|\mathbf{r}-\mathbf{r}'|}}{|\mathbf{r}-\mathbf{r}'|} \propto h^{(2)}_{0}(k|\mathbf{r}-\mathbf{r}'|) \tag{1.107}$$

and find many other applications in mathematical physics [15].

Neumann Functions

The *Bessel function of the second kind*, the Neumann function, may be defined as

$$\mathrm{N}_{\nu}(z) = \frac{1}{2i}\left[\mathrm{H}^{(1)}_{\nu}(z) - \mathrm{H}^{(2)}_{\nu}(z)\right], \qquad \nu \in \mathbb{C}, \quad \mathrm{Re}\,(\nu) > 0. \tag{1.108}$$

When ν is not an integer, we can express $\mathrm{N}_{\nu}(z)$ in terms of $\mathrm{J}_{\nu}(z)$ and $\mathrm{J}_{-\nu}(z)$ in the form

$$\mathrm{N}_{\nu}(z) = \cot(\pi\nu)\mathrm{J}_{\nu}(z) - \csc(\pi\nu)\mathrm{J}_{-\nu}(z).$$

The asymptotic behaviour for $|z| \to \infty$ is as follows [12]:

$$\mathrm{N}_{\nu}(z) = \left(\frac{2}{\pi z}\right)^{1/2}\left[\sin\left(z - \frac{\nu\pi}{2} - \frac{\pi}{4}\right) + O\left(\frac{1}{z}\right)\right], \qquad -\frac{\pi}{2} < \arg(z) < \frac{\pi}{2}.$$

It is possible to compare the Bessel functions already introduced with certain one-dimensional harmonic functions, namely:

$$\begin{aligned}
\mathrm{J}_\nu(z) &\iff \cos(z) \\
\mathrm{N}_\nu(z) &\iff \sin(z) \\
\mathrm{H}^{(1)}_\nu(z) &\iff e^{iz} \\
\mathrm{H}^{(2)}_\nu(z) &\iff e^{-iz}.
\end{aligned}$$

When $\nu = n \in \mathbb{N}$, the following expression holds:

$$\mathrm{N}_n(z) = \frac{1}{\pi}\left[\frac{d}{d\nu}\mathrm{J}_\nu(z) - (-1)^n \frac{d}{d\nu}\mathrm{J}_{-\nu}(z)\right]_{\nu=n}, \tag{1.109}$$

which gives an expression for the Neumann function around $z = 0$ [12]:

$$\begin{aligned}
\mathrm{N}_n(z) = \frac{1}{\pi}\left[2\ln\left(\frac{z}{2}\right) + \gamma - \psi(n+1)\right]\mathrm{J}_n(z) - \frac{1}{\pi}\sum_{m=0}^{n-1}\frac{(n-m-1)!}{m!\,(z/2)^{n-2m}} \\
-\frac{1}{\pi}\sum_{m=1}^{\infty}(-1)^m\frac{(z/2)^{n+2m}}{m!\,(n+m)!}\sum_{s=1}^{m}\left[\frac{1}{s}+\frac{1}{s+n}\right],
\end{aligned}$$

where γ is the Euler–Mascheroni constant, defined in Eq. (1.66). For $n = 0$, we have the special result

$$\mathrm{N}_0(z) \approx \frac{2}{\pi}\left[\ln\left(\frac{z}{2}\right) + \gamma\right], \qquad |z| \to 0,$$

while for $n > 0$, we have:

$$\mathrm{N}_n(z) \approx -\frac{1}{\pi(n-1)!}\left(\frac{2}{z}\right)^n, \qquad |z| \to 0.$$

Modified Bessel Functions

We consider now the modified Bessel functions $\mathrm{I}_\nu(z)$ and $\mathrm{K}_\nu(z)$, which are solutions frequently encountered in physical problems, such as the diffusion problems we will treat in this book. The *modified Bessel* function is the solution to the modified Bessel differential equation, obtained by letting $z \to iz$ in Eq. (1.90):

$$y''(z) + \frac{1}{z}y'(z) - \left(1 + \frac{\nu^2}{z^2}\right)y(z) = 0. \tag{1.110}$$

It is defined as

$$\mathrm{I}_\nu(z) = \sum_{k=0}^{\infty}\frac{(z/2)^{2k+\nu}}{k!\,\Gamma(\nu+k+1)}, \qquad z \in \mathbb{C}\backslash(-\infty, 0], \tag{1.111}$$

and may be expressed via the Bessel function of the first kind as

$$\mathrm{I}_\nu(z) = e^{-\nu\pi i/2}\mathrm{J}_\nu(e^{\pi i/2}z).$$

A useful integral representation has the form:

$$\mathrm{I}_\nu(z) = \frac{(z/2)^\nu}{\sqrt{\pi}\Gamma(\nu+1/2)}\int_{-1}^{1}(1-t^2)^{\nu-1/2}\cosh(zt)\,dt, \quad \mathrm{Re}\,(\nu) > -\frac{1}{2},$$

or, expressing it in terms of a contour integral, in the form

$$\mathrm{I}_\nu(z) = \frac{1}{2\pi i}\int_{-i\pi+\infty}^{i\pi+\gamma} e^{z[\cosh(t)-\nu t]}\,dt, \qquad \nu \in \mathbb{C}; \quad |\arg(z)| < \frac{\pi}{2}.$$

Some asymptotic behaviours of $\mathrm{I}_\nu(z)$ are the following [10]: when $z \to 0$, we have

$$\mathrm{I}_\nu(z) = \frac{1}{\Gamma(\nu+1)}\left(\frac{z}{2}\right)^\nu[1+O(z)], \qquad \nu \in \mathbb{C}\backslash\nu \neq -1,-2,-3,\ldots.$$

When $|z| \to \infty$, we have

$$\mathrm{I}_\nu(z) = \frac{1}{\sqrt{2\pi z}}\left\{e^z\left[1+O\left(\frac{1}{z}\right)\right] + ie^{-z+\nu\pi i}\left[1+O\left(\frac{1}{z}\right)\right]\right\},$$

for $-\pi/2 < \arg(z) < 3\pi/2$.

Finally, the *modified Bessel function of the third kind*, or the *Macdonald function*, $\mathrm{K}_\nu(z)$, may be defined as

$$\mathrm{K}_\nu(z) = \frac{\pi}{2}i^{\nu+1}\mathrm{H}_\nu^{(1)}(iz) = \frac{\pi}{2}i^{\nu+1}\left[\mathrm{J}_\nu(iz) + i\mathrm{N}_\nu(iz)\right], \tag{1.112}$$

which can be transformed as

$$\mathrm{K}_\nu(z) = \frac{\pi}{2\sin(\pi\nu)}\left[\cos(\pi\nu)\mathrm{I}_{-\nu}(z) - \mathrm{I}_\nu(z)\right], \qquad \nu \in \mathbb{C}\backslash\mathbb{Z}. \tag{1.113}$$

There is also an integral representation in the form

$$\mathrm{K}_\nu(z) = \frac{1}{2}\int_0^\infty t^{-\nu-1}\,e^{-z/2[t+1/t]}\,dt, \qquad \mathrm{Re}\,(z) > 0,$$

or, when $0 < \mathrm{Re}\,(\nu) < 1/2$, in the form

$$\mathrm{K}_\nu(z) = \frac{\sqrt{\pi}}{\Gamma(-\nu+1/2)}\left(\frac{2}{z}\right)^\nu\int_1^\infty (t^2-1)^{-\nu-1/2}\,e^{-zt}\,dt.$$

Asymptotic representations for $K_\nu(z)$ are [10]:

$$K_\nu(z) = \frac{\Gamma(\nu)}{2}\left(\frac{2}{z}\right)^{\nu}[1+O(z)], \quad z\to 0, \quad \mathrm{Re}\,(\nu)>0,$$

and

$$K_\nu(z) = \left(\frac{\pi}{2z}\right)^{1/2} e^{-z}\left[1+O\left(\frac{1}{z}\right)\right], \qquad z\to\infty, \quad |\arg(z)| < \frac{3\pi}{2}.$$

The general solution of the modified Bessel equation (1.110) may be written as

$$y(z) = A\,I_\nu(z) + B\,K_\nu(z), \tag{1.114}$$

where A and B are constants.

1.2.3 Mittag-Leffler Functions

The classical and the generalised Mittag-Leffler functions interpolate between a purely exponential law and a power-law-like behaviour and their integral transforms are known. For this reason, they also play an important role in solving some differential equations because they can be used to identify inverse transforms.

Classical Mittag-Leffler Functions

The Mittag-Leffler function was introduced in the beginning of the twentieth century by the Swedish mathematician Magnus Gösta Mittag-Leffler (1846–1927) [16–20]. In the fractional calculus, it naturally occurs whenever one is dealing with fractional integral equations and fractional generalisation of the kinetic equation [21]. It also arises in the investigation of random walks, Lévy flights, and anomalous transport. The *one-parameter* Mittag-Leffler function is defined as

$$E_\alpha(z) = \sum_{k=0}^{\infty}\frac{z^k}{\Gamma(\alpha k+1)}, \quad z\in\mathbb{C}, \quad \alpha\in\mathbb{C}, \quad \mathrm{Re}\,(\alpha)>0. \tag{1.115}$$

For $\alpha = 1$, it is simply the exponential function:

$$E_1(z) = \sum_{k=0}^{\infty}\frac{z^k}{\Gamma(k+1)} = \sum_{k=0}^{\infty}\frac{z^k}{k!} = e^z.$$

For $\alpha = 2$, we have:

$$E_2(z) = \sum_{k=0}^{\infty}\frac{z^k}{\Gamma(2k+1)} = \sum_{k=0}^{\infty}\frac{\sqrt{z}^{2k}}{2k!} = \cosh(\sqrt{z}),$$

i.e., $\mathrm{E}_2(z^2) = \cosh(z)$. For $\alpha = 1/2$, we have:

$$\mathrm{E}_{1/2}(z) = \sum_{k=0}^{\infty} \frac{z^k}{\Gamma(k/2+1)} = e^{z^2}\,\mathrm{erfc}(-z),$$

where $\mathrm{erfc}(-z)$ is the complementary error function, defined as

$$\mathrm{erfc}(-z) = \frac{2}{\sqrt{\pi}} \int_{z}^{\infty} e^{-t^2}\,dt = 1 - \mathrm{erf}(z). \tag{1.116}$$

This particular case has the asymptotic estimate

$$\mathrm{E}_{1/2}(z) \approx 2e^{z^2}, \qquad |z| \to \infty, \quad |\arg(z)| < \frac{\pi}{4},$$

and may be obtained from the general expression for $n \in \mathbb{N}$:

$$\mathrm{E}_{1/n}(z) = e^{z^n} \left\{ 1 + n \int_0^z e^{-t^n} \left[\sum_{k=1}^{n-1} \frac{t^{k-1}}{\Gamma(k/n)} \right] dt \right\}, \qquad n \neq 1.$$

For $\alpha > 0$ and $z \in \mathbb{C}$, the following relation may be proved [2]:

$$\mathrm{E}_\alpha(z) = \frac{1}{2\pi i} \int_{\gamma-i\infty}^{\gamma+i\infty} \frac{\Gamma(s)\Gamma(1-s)}{\Gamma(1-\alpha s)} (-z)^{-s}\,ds, \qquad |\arg(z)| < \pi, \tag{1.117}$$

where the path of integration separates all the poles at $s = -k$, with $k \in \mathbb{N}_0$, to the left and all the poles at $s = n+1$, with $n \in \mathbb{N}_0$, to the right. The proof of this lemma uses the Mellin–Barnes contour integral. Indeed, Eq. (1.117) permits us to obtain the Mellin transform of the Mittag-Lefller function (1.115):

$$\mathcal{M}\{\mathrm{E}_\alpha(-t); s\} = \frac{\Gamma(s)\Gamma(1-s)}{\Gamma(1-\alpha s)}, \qquad 0 < \mathrm{Re}\,(s) < 1. \tag{1.118}$$

The asymptotic behaviour of $\mathrm{E}_\alpha(z)$ as $|z| \to \infty$ may be expressed as follows [22]:

$$\mathrm{E}_\alpha(z) = -\sum_{n=1}^{N-1} \frac{z^{-n}}{\Gamma(1-\alpha n)} + O\left(|z|^{-N}\right), \qquad |\arg(-z)| < \left(1 - \frac{1}{2}\alpha\right)\pi. \tag{1.119}$$

The basic properties of the *two-parameter* function of the Mittag-Leffler type were investigated by Agarwal [23], Humbert and Agarwal [24, 25], and Dzherbashyan [26], but first appeared in a paper by Wiman [19]. It is a generalisation of (1.115), in the form:

$$\mathrm{E}_{\alpha,\beta}(z) = \sum_{k=0}^{\infty} \frac{z^k}{\Gamma(\alpha k + \beta)}, \qquad z, \beta \in \mathbb{C}, \quad \mathrm{Re}\,(\alpha) > 0. \tag{1.120}$$

When $\beta = 1$, it coincides with the Mittag-Leffler function (1.115):

$$\mathrm{E}_{\alpha,1}(z) = \sum_{k=0}^{\infty} \frac{z^k}{\Gamma(\alpha k + 1)} = \mathrm{E}_{\alpha}(z), \qquad z \in \mathbb{C}, \quad \mathrm{Re}\,(\alpha) > 0.$$

Some particular cases are listed below.

(a) $\alpha = \beta = 1$:

$$\mathrm{E}_{1,1}(z) = \sum_{k=0}^{\infty} \frac{z^k}{\Gamma(k+1)} = e^z;$$

(b) $\alpha = 1, \beta = 2$:

$$\mathrm{E}_{1,2}(z) = \sum_{k=0}^{\infty} \frac{z^k}{(k+1)!} = \frac{1}{z} \sum_{k=0}^{\infty} \frac{z^{k+1}}{(k+1)!} = \frac{e^z - 1}{z};$$

(c) $\alpha = 1, \beta = 3$:

$$\mathrm{E}_{1,3}(z) = \sum_{k=0}^{\infty} \frac{z^k}{(k+2)!} = \frac{1}{z^2} \sum_{k=0}^{\infty} \frac{z^{k+2}}{(k+2)!} = \frac{e^z - 1 - z}{z};$$

(d) $\alpha = 2, \beta = 1, z \to z^2$:

$$\mathrm{E}_{2,1}(z^2) = \sum_{k=0}^{\infty} \frac{(z^2)^k}{\Gamma(2k+1)} = \sum_{k=0}^{\infty} \frac{z^{2k}}{(2k)!} = \cosh(z); \tag{1.121}$$

(e) $\alpha = 2, \beta = 2, z \to z^2$:

$$\mathrm{E}_{2,2}(z^2) = \sum_{k=0}^{\infty} \frac{(z^2)^k}{\Gamma(2k+2)} = \frac{1}{z} \sum_{k=0}^{\infty} \frac{z^{2k+1}}{(2k+1)!} = \frac{\sinh(z)}{z}; \tag{1.122}$$

(f) $\alpha = 2, \beta = 1, z \to -z^2$:

$$\mathrm{E}_{2,1}(-z^2) = \sum_{k=0}^{\infty} (-1)^k \frac{z^{2k}}{(2k)!} = \cos(z);$$

(g) $\alpha = 2, \beta = 2, z \to -z^2$:

$$\mathrm{E}_{2,2}(-z^2) = \sum_{k=0}^{\infty} (-1)^k \frac{z^{2k+1}}{(2k+1)!} = \frac{\sin(z)}{z}.$$

The function $\mathrm{E}_{\alpha,\beta}(z)$ has the integral representation

$$\mathrm{E}_{\alpha,\beta}(z) = \frac{1}{2\pi i} \int_C \frac{t^{\alpha-\beta}\, e^t}{t^\alpha - z}\, dt, \tag{1.123}$$

which is a generalisation of (1.117) with the same path C. Notice that the function $\mathrm{E}_{\alpha,\beta}(z)$, similarly to what happens with $\mathrm{E}_\alpha(z)$, presents poles at $s = -k$, with $k \in \mathbb{N}_0$, and poles at $s = n + 1$, with $n \in \mathbb{N}_0$. Likewise, it is possible to show that the Mellin transform of the Mittag-Leffler function $\mathrm{E}_{\alpha,\beta}$ is given by

$$\mathcal{M}\{\mathrm{E}_{\alpha,\beta}(-t); s\} = \frac{\Gamma(s)\Gamma(1-s)}{\Gamma(\beta - \alpha s)}, \qquad 0 < \mathrm{Re}\,(\alpha) < 1. \tag{1.124}$$

The asymptotic behaviour of $\mathrm{E}_{\alpha,\beta}(z)$ for $|z| \to \infty$ is

$$\mathrm{E}_{\alpha,\beta}(z) = -\sum_{k=1}^{N} \frac{1}{\Gamma(\beta - \alpha k)} \frac{1}{z^k} + O\left(\frac{1}{z^{N+1}}\right), \tag{1.125}$$

for $\mu \leq |\arg(z)| \leq \pi$, where $\pi\alpha/2 < \mu < \min[\pi, \pi\alpha]$ [2]. The Mittag-Leffler functions of one and two parameters are perhaps the most commonly used expressions for special functions in this book. For completeness, we briefly introduce another generalisation to the Mittag-Leffler function, which contains three-parameters, α, β, and γ, and has as particular cases the other two functions already defined.

Generalised Mittag-Leffler Functions

A generalisation of the Mittag-Lefller function involving three parameters $\alpha, \beta, \gamma \in \mathbb{C}$ may be given by

$$\mathrm{E}^{\gamma}_{\alpha,\beta}(z) = \frac{1}{\Gamma(\gamma)} \sum_{k=0}^{\infty} \frac{\Gamma(\gamma + k)}{\Gamma(k+1)} \frac{z^k}{\Gamma(\alpha k + \beta)}, \qquad z \in \mathbb{C}, \quad \mathrm{Re}\,(\alpha) > 0. \tag{1.126}$$

This function was introduced by Pabhakar [27] and its properties have been also studied by Kilbas et al. [28]. For $\gamma = 1$, it coincides with the two-parameter Mittag-Leffler function defined in (1.120):

$$\mathrm{E}^{1}_{\alpha,\beta}(z) = \frac{1}{\Gamma(1)} \sum_{k=0}^{\infty} \frac{\Gamma(1 + k)}{\Gamma(k+1)} \frac{z^k}{\Gamma(\alpha k + \beta)} = \mathrm{E}_{\alpha,\beta}(z). \tag{1.127}$$

For $\alpha = 1$, it is connected with the Kummer confluent hypergeometric function ${}_1F_1(\gamma; \beta; z)$, defined as:

$${}_1F_1(\gamma; \beta; z) = \sum_{k=0}^{\infty} \frac{\Gamma(\gamma + k)\Gamma(\beta)}{\Gamma(\gamma)\Gamma(\beta + k)} \frac{z^k}{\Gamma(k+1)} = \Gamma(\beta)\mathrm{E}^{\gamma}_{1,\beta}. \tag{1.128}$$

When $\alpha > 0$, it is possible to express $\mathrm{E}^{\gamma}_{\alpha,\beta}(z)$ in terms of a Mellin–Barnes contour integral as done previously for $\mathrm{E}_{\alpha,\beta}(z)$ and $\mathrm{E}_\alpha(z)$, as

$$\mathrm{E}^{\gamma}_{\alpha,\beta}(z) = \frac{1}{2\pi i \Gamma(\gamma)} \int_{c-i\infty}^{c+i\infty} \frac{\Gamma(s)\Gamma(\gamma - s)}{\Gamma(\beta - \alpha s)} (-z)^{-s}\, ds, \qquad \alpha > 0, \tag{1.129}$$

for $\mathrm{Re}\, z > 0$, with $|\arg(-z)| < \pi$, where c in the contour is such that $0 < c < \mathrm{Re}\,(\gamma)$ and it is assumed that the poles of $\Gamma(s)$ and $\Gamma(\gamma - s)$ are separated by the contour [29]. Its Mellin transform may be given by

$$\mathcal{M}\{\mathrm{E}^{\gamma}_{\alpha,\beta}(-z);s\} = \frac{1}{\Gamma(\gamma)}\frac{\Gamma(s)\Gamma(\gamma - s)}{\Gamma(\beta - \alpha s)}, \qquad 0 < \mathrm{Re}\,(s) < \mathrm{Re}\,(\gamma). \tag{1.130}$$

The function $\mathrm{E}^{\gamma}_{\alpha,\beta}(z)$ is a special case of the Wright's generalised hypergeometric function as well as the H-function of Fox that will be presented in the next sections.

Some useful results may be established for $\mathrm{E}^{\gamma}_{\alpha,\beta}(z)$ when $\alpha, \beta, \gamma, a \in \mathbb{C}$, $\mathrm{Re}\,(\alpha) > 0$, $\mathrm{Re}\,(\beta) > n$, $\mathrm{Re}\,(\gamma) > 0$, where $n \in \mathbb{N}$. These are [29]

$$\frac{d^n}{dz^n}\left[z^{\beta-1}\mathrm{E}^{\gamma}_{\alpha,\beta}(az^{\alpha})\right] = z^{\beta-n-1}\mathrm{E}^{\gamma}_{\alpha,\beta-n}(az^{\alpha})$$

and, in particular,

$$\frac{d^n}{dz^n}\left[z^{\beta-1}\mathrm{E}_{\alpha,\beta}(az^{\alpha})\right] = z^{\beta-n-1}\mathrm{E}_{\alpha,\beta-n}(az^{\alpha}).$$

Finally, we present a Laplace transform formula for $\mathrm{E}^{\gamma}_{\alpha,\beta}$:

$$\mathcal{L}\{t^{\beta-1}\mathrm{E}^{\gamma}_{\alpha,\beta}(at^{\alpha});s\} = s^{-\beta}(1 - as^{-\alpha})^{-\gamma}, \quad s \in \mathbb{C}, \quad \mathrm{Re}\,(s) > 0, \tag{1.131}$$

for $a \in \mathbb{C}$, $\mathrm{Re}\,(\alpha) > 0$, $\mathrm{Re}\,(\beta) > 0$, $\mathrm{Re}\,(\gamma) > 0$, and $|as^{-\alpha}| < 1$. For $\gamma = 1$, this formula reduces to

$$\mathcal{L}\{t^{\beta-1}\mathrm{E}_{\alpha,\beta}(at^{\alpha});s\} = s^{-\beta}(1 - as^{-\alpha})^{-1}.$$

We underline that the three-parameter Mittag-Leffler function may be viewed as a special case of the H-function of Fox [30] to be introduced later (see Section 1.2.5) as

$$\mathrm{E}^{\gamma}_{\alpha,\beta}(-z) = \frac{1}{\Gamma(\gamma)}\mathrm{H}^{1,1}_{1,2}\left[z\,\middle|\,\begin{matrix}(1-\gamma,1)\\(0,1),(1-\beta,\alpha)\end{matrix}\right], \quad \mathrm{Re}\,(\gamma) > 0. \tag{1.132}$$

1.2.4 Wright Functions

The Wright function (or the Fox–Wright function) $W(\alpha, \beta; z)$ is defined as

$$W(\alpha,\beta;z) = \sum_{k=0}^{\infty}\frac{z^k}{k!\,\Gamma(\alpha k + \beta)}, \qquad z, \alpha, \beta \in \mathbb{C}. \tag{1.133}$$

This function was named after the British mathematician Edward Maitland Wright (1906–2005), who introduced it [31] and investigated its asymptotic behaviour at infinity [32, 33]. The definition of this function was originally connected with the investigations of Wright about the asymptotic behaviour of partitions, but recently it has appeared in the investigations of partial differential equations of fractional order [34]. If $\alpha > -1$, the series in (1.133) is absolutely convergent for all $z \in \mathbb{C}$ and is also an entire function [2]. When $\alpha = 0$, and $\beta = 1$, we obtain

$$W(0, 1; z) = e^z.$$

When $\alpha = 1$ and $\beta = \nu + 1$, the Wright function may be expressed in terms of the Bessel functions J_ν and I_ν, respectively, as

$$W\left(1, \nu + 1; -\frac{z^2}{4}\right) = \left(\frac{2}{z}\right)^\nu \mathrm{J}_\nu(z) \quad \text{and} \quad W\left(1, \nu + 1; \frac{z^2}{4}\right) = \left(\frac{2}{z}\right)^\nu \mathrm{I}_\nu(z).$$

The time-fractional diffusion equation in the Caputo sense (see Section 2.3) may be studied by using two functions originally introduced by F. Mainardi and named $F_\nu(z)$ and $M_\nu(z)$, for $0 < \nu < 1$, which are interrelated through $F_\nu(z) = \nu z M_\nu(z)$ [35]. They are represented, respectively, as

$$F_\nu(z) = \sum_{n=1}^{\infty} \frac{(-z)^n}{n!\,\Gamma(\nu n)} = \frac{1}{\pi} \sum_{n=1}^{\infty} \frac{(-z)^{n-1}}{n!} \Gamma(\nu n + 1) \sin(\pi \nu n) \tag{1.134}$$

and

$$M_\nu(z) = \sum_{n=1}^{\infty} \frac{(-z)^n}{n!\,\Gamma[-\nu n + (1 - n)]} = \frac{1}{\pi} \sum_{n=1}^{\infty} \frac{(-z)^{n-1}}{(n-1)!} \Gamma(\nu n) \sin(\pi \nu n). \tag{1.135}$$

The M-Wright function, M_ν, is also referred to as the Mainardi function and plays an important role in studying stochastic processes. It can be also related to the Mittag-Leffler function of one-parameter by means of the Laplace integral transform as [36]:

$$\int_0^\infty e^{st} M_\alpha(t)\, dt = \sum_{n=0}^{\infty} \frac{(-s)^n}{\Gamma(\alpha n + 1)} = \mathrm{E}_\alpha(-s), \quad 0 < \alpha < 1. \tag{1.136}$$

The Wright function can also be represented in terms of the Mellin–Barnes contour integral

$$W(\alpha, \beta; z) = \frac{1}{2\pi i} \int_C \frac{\Gamma(s)}{\Gamma(\beta - \alpha s)} (-z)^{-s}\, ds, \tag{1.137}$$

where C is the Hankel contour, i.e., the path of integration that separates all the poles at $s = -k$ to the left (see Fig. 1.1). This representation permits us to obtain the Mellin transform of (1.133) as

$$\mathcal{M}\{W(\alpha, \beta; z)\} = \frac{\Gamma(s)}{\Gamma(\beta - \alpha s)}, \qquad \mathrm{Re}\,(\alpha) > 0. \tag{1.138}$$

The Laplace transform of (1.133) is given in terms of the Mittag-Leffler function (1.120), i.e.,

$$\mathcal{L}\{W(\alpha,\beta);s\} = \frac{1}{s}\mathrm{E}_{\alpha,\beta}\left(\frac{1}{s}\right), \qquad \alpha > -1, \quad \beta \in \mathbb{C}, \quad \mathrm{Re}\,(s) > 0. \tag{1.139}$$

The asymptotic behaviour of $W(\alpha,\beta;z)$ at infinity is given by

$$W(\alpha,\beta;z) = a_0(\alpha z)^{(1-\beta)/(1+\alpha)} \exp\left[\left(1+\frac{1}{\alpha}\right)(\alpha z)^{1/(1+\alpha)}\right] \times \left\{1 + O\left[\left(\frac{1}{z}\right)\right]^{1/(1+\alpha)}\right\}, \quad z \to \infty, \tag{1.140}$$

for $z \in \mathbb{C}$, $|\arg(z)| \le \pi - \varepsilon\,(0 < \varepsilon < \pi)$, and $a_0 = [2\pi(\alpha+1)]^{-1/2}$. For a real $\alpha > -1$, the Wright function is a particular case of the H-function of Fox, which will be discussed in the next section, as

$$W(\alpha,\beta;z) = \mathrm{H}^{1\,0}_{0\,2}\left[-z\,\middle|_{(0,1),(1-\beta,\alpha)}\right], \tag{1.141}$$

and can be also represented in terms of the generalised hypergeometric function [37]. When $\alpha = \mu$, $\beta = \nu + 1$, and $z \to -z$, the function $W(\alpha,\beta;z)$ becomes

$$\mathrm{J}^{\mu}_{\nu}(z) = W(\alpha,\nu+1;z) = \sum_{k=0}^{\infty} \frac{1}{\Gamma(\mu k + \nu + 1)} \frac{(-z)^k}{k!},$$

which is known as the Bessel–Wright function, or the Wright generalised Bessel function [2]. Finally, we may notice that for $\alpha = \beta = -1/2$, and $z \to -z$, we have

$$W\left(-\frac{1}{2},-\frac{1}{2};-z\right) = \frac{1}{\sqrt{\pi}}e^{-z^2/4},$$

which, together with previous results, suggest that the Wright function is a generalisation of the exponential and the Bessel functions.

1.2.5 The H-Function of Fox

The H-function of Fox – commonly referred to as the Fox function, H-function, generalised Mellin–Barnes, or generalised Meijer G-function – was introduced in 1961 by Charles Fox (1897–1977) [38]. It is of extreme importance in fractional calculus because it includes almost all the special functions of applied mathematics as particular cases [39]. It involves Mellin–Barnes integrals, which are generalisations of the Meijer G-function, and may be written as [30]:

$$\begin{aligned}\mathrm{H}^{m,n}_{p,q}(z) &= \mathrm{H}^{m,n}_{p,q}\left[z\,\middle|\,\begin{matrix}(a_1,A_1),(a_2,A_2),\ldots,(a_p,A_p)\\(b_1,B_1),(b_2,B_2),\ldots,(b_p,B_p)\end{matrix}\right]\\ &= \frac{1}{2\pi i}\int\limits_L ds\, h(s)z^{-s},\end{aligned} \tag{1.142}$$

in which $z \neq 0$, and

$$z^{-s} = \exp\left(-s\left[\ln(|z|) + i\arg(z)\right]\right).$$

The kernel in (1.142) is given by [30]:

$$h(s) = \frac{h_m(s)\,h_n(s)}{h_p(s)\,h_q(s)}, \tag{1.143}$$

where

$$h_m(s) = \prod_{j=1}^{m}\Gamma(b_j + B_j s), \qquad h_n(s) = \prod_{j=1}^{n}\Gamma(1 - a_j - A_j s), \tag{1.144a}$$

and

$$h_p(s) = \prod_{j=n+1}^{p}\Gamma(a_j + A_j s), \qquad h_q(s) = \prod_{j=m+1}^{q}\Gamma(1 - b_j - B_j s). \tag{1.144b}$$

In the preceding equations, $m, n, p, q \in \mathbb{N}$, with $1 \leq m \leq q$, $0 \leq n \leq p$, $(A_j, B_j) \in \mathbb{R}^+$ e $(a_j, b_j) \in \mathbb{C}$. Sometimes, the lower limits in the products are greater than the upper ones; in this case, we found empty products or nullary products, which are by convention equal to the multiplicative identity 1. This means that

$$n = 0 \Leftrightarrow h_n(s) = 1, \quad m = q \Leftrightarrow h_q(s) = 1, \quad \text{and} \quad n = p \Leftrightarrow h_p(s) = 1.$$

Due to the presence of the factor z^{-s} in the integrand of Eq. (1.142), the H-function of Fox is, in general, multivalued, but it can be made single-valued on the Riemann surface of $\log(z)$ by a suitable choice of the branch [40].

The integral in (1.142) has poles because the gamma function is not defined for negative integers. Thus, L is a contour separating these poles, given by

$$\zeta_{j\nu} = -\left(\frac{b_j + \nu}{B_j}\right), \quad j = 1, 2, 3, \ldots, m; \quad \nu = 0, 1, 2, \ldots, \tag{1.145}$$

for the gamma function $\Gamma(b_j + B_j s)$, and the poles

$$\omega_{\lambda k} = \left(\frac{1 - a_\lambda + k}{B_j}\right), \quad \lambda = 1, 2, 3, \ldots, n; \quad k = 0, 1, 2, \ldots, \tag{1.146}$$

for the gamma function $\Gamma(1 - a_\lambda - A_\lambda s)$. These poles do not coincide:

$$A_\lambda(b_j + \nu) \neq B_j(a_\lambda - k - 1).$$

The contour L indicated in (1.142) is an infinite contour which separates all the poles at $s = \zeta_{j\nu}$ in (1.145) to the left and all the poles at $s = \omega_{\lambda k}$ in (1.146) to the right of L. The contour L may have one of the following three forms [2, 30]:

(i) $L = L_{-\infty}$: it is a left loop starting at $-\infty + i\varphi_1$ and terminating at $-\infty + i\varphi_2$, with $-\infty < \varphi_1 < \varphi_2 < \infty$. This loop embodies all the poles of $\Gamma(b_j + B_j s)$ with $j = 1, 2, \ldots, m$, because it follows the positive direction, and does not embody the poles of $\Gamma(1 - a_\lambda - A_\lambda s)$, with $\lambda = 1, 2, \ldots, n$. The integral converges for any z if $\mu > 0$ and $z \neq 0$, or $\mu = 0$ and $0 < |z| < \beta$. It also converges if $\mu = 0$, $|z| = \beta$, and $\mathbb{R}(\delta) < -1$, in which

$$\beta = \left\{\prod_{j=1}^{p}(A_j)^{-A_j}\right\}\left\{\prod_{j=1}^{q}(B_j)^{B_j}\right\},$$

$$\mu = \sum_{j=1}^{q} B_j - \sum_{j=1}^{p} A_j$$

and

$$\delta = \sum_{j=1}^{q} b_j - \sum_{j=1}^{p} a_j + \frac{p-q}{2};$$

(ii) $L = L_{\infty}$: it is a left loop starting at $\infty + i\varphi_1$ and terminating at $\infty + i\varphi_2$, with $-\infty < \varphi_1 < \varphi_2 < \infty$. It embodies all the poles of $\Gamma(1 - a_\lambda - A_\lambda s)$, with $\lambda = 1, 2, \ldots, n$, because it follows the negative direction, and does not embody the poles of $\Gamma(b_j + B_j s)$, with $j = 1, 2, \ldots, m$. The integral converges for any z, if $\mu > 0$ and $z \neq 0$, or $\mu = 0$ e $|z| > \beta$;

(iii) $L = L_{i\xi\infty}$: it is a contour starting at the point $\xi - i\infty$ and terminating at the point $\xi + i\infty$, in which $\xi \in \mathbb{R}$, such that all the poles of $\Gamma(b_j + B_j s)$, with $j = 1, 2, \ldots, m$ are separated from the poles of $\Gamma(1 - a_\lambda - A_\lambda s)$ with $\lambda = 1, 2, \ldots, n$. The integral converges if

$$\alpha > 0, \quad |\arg(z)| < \frac{1}{2}\pi\alpha, \quad \text{and} \quad a \neq 0.$$

Depending on the choice of the contour, it will embody some poles of the kernel in (1.142), in such a way that the H-function of Fox may be written in terms of a series [30, 40–42].

When the poles of $\prod_{j=1}^{m}\Gamma(b_j - sB_j)$ are simple – notice that the gamma function is undetermined when $\nu \in \mathbb{N}$ – from Eq. (1.145), we have:

$$s_1 = \frac{b_h + \nu}{B_h}, \quad h = 1, 2, 3, \ldots, m; \quad \nu = 0, 1, 2, \ldots, \tag{1.147}$$

in which $B_h(b_j + \lambda) \neq B_j(b_h + \nu)$, for $j \neq h$, with $j, h = 1, 2, \ldots, m$ and $\lambda, \nu = 0, 1, 2, \ldots$. Using the residue theorem together with the values of the poles (1.147), the H-function of Fox, defined in (1.142), may be rewritten as

$$\mathrm{H}_{p,q}^{m,n}(z) = \sum_{h=1}^{m}\sum_{\nu=0}^{\infty} \frac{h_m(s_1)\, h_n(s_1)}{h_p(s_1)\, h_q(s_1)} \frac{(-1)^{\nu} z^{(b_h+\nu)/B_h}}{\nu!\, B_h}, \tag{1.148}$$

which is defined for all $z \neq 0$, with

$$h_m(s_1) = \prod_{j=1}^{m}\Gamma(b_j + B_j s_1) = \prod_{\substack{j=1 \\ j\neq h}}^{m}\Gamma(b_j - B_j(b_h + \nu)/B_h),$$

$$h_n(s_1) = \prod_{j=1}^{n}\Gamma(1 - a_j - A_j s_1) = \prod_{j=1}^{n}\Gamma(1 - a_j + A_j(b_h + \nu)/B_h),$$

$$h_p(s_1) = \prod_{j=n+1}^{p}\Gamma(a_j + A_j s_1) = \prod_{j=n+1}^{p}\Gamma(a_j - A_j(b_h + \nu)/B_h), \quad \text{and}$$

$$h_q(s_1) = \prod_{j=m+1}^{q}\Gamma(1 - b_j - B_j s_1) = \prod_{j=m+1}^{q}\Gamma(1 - b_j + B_j(b_h + \nu)/B_h).$$

In a similar manner, if the poles $\prod_{j=1}^{n}\Gamma(1 - b_j + sB_j)$ are simple, we have:

$$s_2 = \frac{a_h - 1 - \nu}{A_h}, \quad h = 1, 2, 3, \ldots, m, \quad \nu = 0, 1, 2, \ldots, \tag{1.149}$$

in which $A_h(1 - a_j + \nu) \neq A_j(1 - a_h + \lambda)$ for $j \neq h$, with $j, h = 1, 2, \ldots, n$ and $\lambda, \nu = 0, 1, 2, \ldots$. By invoking again the residue theorem together with the values of the poles (1.149), the H-function of Fox (1.142) becomes

$$\mathrm{H}_{p,q}^{m,n}(z) = \sum_{h=1}^{n}\sum_{\nu=0}^{\infty} \frac{h_m(s_2)\, h_n(s_2)}{h_p(s_2)\, h_q(s_2)} \frac{(-1)^{\nu}\, (1/z)^{(1-a_h+\nu)/A_h}}{\nu!\, A_h}, \tag{1.150}$$

which is defined for all $z \neq 0$, with

$$h_m(s_2) = \prod_{j=1}^{m} \Gamma(b_j + B_j s_2) = \prod_{j=1}^{m} \Gamma(b_j + B_j(1 - a_h + \nu)/A_h),$$

$$h_n(s_2) = \prod_{j=1}^{n} \Gamma(1 - a_j - A_j s_2) = \prod_{\substack{j=1 \\ j\neq h}}^{n} \Gamma(1 - a_j - A_j(1 - a_h + \nu)/A_h),$$

$$h_p(s_2) = \prod_{j=n+1}^{p} \Gamma(a_j + A_j s_2) = \prod_{j=n+1}^{p} \Gamma(a_j + A_j(1 - a_h + \nu)/A_h), \quad \text{and}$$

$$h_q(s_2) = \prod_{j=m+1}^{q} \Gamma(1 - b_j - B_j s_2) = \prod_{j=m+1}^{q} \Gamma(1 - b_j - B_j(1 - a_h + \nu)/A_h).$$

As pointed out before, the H-function of Fox contains as particular cases most of the special functions of applied mathematics, but it does not contain some of importance, like, for instance, the Riemann zeta function and the polylogarithm. In deriving certain Feynman integrals, a generalisation of the H-function which, besides containing the mentioned special functions, contains also the exact partition function of the Gaussian model from statistical mechanics, was recently investigated [43, 44]. For this reason, some special cases of the H-function of Fox may be now considered here. First, we show that it reduces to the generalised Mittag-Leffler function (1.126). To do this, let us rewrite the H-function of Fox in the form:

$$\mathrm{H}_{1,2}^{1,1}(-z) = \mathrm{H}_{1,2}^{1,1}\left[-z \middle| \begin{matrix} (1-\gamma, 1) \\ (0,1), (1-\beta, \alpha) \end{matrix} \right], \tag{1.151}$$

which permits us to identify

$$m = n = 1, \quad a_1 = 1 - \gamma, \quad b_1 = 0, \quad b_2 = 1 - \beta,$$
$$p = 1, \quad q + 2, \quad A_1 = 1, \quad B_1 = 1, \quad B_2 = \alpha.$$

By using (1.148), we have

$$\mathrm{H}_{1,2}^{1,1}(-z) = \sum_{h=1}^{1} \sum_{\nu=0}^{\infty} \frac{h_{m=1}(s_1)\, h_{n=1}(s_1)}{h_{p=1}(s_1)\, h_{q=2}(s_1)} \frac{(-1)^{\nu}(-z)^{(b_h+\nu)/B_h}}{\nu!\, B_h}.$$

Since the lower index of the product is greater than the upper one, it is made equal to the identity 1; thus,

$$h_{m=1}(s_1) = \prod_{j=1}^{1} \Gamma(b_j + B_j s_1) = \prod_{\substack{j=1 \\ j\neq h}}^{1} \Gamma(b_j - B_j(b_h + \nu)/B_h) = 1$$

and

$$h_{p=1}(s_1) = \prod_{j=n+1}^{1} \Gamma(a_j + A_j s_1) = \prod_{j=1+1}^{1} \Gamma(a_j - A_j(b_h + \nu)/B_h) = 1.$$

Thus, we have

$$\mathrm{H}_{1,2}^{1,1}(-z) = \sum_{\nu=0}^{\infty} \frac{1 \times \prod_{j=1}^{1} \Gamma(1 - a_j + A_j(b_1 + \nu)/B_1)}{\prod_{j=2}^{2} \Gamma(1 - b_j + B_j(b_1 + \nu)/B_1) \times 1} \frac{(-1)^{\nu}(-z)^{(b_1+\nu)/B_1}}{\nu!\, B_1}$$

$$= \sum_{\nu=0}^{\infty} \frac{\Gamma(1 - a_1 + A_1(b_1 + \nu)/B_1)}{\Gamma(1 - b_2 + B_2(b_1 + \nu)/B_1)} \frac{(-1)^{\nu}(-z)^{(b_1+\nu)/B_1}}{\nu!\, B_1}.$$

By substituting the coefficients, we obtain:

$$\mathrm{H}_{1,2}^{1,1}(-z) = \sum_{\nu=0}^{\infty} \frac{\Gamma(1 - (1-\gamma) + 1(0+\nu)/1)}{\Gamma(1 - (1-\beta) + \alpha(0+\nu)/1)} \frac{(-1)^{\nu}(-z)^{(0+\nu)/1}}{\nu!\, 1}$$

$$= \sum_{\nu=0}^{\infty} \frac{\Gamma(\gamma + \nu)}{\Gamma(\alpha\nu + \beta)} \frac{(-1)^{2\nu} z^{\nu}}{\nu!}.$$

Since $(-1)^{2\nu} = 1$, $\nu! = \Gamma(\nu + 1)$, by multiplying and dividing the preceding expression by $\Gamma(\gamma)$, we obtain

$$\mathrm{H}_{1,2}^{1,1}(-z) = \mathrm{H}_{1,2}^{1,1}\left[-z \left| \begin{matrix} (1-\gamma, 1) \\ (0,1), (1-\beta, \alpha) \end{matrix} \right. \right]$$

$$= \Gamma(\gamma) \frac{1}{\Gamma(\gamma)} \sum_{\nu=0}^{\infty} \frac{\Gamma(\gamma + \nu)}{\Gamma(\nu + 1)} \frac{z^{\nu}}{\Gamma(\alpha\nu + \beta)}. \tag{1.152}$$

By comparing it with (1.126), we finally have

$$\mathrm{H}_{1,2}^{1,1}(-z) = \mathrm{H}_{1,2}^{1,1}\left[-z \left| \begin{matrix} (1-\gamma, 1) \\ (0,1), (1-\beta, \alpha) \end{matrix} \right. \right] = \Gamma(\gamma) \mathrm{E}_{\alpha,\beta}^{\gamma}(z). \tag{1.153}$$

As pointed out before, a large number of special functions are particular cases of the H-function. In the first place, the G-function; indeed, for $A_i = 1$ and $B_i = 1$, we obtain the Meijer's functions $G_{p,q}^{m,n}(z)$ [2]:

$$\mathrm{H}_{p,q}^{m,n}(z) = \mathrm{H}_{p,q}^{m,n}\left[z \left| \begin{matrix} (a_1, 1), (a_2, 1), \cdots, (a_p, 1) \\ (b_1, 1), (b_2, 1), \cdots, (b_q, 1) \end{matrix} \right. \right] = \mathrm{G}_{p,q}^{m,n}\left[z \left| \begin{matrix} a_1, \cdots, a_p \\ b_1, \cdots, b_q \end{matrix} \right. \right]$$

$$= \frac{1}{2\pi i} \int_L ds\, h_{\mathrm{M}}(s) z^{-s}, \tag{1.154}$$

where

$$h_{\mathrm{M}}(s) = \frac{\prod_{j=1}^{m} \Gamma(b_j + s) \prod_{j=1}^{n} \Gamma(1 - a_j - s)}{\prod_{j=m+1}^{p} \Gamma(a_j + s) \prod_{j=m+1}^{q} \Gamma(1 - b_j - s)}. \qquad (1.155)$$

It is also possible to obtain special cases of it as, for instance, Bessel functions, Legendre polynomials, Whitakker and Struve functions, and the ordinary generalised hypergeometric functions, among others [39]. For instance, the generalised hypergeometric function may be written as [2]:

$$\begin{aligned} {}_pF_q(a_1, \ldots, a_p; b_1, \ldots, b_q; z) &= \\ &= \frac{\prod_{j=1}^{q} \Gamma(b_j)}{\prod_{j=1}^{p} \Gamma(a_j)} \mathrm{H}^{1,p}_{p,q+1}\left[-z \middle| \begin{matrix} (1-a_1, 1), \ldots, (1-a_p, 1) \\ (0,1), (1-b_1, 1), \ldots, (1-b_q, 1) \end{matrix} \right] \\ &= \frac{\prod_{j=1}^{q} \Gamma(b_j)}{\prod_{j=1}^{p} \Gamma(a_j)} \mathrm{G}^{1,p}_{p,q+1}\left[-z \middle| \begin{matrix} 1-a_1, \ldots, 1-a_p \\ 0, 1-b_1, \ldots, 1-b_q \end{matrix} \right]; \end{aligned}$$

the Bessel function of the first kind, Eq. (1.91), is obtained as

$$\mathrm{J}_\nu(z) = 2^{-\nu}\sqrt{\pi}\,\mathrm{H}^{1,0}_{1,2}\left[z \middle| \begin{matrix} ((1+\nu)/2, 1/2) \\ (\nu, 1), (-\nu/2, 1/2) \end{matrix} \right], \qquad (1.156)$$

whereas the Macdonald function, Eq. (1.112), is

$$\mathrm{K}_\nu(z) = 2^{-\nu-1}\sqrt{\pi}\,\mathrm{H}^{2,0}_{1,2}\left[z \middle| \begin{matrix} ((1+\nu)/2, 1/2) \\ (\nu, 1), (-\nu/2, 1/2) \end{matrix} \right]. \qquad (1.157)$$

It is thus evident that the H-function of Fox embodies a class of functions that are also greater than the family of Mittag-Leffler functions. As shown before, from (1.126), we obtain the two-parameter Mittag-Leffler function if $\gamma = 1$:

$$\mathrm{H}^{1,1}_{1,2}(-z) = \mathrm{H}^{1,1}_{1,2}\left[-z \middle| \begin{matrix} (0,1) \\ (0,1), (1-\beta, \alpha) \end{matrix} \right] = \mathrm{E}^1_{\alpha,\beta}(z) = \mathrm{E}_{\alpha,\beta}(z). \qquad (1.158)$$

In addition, this function may also be reduced to the Mittag-Leffler function if $\beta = 1$, i.e.,

$$\mathrm{H}^{1,1}_{1,2}(-z) = \mathrm{H}^{1,1}_{1,2}\left[-z \middle| \begin{matrix} (0,1) \\ (0,1), (0, \alpha) \end{matrix} \right] = \mathrm{E}^1_{\alpha,1}(z) = \mathrm{E}_\alpha(z). \qquad (1.159)$$

It is also possible to obtain the exponential function as a limiting case for $\alpha = 1$ in this latter case:

$$\mathrm{H}_{1,2}^{1,1}(-z) = \mathrm{H}_{1,2}^{1,1}\left[-z\left|\begin{matrix}(0,1)\\(0,1),(0,1)\end{matrix}\right.\right] = \mathrm{E}_1(z) = e^z. \tag{1.160}$$

We notice that it is also possible to write the exponential function in terms of the H-function. Without entering into details, we can easily prove that:

$$\mathrm{H}_{0,1}^{1,0}(-z) = \mathrm{H}_{0,1}^{1,0}\left[-z\left|\begin{matrix}(0,0)\\(0,1)\end{matrix}\right.\right] = e^z. \tag{1.161}$$

In general, the coefficients $(a_1, A_1) = (0,0)$ may be omitted. Thus, we can write

$$\mathrm{H}_{1,2}^{1,1}\left[-z\left|\begin{matrix}(0,1)\\(0,1),(0,1)\end{matrix}\right.\right] = \mathrm{H}_{0,1}^{1,0}\left[-z\left|\begin{matrix}(0,0)\\(0,1)\end{matrix}\right.\right] = e^z. \tag{1.162}$$

Some reduction formulas hold for the H-function. In particular,

$$\begin{aligned}&\mathrm{H}_{p,q}^{m,n}\left[z\left|\begin{matrix}(a_1,A_1),\ldots,(a_p,A_p)\\(b_1,B_1),\ldots,(b_{q-1},B_{q-1}),(a_1,A_1)\end{matrix}\right.\right]\\ &= \mathrm{H}_{p-1,q-1}^{m,n-1}\left[z\left|\begin{matrix}(a_2,A_2),\ldots,(a_p,A_p)\\(b_1,B_1),\ldots,(b_{q-1},B_{q-1})\end{matrix}\right.\right],\end{aligned} \tag{1.163a}$$

if $n \geq 1$ and $q > m$, and

$$\begin{aligned}&\mathrm{H}_{p,q}^{m,n}\left[z\left|\begin{matrix}(a_1,A_1),\ldots,(a_{p-1},A_{p-1}),(b_1,B_1)\\(b_1,B_1),\ldots,(b_q,B_q)\end{matrix}\right.\right]\\ &= \mathrm{H}_{p-1,q-1}^{m-1,n}\left[z\left|\begin{matrix}(a_1,A_1),\ldots,(a_{p-1},A_{p-1})\\(b_2,B_2),\ldots,(b_q,B_q)\end{matrix}\right.\right],\end{aligned} \tag{1.163b}$$

if $m \geq 1$ and $p > n$. Other useful properties are

$$\mathrm{H}_{p,q}^{m,n}\left[z\left|\begin{matrix}(a_p,A_p)\\(b_q,B_q)\end{matrix}\right.\right] = \mathrm{H}_{q,p}^{n,m}\left[\frac{1}{z}\left|\begin{matrix}(1-b_q,B_q)\\(1-a_p,A_p)\end{matrix}\right.\right], \tag{1.164}$$

$$\mathrm{H}_{p,q}^{m,n}\left[z\left|\begin{matrix}(a_p,A_p)\\(b_q,B_q)\end{matrix}\right.\right] = k\,\mathrm{H}_{p,q}^{m,n}\left[z^k\left|\begin{matrix}(a_p,kA_p)\\(b_q,kB_q)\end{matrix}\right.\right], \tag{1.165}$$

and

$$z^k\mathrm{H}_{p,q}^{m,n}\left[z\left|\begin{matrix}(a_p,A_p)\\(b_q,B_q)\end{matrix}\right.\right] = \mathrm{H}_{p,q}^{m,n}\left[z\left|\begin{matrix}(a_p+kA_p,A_p)\\(b_q+kB_p,B_q)\end{matrix}\right.\right], \tag{1.166}$$

with $k \in \mathbb{C}$.

The asymptotic behaviour of the H-function of Fox, when $n = 0$, is given by [45]

$$\mathrm{H}_{p,q}^{m,0}\left[z\left|\begin{matrix}(a_j,A_j)\\(b_j,B_j)\end{matrix}\right.\right] \sim Fz^{\gamma/\mu}\exp\left[-\mu(z/\beta')^{1/\mu}\right] \tag{1.167}$$

for large $|z|$, uniformly on every closed sector contained in $|\arg(z)| < \delta\pi/2$, where $\gamma = \delta + 1/2$

$$\beta' = \prod_{j=1}^{p} A_j^{A_j} \prod_{j=1}^{q} B_j^{-B_j}, \tag{1.168}$$

$$F = (2\pi)^{(q-p-1)/2} \beta'^{\gamma/\mu} \mu^{-1/2} \prod_{j=1}^{p} A_j^{1/2-a_j} \prod_{j=1}^{P} B_j^{b_j-1/2}. \tag{1.169}$$

It is worth mentioning that a similar behaviour may be found in several other contexts, in particular, in those situations related to diffusion on fractals [41].

The asymptotic behaviour for $n \neq 1$ is different from the previous one and it may be characterised by power laws, which can be connected with Lévy distributions. For example, let us consider the following H-function of Fox:

$$\mathrm{H}_{2,2}^{1,1}\left[z \,\middle|\, \begin{matrix} \left(1,\frac{1}{\mu}\right),\left(1,\frac{1}{2}\right) \\ (1,1),\left(1,\frac{1}{2}\right) \end{matrix}\right], \tag{1.170}$$

which can be used to represent a symmetric Lévy distribution [41], i.e.,

$$L_\mu(z) = \frac{1}{\mu|z|} \mathrm{H}_{2,2}^{1,1}\left[|z| \,\middle|\, \begin{matrix} \left(1,\frac{1}{\mu}\right),\left(1,\frac{1}{2}\right) \\ (1,1),\left(1,\frac{1}{2}\right) \end{matrix}\right]. \tag{1.171}$$

The asymptotic behaviour of this H-function of Fox is given by

$$\mathrm{H}_{2,2}^{1,1}\left[z \,\middle|\, \begin{matrix} \left(1,\frac{1}{\mu}\right),\left(1,\frac{1}{2}\right) \\ (1,1),\left(1,\frac{1}{2}\right) \end{matrix}\right] \sim \sum_{n=1}^{\infty} \frac{\Gamma(1-\mu n)}{\Gamma(1+n)\,\Gamma\left(-\frac{\mu}{2}n\right)\Gamma\left(1+\frac{\mu}{2}n\right)} \left(-\frac{1}{z^\mu}\right)^n, \tag{1.172}$$

for large $|z|$, which implies that $L_\mu(z) \sim 1/|z|^{1+\mu}$.

The fundamental solutions of the Cauchy problem for the space-time fractional diffusion equation may be expressed in terms of appropriate H-functions, based on their Mellin-Barnes integral representation [40]. In Chapter 4, we shall investigate in detail the solutions of the fractional diffusion equations in terms of these functions, and several limiting situations will also be discussed.

1.3 Integral Transforms of Special Functions

Some useful formulas of the Mittag-Leffler and H-function of Fox have been established involving integral transforms [30, 46]. For the two-parameter Mittag-Leffler function (1.120), by considering its k-derivative, given by

$$\mathrm{E}_{\alpha,\beta}^{(k)}(z) = \frac{d^k}{dz^k}\mathrm{E}_{\alpha,\beta}(z),$$

we have

$$\mathfrak{L}\left\{z^{\alpha k+\beta-1}\mathrm{E}_{\alpha,\beta}^{(\mathrm{k})}(\pm\xi z^{\alpha});s\right\} = \int_0^\infty dz\, z^{\alpha k+\beta-1}\mathrm{E}_{\alpha,\beta}^{(\mathrm{k})}(\pm\xi z^{\alpha})e^{-sz} = \frac{k!\, s^{\alpha-\beta}}{(s^\alpha \mp \xi)^{k+1}}. \tag{1.173}$$

In this way as well, we obtain

$$\mathfrak{L}^{-1}\left\{\frac{k!\, s^{\alpha-\beta}}{(s^\alpha \mp \xi)^{k+1}};z\right\} = z^{\alpha k+\beta-1}\mathrm{E}_{\alpha,\beta}^{(\mathrm{k})}(\pm\xi z^{\alpha}), \qquad \mathrm{Re}\,(s) > |\xi|^{1/\alpha}. \tag{1.174}$$

If $k = 0$, i.e., if there is not a derivative operation, it is possible to show that [46]

$$\mathfrak{L}\left\{z^{(\beta-1)}\mathrm{E}_{\alpha,\beta}(\pm\xi z^{\alpha});s\right\} = \int_0^\infty dz\, z^{(\beta-1)}\mathrm{E}_{\alpha,\beta}(\pm\xi z^{\alpha})e^{-sz} = \frac{s^{\alpha-\beta}}{s^\alpha \mp \xi}, \tag{1.175}$$

with $s, \alpha, \beta \in \mathbb{C}$, $\mathrm{Re}\,(\alpha) > 0$, and $|\xi/s^\alpha| < 1$. Thus, we have

$$\mathfrak{L}^{-1}\left\{\frac{s^{\alpha-\beta}}{s^\alpha \mp \xi};z\right\} = z^{(\beta-1)}\mathrm{E}_{\alpha,\beta}(\pm\xi z^{\alpha}). \tag{1.176}$$

If $s = 1$ in (1.175), we may write

$$\int_0^\infty dz\, z^{(\beta-1)}\mathrm{E}_{\alpha,\beta}(\pm\xi z^{\beta})e^{-z} = \frac{1}{1 \mp \xi}, \qquad |\xi| < 1. \tag{1.177}$$

Finally, let us consider some useful properties of the integral transforms of the H-function of Fox. The first one is the Laplace transform, given by [30]

$$\mathfrak{L}\left\{z^{\rho-1}\mathrm{H}_{p+1,q}^{m,n}\left[\xi z^{-\sigma}\left|\begin{matrix}(a_p,A_p),(\rho,\sigma)\\(b_q,B_q)\end{matrix}\right.\right];s\right\} = s^{-\rho}\mathrm{H}_{p,q}^{m,n}\left[\xi s^{\sigma}\left|\begin{matrix}(a_p,A_p)\\(b_q,B_q)\end{matrix}\right.\right], \tag{1.178}$$

which may be rewritten using the properties (1.163a) to (1.166) of the H-function of Fox. The inverse is given by

$$\mathfrak{L}^{-1}\left\{s^{-\rho}\mathrm{H}_{p,q}^{m,n}\left[\xi s^{\sigma}\left|\begin{matrix}(a_p,A_p)\\(b_q,B_q)\end{matrix}\right.\right];z\right\} = z^{\rho-1}\mathrm{H}_{p+1,q}^{m,n}\left[\xi z^{-\sigma}\left|\begin{matrix}(a_p,A_p),(\rho,\sigma)\\(b_q,B_q)\end{matrix}\right.\right]. \tag{1.179}$$

Another useful property is connected with the cosine Fourier transform, defined in Eq. (1.9), written as [41]

$$\int_0^\infty dz \cos(kz)\, \mathrm{H}^{m,n}_{p,q}\left[z \left| \begin{matrix}(a_p, A_p)\\(b_q, B_q)\end{matrix}\right.\right] = \frac{\pi}{k}\mathrm{H}^{n+1,m}_{q+1,p+2}\left[k \left| \begin{matrix}(1-b_q, B_q), \left(1, \frac{1}{2}\right)\\(1,1), (1-a_p, A_p), \left(1, \frac{1}{2}\right)\end{matrix}\right.\right] \tag{1.180}$$

or

$$\mathfrak{F}_c\left\{\mathrm{H}^{m,n}_{p,q}\left[z \left| \begin{matrix}(a_p, A_p)\\(b_q, B_q)\end{matrix}\right.\right]; k\right\} = \frac{\sqrt{\pi}}{k}\mathrm{H}^{n+1,m}_{q+1,p+2}\left[k \left| \begin{matrix}(1-b_q, B_q), \left(1, \frac{1}{2}\right)\\(1,1), (1-a_p, A_p), \left(1, \frac{1}{2}\right)\end{matrix}\right.\right], \tag{1.181}$$

which may also be rewritten using the properties (1.163a) to (1.166) of the H-function of Fox, discussed in Section 1.2.5.

The Mellin-cosine transform of the H-function of Fox is defined as [30]

$$\int_0^\infty k^{\rho-1}\cos(kx)\mathrm{H}^{m,n}_{p,q}\left[ak^\delta \left| \begin{matrix}(a_p, A_p)\\(b_q, B_q)\end{matrix}\right.\right]\mathrm{d}k =$$

$$= \frac{\pi}{x^\rho}\mathrm{H}^{n+1,m}_{q+1,p+2}\left[\frac{x^\delta}{a} \left| \begin{matrix}(1-b_q, B_q), \left(\frac{1+\rho}{2}, \frac{\delta}{2}\right)\\(\rho, \delta), (1-a_p, A_p), \left(\frac{1+\rho}{2}, \frac{\delta}{2}\right)\end{matrix}\right.\right],$$

where $x^\delta > 0$, $|\arg(a)| < \pi\theta/2$, with $\theta > 0$,

$$\mathrm{Re}\left[\rho + \delta \min_{1\le j\le m}\left(\frac{b_j}{B_j}\right)\right] > 1,$$

and

$$\mathrm{Re}\left[\rho + \delta \max_{1\le j\le n}\left(\frac{a_j - 1}{A_j}\right)\right] < \frac{3}{2},$$

in which

$$\theta = \sum_{j=1}^{n} A_j - \sum_{j=n+1}^{p} A_j + \sum_{j=1}^{m} B_j - \sum_{j=m+1}^{q} B_j.$$

The Mellin transform of the H-function of Fox is [30]

$$\int_0^\infty x^{\xi-1}\mathrm{H}^{m,n}_{p,q}\left[ax \left| \begin{matrix}(a_p, A_p)\\(b_q, B_q)\end{matrix}\right.\right]\mathrm{d}x = a^{-\xi}h(-\xi), \tag{1.182}$$

where $h(-\xi)$ is the kernel defined in Eq. (1.143).

This completes this first chapter, which is intended solely to provide general guidance on mathematical tools of interest for the analysis and solutions to the problems treated in this book. More detailed information may be found in the References at the end of the book.

2

A Survey of Fractional Calculus

The starting point of fractional calculus could be represented by the search for a meaning to the expression

$$\frac{d^{1/2}f(x)}{dx^{1/2}} \tag{2.1}$$

or to any expression of the kind $d^{\nu}f/dx^{\nu}$, where ν is any number (positive or negative, real or complex). The proposition of this question and the several tentative answers to it form the historical pathway to the origins of fractional calculus. This chapter starts with a brief historical account of the evolution of the concepts of differentiation and integration of arbitrary order, ranging from their origins until the consolidation of the field at the beginning of the twentieth century. The historical survey is mainly concerned with the definitions connected with the physical applications treated in this book.

2.1 The Origins of Fractional Calculus

There is a famous excerpt of the correspondence between Gottfried Wilhelm Leibniz (1646–1716) and the French mathematician Guillaume François Antoine, marquis de L'Hôpital (1661–1704), in a letter dated September 30, 1695 – now recognised as the exact birthday of fractional calculus – in which the former answers the question of the latter asking for the meaning of a derivative of order $\nu = 1/2$, as pointed out before [47]:

You can see here, sir, that one can express a term like $d^{1/2}\overline{x}y$ or $d^{1:2}\overline{x}y$ by an infinite series, even though it seems to be far from the geometry, which usually only considers the differences of positive integer exponents or the negatives with respect to sums, but not yet those, whose exponents are fractional. It is true that it is still to show that it is this series for $d^{1:2}x$; but not only this can be explained in a way. Because the ordinates x are expressed in a geometric series, such that by choosing a constant $d\beta$ it follows that $dx = xd\beta : a$, or (taking a as unit) $dx = xd\beta$, meaning ddx would be $x \cdot \overline{d\beta}^2$, and d^3x would be $= x \cdot \overline{d\beta}^3$ etc. and $d^e x = x \cdot \overline{d\beta}^e$. And thus by this artifice the differential exponent has been changed by the exponents and by replacing $d\beta$ with $\overline{dx} : x$, yielding $d^e x = \overline{\overline{dx} : x}^e \cdot x$. Thus it follows that $d^{1/2}x$ will be equal to $x\sqrt[2]{dx : x}$. It seems like one day very useful consequences will be drawn from this paradox, since there are little paradoxes without usefulness.

In the recent literature dedicated to the fractional calculus, the emphasis of the authors lies on the last phrase. Indeed, the first useful consequence drawn from the paradox was the birth of fractional calculus, which is the name (actually a misnomer) for the theory of integrations and derivatives of arbitrary order. It unifies and generalises the notions of integer order differentiation and n-fold integration [46].

Leonhard Euler (1707–1783), in a paper in 1738, mentioned interpolating between orders of a derivative to face the problem of a fractional order [48]:

To round off this discussion, let me add something which certainly is more curious than useful. It is known that $d^n x$ denotes the differential of x of order n, and if p denotes any function of x and dx is taken to be constant, then $d^n p$ is homogeneous with dx^n; but whenever n is a positive integer number, the ratio of $d^n p$ to dx^n can be expressed algebraically; for example, if $n = 2$ and $p = x^3$, then $d^2(x^3)$ is to dx^2 as $6x$ to 1. We now ask, if n is a fractional number, what the value of that ratio should be. It is easy to understand the difficulty in this case; for if n is the positive integer number, d^n can be found by continued differentiation; such an approach is not available if n is a fractional number. But it will nevertheless be possible to disentangle the matter by using interpolation in progressions, which I have discussed in this essay.

In a few words, even if Euler did not consider this kind of calculation as useful, he observed that the result of the evaluation of

$$\frac{d^n f}{dx^n}, \quad \text{for} \quad f = x^p, \tag{2.2}$$

has a meaning even for noninteger n.

In the work of Joseph-Louis Lagrange (1736–1813), in 1772, dedicated to the law of exponents for differential operators of integer order, he demonstrates that [49]

$$\frac{d^\mu}{dx^\mu}\frac{d^\nu}{dx^\nu}u = \frac{d^{\mu+\nu}}{dx^\mu dx^\nu}u. \tag{2.3}$$

This result contributed in an indirect way to the development of fractional calculus, because, under appropriate conditions, it may be applied to arbitrary values of μ and ν.

In 1812, Pierre Simon Laplace (1749–1827) defined a fractional derivative in terms of a definite integral [50]:

$$y_x = \int T(t)t^{-x}dt,$$

where T is a function of t, and the integral is evaluated on a given interval. Defining a variation α of x as

$$\Delta y_x = \int T.dt.t^{-x}\left(\frac{1}{t^\alpha} - 1\right),$$

and, in general,

$$\Delta^i \cdot y_x = \int T.dt.t^{-x}\left(\frac{1}{t^\alpha} - 1\right)^i;$$

by making i negative, the characteristic Δ in the integral changes the integral sign Σ. If we suppose α infinitely small and equal to dx, one has

$$\frac{1}{t} = 1 + dx \log\frac{1}{t};$$

by observing that $\Delta^i y_x$ changes in $d^i y_x$, one has

$$\frac{d^i y_x}{dx^i} = \int Tdtt^{-x}\left(\log\frac{1}{t}\right)^i.$$

In the same manner, by adopting the notation of n. 2,

$$\nabla^i y_x = \int Tdtt^{-x}\left(a + \frac{b}{t} + \cdots + \frac{q}{t^n}\right)^i.$$

Thus, the same analysis which yields the generating functions from the successive derivations of the variables, yields the functions under the sign $\int$, the definite integrals that express theses derivatives. The characteristics ∇^i express, strictly speaking, only a number i of consecutive operations; the consideration of the generating functions reduces these operations to the elevation of a polynomial to its different powers; and the consideration of the definite integrals yields directly the expression $\nabla^i \cdot y_x$, even in the case one supposes i as a fractional number.

Another mention of the fractional exponent in a derivative is made by Laplace later on in the book, where he considers a function y, of s, that can be expressed by means of a definite integral, in the form $\int x^s \varphi dx$; the differences infinitely small and finite of arbitrary order n will be given by [50]:

$$\frac{d^n y_s}{ds^n} = \int x^s \varphi dx(\log x)^n,$$

$$\Delta^n \cdot y_s = \int x^s \cdot \varphi dx \cdot (x-1)^n.$$

If, instead of representing the function of s by means of the integral $\int x^s \varphi dx$ one uses the integral $\int c^{-sx} \cdot \varphi\, dx$, one obtains

$$\frac{d^n y_s}{ds^n} = (-1)^n \int x^n \cdot \varphi dx \cdot c^{-sx}$$

$$\Delta^n \cdot y_s = \int \varphi dx \cdot c^{-sx} \cdot (c^{-x} - 1)^n.$$

To obtain negative integrals, finite or infinitesimally small, it is sufficient to choose n negative in the preceding formulae. One can observe that these formulae are valid in general, whatever the value of n, even if it is fractional.

In 1819, the fractional calculus is mentioned in the book of Sylvestre François Lacroix (1765–1843), where an explicit formula for a fractional derivative is given [51]:

By means of definite integrals, Euler arrived at a remarkable interpolation, the one of the differential functions. As in between the integer powers, by extracting the roots, one can use fractional powers to understand the terms in between the series

$$V,\ dV,\ d^2V,\ d^3V,\dots,d^nV,$$

the differentials of a given function, and to designate these terms by means of a fractional index that marks the position occupied in the series. It is not possible to interpret these quantities by successive differentiation nor to explain the fractional power by repeating multiplication; but the formulae $d^{1/2}V$ and $V^{1/2}$ will be expression formed by analogy, one in the series of differentials, another one in the series of powers.

Let, for example, be $V = v^m$; when n is an integer, one has for any arbitrary m,

$$d^n(v^m) = m(m-1)\dots(m-n+1)v^{m-n}dv^n = \frac{[m]^m}{[m-n]^{m-n}}v^{m-n}dv^n;$$

setting for $[m]^m$ and $[m-n]^{m-n}$ the expression of n. 1160, one finds[1]

$$d^n(v^m) = v^{m-n}dv^n\frac{\int dx\left(l\frac{1}{x}\right)^m}{\int dx\left(l\frac{1}{x}\right)^{m-n}}.$$

This result is susceptible of an immediate verification, by confirming that it gives the known result for the case in which n is a positive integer.

If one puts $m = 1$ and $n = 1/2$, one gets

$$d^{1/2}v = \sqrt{vdv}\frac{\int dx(l1/x)}{\int dx(l1/x)^{1/2}} = \frac{\sqrt{vdv}}{\frac{1}{2}\sqrt{\pi}},$$

and observing that between the limits 0 and 1,

$$\int dx(l1/x) = 1 \qquad \int dx(l1/x)^{1/2} = \left[\frac{1}{2}\right]^{1/2} = \frac{1}{2}\sqrt{\pi},$$

where π is the half-circumference of the circle with radius 1 (1160).

Thus, one obtains the primitive equation for the curve that corresponds to the differential equation

$$yd^{1/2}v = v\sqrt{dy},$$

[1] In today's notation $l(1/x) = \log(1/x)$ and $[m]^m = \int_0^\infty e^{-t}t^m dt = m!$.

over which dv is supposed constant. By means of the preceding value of $d^{1/2}v$, one transforms the preceding equation as

$$\frac{y\sqrt{vdv}}{\frac{1}{2}\sqrt{\pi}} = v\sqrt{dy};$$

and by squaring both sides of its members, one obtains

$$\frac{y^2 dv}{\frac{1}{4}\pi} = vdy,$$

from which one concludes

$$\frac{1}{\frac{1}{4}\pi} l\,v = C - \frac{1}{y}, \quad \text{or} \quad yl\,v = \frac{1}{4}C\pi y - \frac{1}{4}\pi.$$

In today's notation, Lacroix expressed the derivative of order n ($n < m$) in terms of Euler's gamma function (see Section 1.2.1), written as

$$\Gamma(a) = \int_0^\infty t^{a-1}e^{-t}\,dt, \quad \mathrm{Re}\,(a) > 0, \tag{2.4}$$

for $y(x) = x^m$:

$$\frac{d^n}{dx^n}y(x) = \frac{m!}{(m-n)!}x^{m-n} = \frac{\Gamma(m+1)}{\Gamma(m-n+1)}x^{m-n}, \quad m \geq n.$$

For the special case $m = 1$ and $n = 1/2$, Lacroix obtained

$$\frac{d^{1/2}}{dx^{1/2}}x = \frac{\Gamma(2)}{\Gamma(3/2)}x^{1/2} = \frac{2\sqrt{x}}{\sqrt{\pi}}.$$

Consider also another simple example, for which $n = 1/2$ and $m = 0$; the fractional derivative of a constant $y = x^0 = 1$ is

$$\frac{d^{1/2}}{dx^{1/2}}x^0 = \frac{\Gamma(1)}{\Gamma(1/2)}x^{-1/2} = \frac{1}{\sqrt{\pi x}}, \tag{2.5}$$

is not zero as in the usual calculus! The Lacroix results are a particular case of the Riemann–Liouville fractional derivative, which will be defined in Eq. (2.8):

$${}_0\mathrm{D}_x^\alpha x^a = \frac{\Gamma(a+1)}{\Gamma(a-\alpha+1)}x^{a-\alpha}.$$

For $\alpha = 1/2$, this definition coincides with the example discussed before.

In 1822, Jean-Baptiste Joseph Fourier (1768–1830) presented an applicable definition of fractional operator starting from the differential coefficient of arbitrary order [52], written as[2]

$$\frac{d^i}{dx^i} fx = \frac{d^i}{dx^i} f(x) \quad \text{and} \quad \int^i dx^i \cdot fx = \int^i dx^i \cdot f(x).$$

By recognising that $\cos \cdot (r + i\frac{\pi}{2})$ may be written as

$$\cos \cdot \cos \cdot \left(i\frac{\pi}{2}\right) - \sin \cdot r \cdot \sin \cdot \left(i\frac{\pi}{2}\right)$$

which successively becomes

$$-\sin \cdot r, -\cos \cdot r, +\sin \cdot r, +\cos \cdot r, -\sin \cdot r, \ldots, \text{etc.}$$

if the corresponding values of i are $1, 2, 3, 4, 5$, etc. ... The same results come in the same order, when the value of i increases.

Now the second part of the equation

$$f(x) = \frac{1}{2\pi} \int d\alpha f(\alpha) \int dp \cos \cdot (px - p\alpha)$$

may be rewritten by putting the factor p^i in front of the cosine, and to add the term $i\pi/2$ under the cosine. Thus, we have

$$\frac{d^i}{dx^i} f(x) = \frac{1}{2\pi} \int_{-\infty}^{\infty} d\alpha f(\alpha) \int_{-\infty}^{\infty} dp\, p^i \cos(px - p\alpha + i\pi/2).$$

The number i, which enters in the second member, can be viewed as an arbitrary quantity, positive or negative.

Notice that this definition of fractional derivative is not restricted to a power function, as the one of Lacroix; it applies to any "well behaved" function [53].

A relevant part of the history of fractional calculus began with the papers of Abel and Liouville. The first specific problem in which derivatives of arbitrary order are used is the tautochrone problem, which consists in the determination of a curve in the (x, y) plane such that the time required for a particle to slide down the curve to its lowest point under gravity is independent of its initial position (x_0, y_0) on the curve. Niels Henrik Abel (1802–1829), in 1823, used the mathematical tool to solve an integral equation arising in this problem [54]. According to his approach [55], the time the particle needs to reach the lowest point of the curve, which is a constant, is given by [56]

$$k = \int_0^x dt (x - t)^{-1/2} f(t)\, dt,$$

[2] We used today's notation such that $fx = f(x)$.

where $f(t)$ is an unknown function. By multiplying both sides of the integral equation by $1/\Gamma(1/2)$, we obtain:

$$\frac{k}{\Gamma(1/2)} = \frac{1}{\Gamma(1/2)} \int_0^x (x-t)^{-1/2} f(t)\, dt,$$

in which the right-hand side is a fractional integral of order 1/2, i.e.,

$$\frac{k}{\Gamma(1/2)} = \frac{d^{-1/2}}{dx^{-1/2}} f(x).$$

This expression can be inverted to yield

$$\frac{1}{\Gamma(1/2)} \frac{d^{1/2}}{dx^{1/2}} k = f(x).$$

Since k is a constant, as we have demonstrated before, we obtain

$$f(x) = \frac{k}{\pi\sqrt{x}}.$$

In summary, to solve the tautochrone problem we have to determine $f(t)$, which consists in computing the fractional derivative of the constant k. In general, however, Abel provided the solution for the integral equation

$$k(x) = \int_a^x \frac{f(t)}{(x-t)^\alpha}\, dt, \quad x > a, \quad 0 < \alpha < 1.$$

Abel's solution of the problem attracted the attention of Joseph Liouville (1809–1882), who carried out a study of fractional calculus in a series of papers and extended the expression for the derivative of integral order $D^n e^{ax}$, with $n \in \mathbb{N}$, to derivatives of arbitrary order $D^\alpha e^{ax} = a^\alpha e^{ax}$. This definition introduced by Liouville of a derivative of order α involved an infinite series in the form

$$f(x) = \sum_0^\infty c_n e^{a_n x},$$

such that

$$D^\alpha f(x) = \sum_0^\infty c_n a_n^\alpha e^{a_n x},$$

which is known as the first formula of Liouville for the fractional derivative [57]. However, this definition is restricted by the fact that the values of α have to be such that to ensure the series converges. There is a second definition, given in the same paper, that succeeded in establishing a definition of a fractional derivative for

$f(x) = x^{-a}$, whenever both x and a are positive. The steps for this definition can be summarised as follows. Consider again the Euler's gamma function rewritten as

$$I = \int_0^\infty u^{a-1} e^{xu}\, du, \quad a > 0, \quad u > 0.$$

By the substitution $t = xu$, we obtain

$$I = x^{-a} \int_0^\infty t^{a-1}\, e^{-t}\, dt = x^{-a}\Gamma(a),$$

where $\Gamma(a)$ is given by Eq. (2.4). The preceding equation yields the integral formula

$$x^{-a} = \frac{1}{\Gamma(a)} \int_0^\infty u^{a-1}\, e^{-xu}\, du.$$

Application of the operator D^α on both sides of the expression yields

$$D^\alpha x^{-a} = \frac{(-1)^\alpha}{\Gamma(a)} \int_0^\infty u^{a+\alpha}\, e^{-xu}\, du.$$

Again, by using $t = xu$, we obtain

$$D^\alpha x^{-a} = \frac{(-1)^\alpha \Gamma(a+\alpha)}{\Gamma(a)} x^{-a-\alpha}.$$

The presence of the term $(-1)^\alpha$ suggests the need to include complex numbers in the theory. However, even this second definition of Liouville is restricted to a class of functions in the form $f(x) = 1/x^a$.

At this point it is convenient to introduce a notation to be used hereafter, which is due to the mathematician Harold Thayer Davis (1892–1974) [58]. If α is a positive real number,

$${}_c\mathrm{D}_x^\alpha f(x)$$

will denote *differentiation* of order α of the function along the x-axis; in the same way, the operator

$${}_c\mathrm{D}_x^{-\alpha} f(x)$$

will denote the *integration* of arbitrary order α of the function $f(x)$ along the x-axis. The meaning of the label c, which is very important and stands for a lower integration limit, will be discussed later.

In 1847, Georg Friedrich Bernhard Riemann (1826–1866), when searching for a generalisation of the Taylor series, deduced an expression for the fractional integral of order α of a given function $f(x)$ as

$$
{}_c\mathrm{D}_x^{-\alpha} f(x) = \frac{1}{\Gamma(\alpha)} \int_c^x (x-t)^{\alpha-1} f(t)\, dt + \psi(x), \tag{2.6}
$$

in which $\psi(x)$ was added in view of the ambiguity of the lower limit of integration, c. The addition of exponents in the operators

$$
{}_c\mathrm{D}_x^{-\mu}\, {}_c\mathrm{D}_x^{-\alpha} = {}_c\, \mathrm{D}_x^{-\mu-\alpha} f(x)
$$

is valid only when c is the same. The complementary function $\psi(x)$ is added just for the case in which these limits are not the same, i.e., ${}_c\mathrm{D}_x^{-\mu}\, {}_{c'}\mathrm{D}_x^{-\alpha}$. The idea of a complementary function comes from the work of Liouville [59]. According to him, the ordinary differential equation

$$
\frac{d^n y}{dx^n} = 0
$$

has the complementary solution

$$
y_c = c_0 + c_1 x + c_2 x^2 + \cdots + c_{n-1} x^{n-1},
$$

when no initial conditions are specified. In the same way as well, a complementary solution, y_c, exists when the differential equation is changed from $n \to \alpha$, for α arbitrary. This difficulty arises because the different definitions of fractional derivatives available to Liouville may require different complementary functions. The addition of this complementary function was not satisfactory for Riemann, whose paper was not published during his life time; it was published posthumously, ten years after his death [60]. Anyway, the existence of this complementary function gave rise to a "longstanding controversy" in the young field of the mathematics dedicated to fractional calculus [56].

In 1869, Nikolay Yakovlevich Sonin (1849–1915) made an important step towards solving the problem with the ambiguities in the definition of a fractional operator [61] by using the Cauchy integral formula for derivatives of integer order, namely

$$
D^n f(x) = f^{(n)}(x) = \frac{n!}{2\pi i} \int_C \frac{f(t)}{(t-z)^{n+1}}\, dt, \qquad n \in \mathbb{N}.
$$

Since n is an integer, the integrand has a pole of order n. The generalisation for arbitrary order may be done by changing $n \in \mathbb{N}$ to $\alpha \in \mathbb{R}$, and using again Euler's gamma function $\alpha! = \Gamma(\alpha+1)$. In this case, however, the integrand has not a pole when α is fractional but instead a branch point and, thus, the appropriate contour would require a branch cut [56].

To obtain a useful formula for the fractional integration one can consider that since the function $f(x)$ may be written in terms of a power series as $f(x) = \sum_n a_n x^n$, the integral of a function $f(x)$ may be written in terms of one or more integration steps, by appropriately changing the interval of integration. For instance, the function $f(x)$ may be integrated twice in the interval (c, x) to yield

$$f^{(-2)} = \int_c^x dx_1 \int_c^{x_1} f(t)\, dt = \int_c^x f(t)\, dt \int_t^x dx_1$$
$$= \int_c^x f(t)\,(x-t)\, dt.$$

Analogously, the function may be integrated three times:

$$f^{(-3)} = \int_c^x dx_2 \int_c^{x_2} dx_1 \int_c^{x_1} f(t)\, dt$$
$$= \int_c^x f(t)\, \frac{(x-t)^2}{2}\, dt.$$

If the integration is performed n times, we easily obtain

$$f^{(-n)} = \int_c^x f(t)\, \frac{(x-t)^{n-1}}{(n-1)!}\, dt = \frac{1}{\Gamma(n)} \int_c^x \frac{f(t)}{(x-t)^{-n+1}} dt,$$

using the Euler's gamma function. A general expression is obtained in the form

$$f^{(-\alpha)} = \frac{1}{\Gamma(\alpha)} \int_c^x \frac{f(t)}{(x-t)^{-\alpha+1}}\, dt, \qquad \mathrm{Re}\,(\alpha) > 0,$$

when the integer n is replaced by an arbitrary α. By using the notation already introduced, we may rewrite

$$_c\mathrm{D}_x^{-\alpha} f(x) = \frac{1}{\Gamma(\alpha)} \int_c^x (x-t)^{\alpha-1} f(t)\, dt. \tag{2.7}$$

This is called the *Riemann–Liouville fractional integral*, because, when $c = 0$, we obtain the definition of Riemann, Eq. (2.6), without the complementary function; when $c \to \infty$, it is possible to recover the first definition of Liouville for the class of functions $f(x) = x^{-a}$.

This definition was made possible after the work of Paul Matthieu Hermann Laurent (1841–1908), who used a contour as an open circuit, called *Laurent loop* [62], instead of a closed circuit as used by Sonin and Aleksey Vasilievich Letnikov (1837–1888), who, in 1872, also utilising the Cauchy integral formula as a starting point, extended the idea of Sonin [63]. Indeed, in the paper of Laurent, a Cauchy's integral formula for complex-valued analytical functions was used. From this perspective, Eq. (2.7) may be understood as an indication that the integration has to be performed up to a point from which the desired result can be obtained by means of conventional operations.

One notices that by replacing $-\alpha$ by α, in Eq. (2.7), it becomes a definition for the fractional derivative operator. However, in this case, the integral

$$\int_c^x (x-t)^{-\alpha-1} f(t)\, dt$$

becomes divergent. Thus to calculate the derivative of order α of the function $f(x)$, continuous in the interval (c, x), one writes $\alpha = k - p$, such that

$$\begin{aligned}
{}_c\mathrm{D}_x^{\alpha} f(x) &= {}_c\mathrm{D}_x^{k-p} f(x) = {}_c\mathrm{D}_x^{k}\, {}_c\mathrm{D}_x^{-p} f(x) \\
&= \frac{d^k}{dx^k}\left[\frac{1}{\Gamma(p)} \int_c^x (x-t)^{p-1} f(t)\, dt\right] \\
&= \frac{1}{\Gamma(k-\alpha)} \frac{d^k}{dx^k}\left[\int_c^x (x-t)^{k-\alpha-1} f(t)\, dt\right] \\
&= \frac{1}{\Gamma(k-\alpha)} \frac{d^k}{dx^k}\left[\int_c^x \frac{f(t)}{(x-t)^{\alpha+1-k}}\, dt\right],
\end{aligned}$$

in which k is the smallest integer greater than α and $0 < p = k - \alpha < 1$, with ${}_c\mathrm{D}_x^k$ being the usual operator d^k/dx^k. Then

$$ {}_c\mathrm{D}_x^{\alpha} f(x) = \frac{1}{\Gamma(k-\alpha)} \frac{d^k}{dx^k}\left[\int_c^x \frac{f(t)}{(x-t)^{\alpha+1-k}}\, dt\right] \tag{2.8}$$

for $\alpha = k - p$ is the today's definition of *Riemann–Liouville fractional derivative*. For $c = 0$ or $c \to \infty$, one can use the definition of the beta function, Eq. (1.77), i.e.,

$$B(z, w) = \int_0^1 (x-t)^{w-1}\, t^{z-1}\, dt = \frac{\Gamma(z)\Gamma(w)}{\Gamma(z+w)},$$

for $\mathrm{Re}\,(z) > 0$ and $\mathrm{Re}\,(w) > 0$, to represent the integral in Eq. (2.8). For $f(x) = x^a$, when $c = 0$, $a > 0$, and $\mathrm{Re}\,(\alpha) > 0$, we obtain

$$ {}_0\mathrm{D}_x^{-\alpha} x^a = \frac{\Gamma(a+1)}{\Gamma(a+\alpha+1)} x^{a+\alpha}$$

and

$$ {}_0\mathrm{D}_x^{\alpha} x^a = \frac{\Gamma(a+1)}{\Gamma(a-\alpha+1)} x^{a-\alpha}, $$

which is the definition of Riemann's fractional derivative of the function $f(x) = x^a$, when a is an arbitrary parameter. On the other hand, if $f(x) = x^{-a}$, with $c \to -\infty$, $a > 0$, and Re $(\alpha) > 0$, we obtain

$$ {}_{-\infty}\mathrm{D}_x^{-\alpha} x^{-a} = \frac{(-1)^{\alpha}\Gamma(a-\alpha)}{\Gamma(a-\alpha+1)} x^{-a+\alpha} \tag{2.9} $$

and

$$ {}_{-\infty}\mathrm{D}_x^{\alpha} x^{-a} = \frac{(-1)^{\alpha}\Gamma(a+\alpha)}{\Gamma(a)} x^{-a-\alpha}, \tag{2.10} $$

which corresponds to the definitions of Liouville. To appreciate the meaning of the preceding definitions, consider the simple case in which $a = 0$, i.e., $f(x) = x^0 = 1$ and $\alpha = 1/2$. This implies $k = 1$ and $p = 1/2$. Using Eq. (2.8), we obtain

$$ \begin{aligned} {}_0\mathrm{D}_x^{1/2} 1 &= {}_0\mathrm{D}_x^{1-1/2} 1 =_0 \mathrm{D}_x^{-1/2} 1 \\ &= \frac{1}{\Gamma(1/2)} \frac{d}{dx} \int_0^x (x-t)^{-1/2}\, 1\, dt \\ &= \frac{1}{\Gamma(1/2)} \frac{d}{dx} (2x^{1/2}) = \frac{1}{\sqrt{\pi x}}, \end{aligned} $$

which coincides with the fractional derivative of Lacroix, given by Eq. (2.5).

2.2 The Grünwald–Letnikov Operator

In the papers of Anton Karl Grünwald (1838–1920) and Aleksey Letnikov (1837–1888), an approach to fractional differentiation based on the limit of a sum was developed, inspired by the original idea of Liouville of using the limit of a difference quotient in which differences of fractional order intervene [64, 65]. The conventional definition of a derivative of a function $f(x)$ is the starting point to introduce the operators. From elementary calculus, we know that the first derivative is defined as

$$ f'(x) = \frac{df}{dx} = \lim_{h\to 0} \frac{f(x) - f(x-h)}{h}, $$

and that the second derivate, in analogous way, is given by

$$ \begin{aligned} f''(x) = \frac{d^2 f}{dx^2} &= \lim_{h\to 0} \frac{f'(x) - f'(x-h)}{h} \\ &= \lim_{h\to 0} \frac{f(x) - 2f(x-h) + f(x-2h)}{h^2}. \end{aligned} $$

This procedure permits us to obtain for the n_{th} derivative the general expression

$$f^{(n)}(x) = \frac{d^n f}{dx^n} = \lim_{h\to 0} \frac{1}{h^n} \sum_{r=0}^{n} \binom{n}{r} f(x - rh),$$

in which the binomial coefficients are defined as

$$\binom{n}{r} = \frac{n!}{r!\,(n-r)!} = \frac{n(n-1)(n-2)\cdots(n-r+1)}{r!}.$$

In general, we may define the quantity

$$f_h^{(p)}(x) = \frac{1}{h^p} \sum_{r=0}^{n} \binom{p}{r} f(x - rh),$$

where $p \leq n$ is an arbitrary integer. If $p > n$, all the coefficients in the numerator vanish. Consequently,

$$\lim_{h\to 0} f_h^{(n)}(x) = f^{(p)}(x) = \frac{d^p f}{dx^p}.$$

If $p < 0$, we may introduce

$$\binom{-p}{r} = (-1)^r \begin{bmatrix} p \\ r \end{bmatrix} = (-1)^r \frac{p(p+1)\cdots(p+r-1)}{r!},$$

and thus

$$f_h^{(-p)}(x) = \frac{1}{h^{-p}} \sum_{r=0}^{n} (-1)^r \begin{bmatrix} p \\ r \end{bmatrix} f(x - rh). \tag{2.11}$$

Notice that in the limit $h \to 0$, $f_h^{(-p)}(x) \to 0$. For this reason, we may put $h = (x - c)/n$, where c is a real constant, and then $h \to 0$, if $n \to \infty$. It is also possible to connect the previous results with the notation already introduced for the operators, namely

$$\lim_{\substack{h\to 0 \\ nh=x-c}} f_h^{(-p)}(x) = {}_c\mathrm{D}_x^{-p} f(x),$$

leading to the general expression [46]:

$$\begin{aligned} \left[{}_c\mathrm{D}_x^{-p} f(x)\right] = {}_c^{\mathrm{GL}}\mathrm{D}_x^{-p} f(x) &= \lim_{\substack{h\to 0 \\ nh=x-c}} h^p \sum_{r=0}^{n} \begin{bmatrix} p \\ r \end{bmatrix} f(x - rh) \\ &= \frac{1}{(p-1)!} \int_c^x (x-t)^{p-1} f(t)\, dt, \end{aligned} \tag{2.12}$$

which permits us to demonstrate that the operator represents p successive integrations of the function $f(x)$. By calculating the first derivative of Eq. (2.12), we obtain

$$\frac{d}{dx}\left[{}_c\mathrm{D}_x^{-p}f(x)\right] = \frac{1}{(p-2)!}\frac{d}{dx}\int_c^x (x-t)^{p-2}f(t)\,dt = {}_c\,\mathrm{D}_x^{-p+1}f(x).$$

By integrating this result in the interval (c, x), we have:

$${}_c\mathrm{D}_x^{-p}f(x) = \int_0^c {}_c\mathrm{D}_x^{-p+1}f(x)\,dx. \tag{2.13}$$

If we now change $p \to (p-1)$, Eq. (2.13) becomes

$${}_c\mathrm{D}_x^{-p+1}f(x) = \int_0^c {}_c\mathrm{D}_x^{-p+2}f(x)\,dx. \tag{2.14}$$

Combining Eqs. (2.13) and (2.14), we deduce that

$$\begin{aligned}
{}_c\mathrm{D}_x^{-p}f(x) &= \int_0^c dx \int_c^x dx\,{}_c\mathrm{D}_x^{-p+2}f(x)\,dx \\
&= \int_0^c dx \int_c^x dx \int_c^x {}_c\mathrm{D}_x^{-p+3}f(x)\,dx \\
&= \underbrace{\int_0^c dx \int_c^x dx \cdots \int_c^x f(x)\,dx}_{p\text{ times}}.
\end{aligned} \tag{2.15}$$

The general expression embodying these results is

$${}_c\mathrm{D}_x^{p}f(x) = \lim_{\substack{h\to 0\\ nh=x-c}} h^{-p}\sum_{r=0}^{n}(-1)^r\binom{p}{r}f(x-r\,h), \tag{2.16}$$

which represents a derivative of integer order m, when $p = m$, and represents also a succession of integrations when $p = -m$. Thus the operator (2.16) represents a fractional derivative for $p > 0$, whereas for $-p$, it represents a fractional integration operator.

These results may be extended to operators of arbitrary order. In particular, the expression (2.16), written for $p = -\alpha$, becomes

$${}_c\mathrm{D}_x^{-\alpha}f(x) = \lim_{\substack{h\to 0\\ nh=x-c}} h^{\alpha}\sum_{k=0}^{n}(-1)^k\begin{bmatrix}\alpha\\ k\end{bmatrix}f(x-k\,h), \quad \alpha > 0. \tag{2.17}$$

If the factorial $\left[\begin{smallmatrix}\alpha\\k\end{smallmatrix}\right]$ is replaced with Euler's gamma function, Eq. (2.17) represents the Grünwald–Letnikov integral operator, written in a discretised form, very useful in numerical calculations. When the limit operation indicated in (2.17) is performed, we obtain

$$
{}_c\mathrm{D}_x^{-\alpha} f(x) = \frac{1}{\Gamma(\alpha)} \int_c^x (x-t)^{\alpha-1} f(t)\, dt, \tag{2.18}
$$

which is nothing but the Riemann–Liouville fractional integral operator defined in (2.7). By assuming that the function $f(x)$ has $m+1$ continuous derivatives, we may write

$$
\begin{aligned}
{}_c\mathrm{D}_x^{-\alpha} f(x) = \sum_{k=0}^{m} \Bigg\{ & \frac{f^{(k)}(c)(x-c)^{\alpha+k}}{\Gamma(\alpha+k+1)} \\
& + \frac{1}{\Gamma(\alpha+k+1)} \int_c^x (x-t)^{\alpha+m} f^{(m+1)}(t)\, dt \Bigg\}.
\end{aligned} \tag{2.19}
$$

It is another form of the Grünwald–Letnikov fractional *integral* operator and yields the asymptotic value of ${}_c\mathrm{D}_x^{-\alpha} f(x)$, when $x = c$. When $p = \alpha > 0$, expression (2.16) becomes

$$
{}_c\mathrm{D}_x^{\alpha} f(x) = \lim_{\substack{h\to 0\\ nh = x-c}} \frac{1}{h^\alpha} \sum_{k=0}^{n} (-1)^k \frac{\Gamma(p+1)}{\Gamma(k+1)\Gamma(p-k+1)} f(x-kh). \tag{2.20}
$$

It is now the Grünwald–Letnikov fractional *derivative* operator, also written in a discretised form, like the integral operator. Similarly, when the limit operation indicated is performed, we may deduce that

$$
\begin{aligned}
{}_c\mathrm{D}_x^{\alpha} f(x) = & \sum_{k=0}^{m} \frac{f^{(k)}(c)(x-c)^{-\alpha+k}}{\Gamma(-\alpha+k+1)} \\
& + \frac{1}{\Gamma(-\alpha+m+1)} \int_c^x (x-t)^{-\alpha+m} f^{(m+1)}(t)\, dt,
\end{aligned} \tag{2.21}
$$

which is valid when the derivatives $f^{(k)}$, with $k = 1, 2, 3, \ldots, m+1$, are continuous in the interval $[c, x]$, for m an integer such that $m > \alpha - 1$. The lowest value of m is established by the inequality $m \le \alpha < m+1$. For the particular cases in which $c = 0$ and $m = 0$, we obtain $0 < \alpha < 1$ and Eq. (2.21) reduces to the simple form:

$$
{}_0\mathrm{D}_x^{\alpha} f(x) = \frac{f(0)x^{-\alpha}}{\Gamma(1-\alpha)} + \frac{1}{\Gamma(1-\alpha)} \int_0^x (x-t)^{-\alpha} f'(t)\, dt. \tag{2.22}
$$

Let us briefly explore the meaning of these operators for the "classical" case of $f(x) = x^a$. From (2.22), we have

$$ {}_0\mathrm{D}_x^\alpha x^a = \frac{1}{\Gamma(1-\alpha)} \int_0^x (x-t)^{-\alpha} \, a t^{a-1} \, dt = \frac{\Gamma(a+1)}{\Gamma(a-\alpha+1)} x^{a-\alpha}. \tag{2.23} $$

Now, if we make $c = 0$ and $\alpha \to -\alpha$ in (2.18), and use the expression for $f(x) = x^a$, we obtain

$$ {}_0\mathrm{D}_x^\alpha x^a = \frac{1}{\Gamma(-\alpha)} \int_0^x (x-t)^{-\alpha-1} \, t^a \, dt = \frac{\Gamma(a+1)}{\Gamma(a-\alpha+1)} x^{a-\alpha}. \tag{2.24} $$

Notice that both results (2.23) and (2.24) coincide under these circumstances; they coincide also with the result (2.9), obtained for the Riemann–Liouville operator in the limit in which this operator reduces to the one of Riemann.

2.3 The Caputo Operator

In 1967, M. Caputo introduced a new definition of a fractional derivative, which is connected with the fractional Riemann–Liouville integral and differential operators, now known as the Caputo fractional derivative [66]. The simplest way to introduce it is to consider certain properties of the operator ${}_c\mathrm{D}_x^\alpha$ on $f(x)$. If $\alpha = n - p$, $\alpha \in \mathbb{R}$, $0 < p < 1$, and $n \in \mathbb{N}$, n being the smallest integer greater than α, we may write

$$ {}_c\mathrm{D}_x^\alpha = {}_c\mathrm{D}_x^{n-p} = {}_c\mathrm{D}_x^n \, {}_c\mathrm{D}_x^{-p} = {}_c\mathrm{D}_x^{-p} \, {}_c\mathrm{D}_x^n = {}_c\mathrm{D}_x^{-p+n}, $$

where we have explored the commutative properties of the operator ${}_c\mathrm{D}_x^\alpha$. Thus, it is possible also to write

$$ {}_c\mathrm{D}_x^\alpha f(x) = {}_c\mathrm{D}_x^{n-p} f(x) = {}_c\mathrm{D}_x^{-p} \, {}_c\mathrm{D}_x^n f(x). $$

Since n is an integer, ${}_c\mathrm{D}_x^n f(x) = f^{(n)}(x)$. Then

$$ {}_c\mathrm{D}_x^\alpha f(x) = {}_c\mathrm{D}_x^{-p} f^{(n)}(x). $$

Now, by using the Riemann–Liouville integral operator defined in Eq. (2.7), we establish that

$$ \begin{aligned} {}_c^{\mathrm{C}}\mathrm{D}_x^\alpha f(x) &= \frac{1}{\Gamma(p)} \int_c^x (x-t)^{p-1} f^{(n)}(t) dt \\ &= \frac{1}{\Gamma(n-\alpha)} \int_c^x \frac{f^{(n)}(t)}{(x-t)^{\alpha+1-n}} dt, \end{aligned} \tag{2.25} $$

which is the *Caputo fractional operator*. This derivative is strongly related to the Riemann–Liouville fractional derivative. We notice that it is a formulation in which the operations of integral and derivative are exchanged with respect to the Riemann–Liouville forms. It has been applied thoroughly in physics because it makes the process of working with differential equations of fractional order similar to the method employed for usual differential equations. This fact is connected with the existence of integration constants in the process of solution. In this formulation, one can specify the initial conditions of fractional differential equations as

$$f^{(k)} = b_k, \qquad k = 0, 1, \ldots, n-1,$$

i.e., the conditions may be written as derivatives of integer order, which makes the physical interpretation more akin to the usual problems.

2.4 The Riesz–Weyl Operator

When the limit $c \to \infty$ or $c \to -\infty$, integrals of the kind

$$\frac{1}{\Gamma(-\alpha)} \int_c^x f(t)(x-t)^{-\alpha-1}\, dt, \quad \mathrm{Re}(\alpha) < 0,$$

become expressions that are often called "Weyl fractional derivatives" and could be written as

$${}_{-\infty}\mathrm{D}^{\alpha}_{x}f(x) = \frac{(-1)^{-\alpha}}{\Gamma(-\alpha)} \int_x^{\infty} f(t)(t-x)^{-\alpha-1}\, dt,$$

for which some suitable branch of $(-1)^{-\alpha}$ must be specified, and

$${}_{-\infty}\mathrm{D}^{\alpha}_{x}f(x) = \frac{1}{\Gamma(-\alpha)} \int_{-\infty}^{x} f(t)(x-t)^{-\alpha-1}\, dt.$$

To obtain another representation of the fractional derivative, we can consider the operators of Riemann–Liouville, Grünwald and Letnikov, and Caputo, in the limit $c \to -\infty$. This leads to

$$\frac{1}{\Gamma(k-\alpha)} \int_{-\infty}^{x} \frac{f^{(k)}(t)}{(x-t)^{\alpha+1-k}}\, dt = {}^{\mathrm{RW}}_{-\infty}\mathrm{D}^{\alpha}_{x}f(x), \tag{2.26}$$

which is known as the *operator of Riesz–Weyl for the fractional derivative* of function $f(x)$. The use of this operator may be helpful in some specific applications

Table 2.1 Table of some fractional derivative operators to be employed in this book

Name	**Operator**
Grünwald–Letnikov	${}_{c}^{\mathrm{GL}}\mathrm{D}_x^{\nu}f(x) = \lim_{\substack{h\to 0\\ nh=x-c}} h^{-p}\sum_{r=0}^{n}(-1)^r\binom{p}{r}f(x-rh)$
Riemann–Liouville	${}_{c}^{\mathrm{RL}}\mathrm{D}_x^{\nu}f(x) = \frac{1}{\Gamma(k-\nu)}\frac{d^k}{dx^k}\int\limits_c^x d\tau\,\frac{f(\tau)}{(x-\tau)^{\nu+1-k}}$
Caputo	${}_{c}^{\mathrm{C}}\mathrm{D}_x^{\nu}f(x) = \frac{1}{\Gamma(k-\nu)}\int\limits_c^x d\tau\,\frac{f^{(k)}(\tau)}{(x-\tau)^{\nu+1-k}}$
Riesz–Weyl	${}_{-\infty}^{\mathrm{RW}}\mathrm{D}_x^{\nu}f(x) = \frac{1}{\Gamma(k-\nu)}\int\limits_{-\infty}^x d\tau\,\frac{f^{(k)}(\tau)}{(x-\tau)^{\nu+1-k}}$

because its Fourier transform is simple, and the initial values are not present in the result. Table 2.1 summarises some useful fractional derivative operators defined in this chapter, which will be used later on in this book.

2.5 Integral Transforms of Fractional Operators

In this section, we shall present some Fourier and Laplace transforms of fractional derivatives arising in problems of fractional reaction or fractional diffusion [30]. As we have shown before, the Riemann–Liouville integral operator is given by Eq. (2.18), here rewritten as

$$ {}_0\mathrm{D}_t^{-\nu}f(t) = \frac{1}{\Gamma(\nu)}\int\limits_0^t d\tau\,(t-\tau)^{\nu-1}f(\tau), \quad \mathrm{Re}(\nu) > 0, \tag{2.27}$$

in which we have chosen $c = 0$ and changed the variable $x \to t$. By applying the Laplace transform and using the convolution theorem, we have

$$\begin{aligned}
\mathfrak{L}\left\{{}_0\mathrm{D}_t^{-\nu}f(t);s\right\} &= \int\limits_0^{\infty}\left[{}_0\mathrm{D}_t^{-\nu}f(t)\right]e^{-st}dt \\
&= \int\limits_0^{\infty}\left[\frac{1}{\Gamma(\nu)}\int\limits_0^t d\tau\,(t-\tau)^{\nu-1}f(\tau)\right]e^{-st}dt \\
&= \int\limits_0^{\infty}\left[\frac{t^{\nu-1}}{\Gamma(\nu)} * f(\tau)\right]e^{-st}dt
\end{aligned} \tag{2.28}$$

$$
\begin{aligned}
&= \mathfrak{L}\left\{\frac{t^{\nu-1}}{\Gamma(\nu)} * f(\tau); s\right\} \\
&= \mathfrak{L}\left\{\frac{t^{\nu-1}}{\Gamma(\nu)}; s\right\} \mathfrak{L}\{f(t); s\} \\
&= s^{-\nu} F(s),
\end{aligned}
$$

in which we have used (1.23) and $s, \nu \in \mathbb{C}$, $\mathrm{Re}(s) > 0$ and $\mathrm{Re}(\nu) > 0$. The Riemann–Liouville fractional derivative operator, for $c = 0$ and $x = t$, may be rewritten as in Eq. (2.8):

$$
{}_{c=0}^{\mathrm{RL}}\mathrm{D}_t^{\nu} f(t) = \frac{1}{\Gamma(k-\nu)} \frac{d^k}{dt^k} \int_0^t d\tau \, \frac{f(\tau)}{(t-\tau)^{\nu+1-k}}, \tag{2.29}
$$

in which $\nu = k - p$, with k being the first integer greater than ν, such that $k - 1 \leq \nu < k$ and $0 < p \leq 1$. By applying the Laplace transform, we have:

$$
\mathfrak{L}\left\{{}_0^{\mathrm{RL}}\mathrm{D}_t^{\nu} f(t); s\right\} = \int_0^\infty dt \left[\frac{1}{\Gamma(k-\nu)} \frac{d^k}{dt^k} \int_0^t d\tau \, \frac{f(\tau)}{(t-\tau)^{\nu+1-k}}\right] e^{-st}. \tag{2.30}
$$

It is possible to identify the Riemann–Liouville integral operator inside the brackets in such a way that one can write

$$
\mathfrak{L}\left\{{}_0^{\mathrm{RL}}\mathrm{D}_t^{\nu} f(t); s\right\} = \int_0^\infty dt \left[\frac{d^k}{dt^k}\, {}_0\mathrm{D}_t^{-p} f(t)\right] e^{-st}. \tag{2.31}
$$

Using the result (1.25) together with (2.28), we have

$$
\begin{aligned}
\mathfrak{L}\left\{{}_0^{\mathrm{RL}}\mathrm{D}_t^{\nu} f(t); s\right\} &= s^k \mathfrak{L}\left\{{}_0\mathrm{D}_t^{-\nu} f(t); s\right\} - \sum_{r=0}^{k-1} s^r \frac{d^{k-r-1}}{dt^{k-r-1}}\, {}_0\mathrm{D}_t^{-p} f(t)\bigg|_{t=0} \\
&= s^k s^{-\nu} F(s) - \sum_{r=0}^{k-1} s^r \frac{d^{k-r-1}}{dt^{k-r-1}}\, {}_0\mathrm{D}_t^{-(k-\nu)} f(t)\bigg|_{t=0} \\
&= s^{k-\nu} F(s) - \sum_{r=0}^{k-1} s^r \frac{d^{\nu-r-1}}{dt^{\nu-r-1}} f(t)\bigg|_{t=0}.
\end{aligned} \tag{2.32}
$$

By changing $r \to r - 1$, we obtain:

$$
\mathfrak{L}\left\{{}_0^{\mathrm{RL}}\mathrm{D}_t^{\nu} f(t); s\right\} = s^{k-\nu} F(s) - \sum_{r=1}^{k} s^{r-1} \frac{d^{\nu-r}}{dt^{\nu-r}} f(t)\bigg|_{t=0}. \tag{2.33}
$$

The same procedure may be applied to the Caputo derivative. For $c = 0$ and $x = t$, from Eq. (2.25) we obtain:

$$ {}_0^{\mathrm{C}}\mathrm{D}_x^{\nu} f(x) = {}_0\mathrm{D}_t^{-(p-k)} f(t) = \frac{1}{\Gamma(k-\nu)} \int_0^x d\tau \, \frac{f^{(k)}(\tau)}{(x-\tau)^{\nu+1-k}}, \tag{2.34} $$

with $\nu = k - p$. By applying the Laplace transform, we obtain:

$$ \begin{aligned} \mathfrak{L}\left\{ {}_0^{\mathrm{C}}\mathrm{D}_t^{\nu} f(t); s \right\} &= \mathfrak{L}\left\{ {}_0\mathrm{D}_t^{-(p-k)} f(t); s \right\} \\ &= \mathfrak{L}\left\{ {}_0\mathrm{D}_t^{-p} \frac{d^k f}{dt^k}; s \right\} \\ &= \int_0^{\infty} dt \left[\frac{1}{\Gamma(p)} \int_0^t d\tau \, (t-\tau)^{p-1} \frac{d^k f}{dt^k} \right] e^{-st}. \end{aligned} \tag{2.35} $$

The term inside the brackets is a convolution. Thus,

$$ \begin{aligned} \mathfrak{L}\left\{ {}_0^{\mathrm{C}}\mathrm{D}_t^{\nu} f(t); s \right\} &= \int_0^{\infty} dt \left[\frac{t^{p-1}}{\Gamma(p)} * \frac{d^k f}{dt^k} \right] e^{-st} \\ &= \mathfrak{L}\left\{ \frac{t^{p-1}}{\Gamma(p)} * \frac{d^k f}{dt^k}; s \right\} \\ &= \mathfrak{L}\left\{ \frac{t^{p-1}}{\Gamma(p)}; s \right\} \mathfrak{L}\left\{ \frac{d^k f}{dt^k}; s \right\}. \end{aligned} \tag{2.36} $$

Using (1.23) and (2.28), we obtain

$$ \begin{aligned} \mathfrak{L}\left\{ {}_0^{\mathrm{C}}\mathrm{D}_t^{\nu} f(t); s \right\} &= s^{-p} \left[s^k F(s) - \sum_{r=0}^{k-1} s^{k-r-1} \left. \frac{d^r}{dt^r} f(t) \right|_{t=0} \right] \\ &= s^{-(k-\nu)} \left[s^k F(s) - \sum_{r=0}^{k-1} s^{k-r-1} \left. \frac{d^r}{dt^r} f(t) \right|_{t=0} \right] \\ &= s^{\nu} F(s) - \sum_{r=0}^{k-1} s^{\nu-r-1} \left. \frac{d^r}{dt^r} f(t) \right|_{t=0}. \end{aligned} \tag{2.37} $$

The Grünwald–Letnikov fractional derivative operator may be rewritten from Eq. (2.19), for $\alpha = -\nu$, as

$$ \begin{aligned} {}_c^{\mathrm{GL}}\mathrm{D}_x^{\nu} f(x) = \sum_{k=0}^{m} \Bigg\{ & \frac{f^{(k)}(x-c)^{-\nu+k}}{\Gamma(-\nu+k+1)} \\ & + \frac{1}{\Gamma(-\nu+m+1)} \int_c^x d\tau \, (x-\tau)^{-\nu+k} f^{(m+1)}(\tau) \Bigg\}. \end{aligned} \tag{2.38} $$

By considering the case in which $c = 0$ and $m = 0$, such that $0 \leq \nu < 1$, we may write:

$$ {}^{\mathrm{GL}}_{\;\;0}\mathrm{D}^{\nu}_{t} f(t) = \frac{f(0)t^{-\nu}}{\Gamma(1-\nu)} + \frac{1}{\Gamma(1-\nu)} \int_0^t d\tau \, (t-\tau)^{-\nu} f'(\tau). \tag{2.39} $$

By applying the Laplace transform, we have:

$$ \begin{aligned} \mathfrak{L}\left\{ {}^{\mathrm{GL}}_{\;\;0}\mathrm{D}^{\nu}_{t} f(t); s \right\} &= \mathfrak{L}\left\{ \frac{f(0)t^{-\nu}}{\Gamma(1-\nu)} + \frac{1}{\Gamma(1-\nu)} \int_0^t d\tau \, (t-\tau)^{-\nu} f'(\tau); s \right\} \\ &= \mathfrak{L}\left\{ \frac{f(0)t^{-\nu}}{\Gamma(1-\nu)}; s \right\} \\ &\quad + \mathfrak{L}\left\{ \frac{1}{\Gamma(1-\nu)} \int_0^t d\tau \, (t-\tau)^{-\nu} f'(\tau); s \right\}. \end{aligned} \tag{2.40} $$

By using the convolution theorem, Eq. (1.21), we obtain

$$ \begin{aligned} \mathfrak{L}\left\{ {}^{\mathrm{GL}}_{\;\;0}\mathrm{D}^{\nu}_{t} f(t); s \right\} &= \mathfrak{L}\left\{ \frac{f(0)t^{-\nu}}{\Gamma(1-\nu)}; s \right\} + \mathfrak{L}\left\{ \frac{t^{-\nu}}{\Gamma(1-\nu)} * f'(t); s \right\} \\ &= \mathfrak{L}\left\{ \frac{f(0)t^{-\nu}}{\Gamma(1-\nu)}; s \right\} + \mathfrak{L}\left\{ \frac{t^{-\nu}}{\Gamma(1-\nu)} * f'(t); s \right\} \\ &= \mathfrak{L}\left\{ \frac{f(0)t^{-\nu}}{\Gamma(1-\nu)}; s \right\} + \mathfrak{L}\left\{ \frac{t^{-\nu}}{\Gamma(1-\nu)}; s \right\} \mathfrak{L}\left\{ f'(t); s \right\}. \end{aligned} \tag{2.41} $$

Now, by using (1.23), with $k = 1 - \nu$ and (1.25), we may write

$$ \begin{aligned} \mathfrak{L}\left\{ {}^{\mathrm{GL}}_{\;\;0}\mathrm{D}^{\nu}_{t} f(t); s \right\} &= f(0)s^{\nu-1} + s^{\nu-1}\left[-f(0) + sF(s) \right] \\ &= s^{\nu} F(s). \end{aligned} \tag{2.42} $$

The Laplace transform of the Grünwald–Letnikov fractional derivative operator of order $\nu > 1$ does not exist in the classical sense [46], because there are noninte-grable functions in the first term of the sum in (2.38).

Regarding the Riesz–Weyl fractional derivative operator, the Fourier transform is the appropriate tool to face it. Indeed, this operator is defined in the limit for which $c \to -\infty$ and the operators ${}^{\mathrm{GL}}_{\;\;c}\mathrm{D}^{\nu}_{x}$, ${}^{\mathrm{RL}}_{\;\;c}\mathrm{D}^{\nu}_{x}$ e ${}^{\mathrm{C}}_{c}\mathrm{D}^{\nu}_{x}$ are all identical, i.e.,

$$ \left. \begin{aligned} &{}^{\mathrm{GL}}_{-\infty}\mathrm{D}^{\nu}_{x} f(x) \\ &{}^{\mathrm{RL}}_{-\infty}\mathrm{D}^{\nu}_{x} f(x) \\ &{}^{\;\mathrm{C}}_{-\infty}\mathrm{D}^{\nu}_{x} f(x) \end{aligned} \right\} = \frac{1}{\Gamma(k-\nu)} \int_{-\infty}^{x} d\tau \, \frac{f^{(k)}(\tau)}{(x-\tau)^{\nu+1-k}} $$

$$ \begin{aligned} &= {}^{\mathrm{RW}}_{-\infty}\mathrm{D}^{\nu-k}_{x} f^{(k)}(x) \\ &= {}^{\mathrm{RW}}_{-\infty}\mathrm{D}^{\nu}_{x} f(x). \end{aligned} \tag{2.43} $$

Let us first calculate the Fourier transform of the Riemann–Liouville fractional integral operator when $c \to -\infty$, i.e., when it can be rewritten as

$$ {}_{-\infty}^{\mathrm{RL}}\mathrm{D}_x^{-\nu} f(x) = \frac{1}{\Gamma(\nu)} \int_{-\infty}^{x} d\tau \; (x-\tau)^{\nu-1} f(\tau), \tag{2.44} $$

where, to assure convergence, $0 < \nu < 1$. Using (1.23), with $t = x$, we have

$$ \mathfrak{L}\left\{ \frac{t^{\nu-1}}{\Gamma(\nu)}; s \right\} = \int_0^{\infty} dt \; \frac{t^{\nu-1}}{\Gamma(\nu)} e^{-st} = s^{-\nu}. \tag{2.45} $$

If we assume now that $s = -ik$, with $k \in \mathbb{R}$, the previous integral converges for [46] $0 < \nu < 1$, and the Fourier transform of the function

$$ h_+(x) = \begin{cases} \dfrac{x^{\nu-1}}{\Gamma(\nu)}, & x > 0 \\ 0, & x \leq 0 \end{cases} \tag{2.46} $$

may be written as

$$ \mathfrak{F}\{h_+(x); k\} = (-ik)^{-\nu}, \qquad 0 < \nu < 1. \tag{2.47} $$

In this way as well the Riemann–Liouville fractional integral operator may be put in the form

$$ {}_{-\infty}^{\mathrm{RL}}\mathrm{D}_x^{-\nu} f(x) = \int_{-\infty}^{0} d\tau \; h_+(x-\tau)^{\nu-1} f(\tau) + \int_0^{x} d\tau \; h_+(x-\tau)^{\nu-1} f(\tau). \tag{2.48} $$

The first integral vanishes because of (2.46) and the second one represents a convolution. Thus, we obtain

$$ {}_{-\infty}^{\mathrm{RL}}\mathrm{D}_x^{-\nu} f(x) = h_+(x) * f(x). \tag{2.49} $$

By applying the Fourier transform now, we have

$$ \begin{aligned} \mathfrak{F}\left\{ {}_{-\infty}^{\mathrm{RL}}\mathrm{D}_x^{-\nu} f(x) \right\} &= \mathfrak{F}\{h_+(x) * f(x); k\} \\ &= \mathfrak{F}\{h_+(x); k\} \, \mathfrak{F}\{f(x); k\} \\ &= (-ik)^{-\nu} F(k). \end{aligned} \tag{2.50} $$

This relation also yields the Fourier transform of the Grünwald–Letnikov integral fractional operator (2.19) when $c \to -\infty$, because, in this case, it coincides with the Fourier transform of the Riemann–Liouville integral fractional operator. Since the Riesz–Weyl operator involves the number k, the Fourier transform may be done for the coordinate ω in the Fourier space, to avoid any confusion. By applying the Fourier transform to the Riesz–Weyl operator, we have

$$\begin{aligned}\mathfrak{F}\left\{{}^{\text{RW}}_{-\infty}\text{D}^{\nu}_{x}f(x);\omega\right\} &= \mathfrak{F}\left\{{}^{\text{RL}}_{-\infty}\text{D}^{-(k-\nu)}_{x}f^{(k)}(x);\omega\right\} \\ &= \mathfrak{F}\left\{\frac{1}{\Gamma(k-\nu)}\int_{-\infty}^{x} d\tau\ (x-\tau)^{(k-\nu)-1}f(\tau);\omega\right\}.\end{aligned} \tag{2.51}$$

This integral converges only for $0 < k - \nu < 1$ or $k - 1 < \nu < 1$. Therefore, we may use (2.46) and (2.48) to write

$${}^{\text{RL}}_{-\infty}\text{D}^{-(k-\nu)}_{x}f(x) = \frac{x^{(k-\nu)-1}}{\Gamma(k-\nu)} * f^{(k)}(x). \tag{2.52}$$

The first term in the convolution is not zero only for $x > 0$. Thus, using the Fourier transform for the kth derivative of a function $f(x)$ (1.7), we obtain

$$\begin{aligned}\mathfrak{F}\left\{{}^{\text{RW}}_{-\infty}\text{D}^{\nu}_{x}f(x);\omega\right\} &= \mathfrak{F}\left\{\frac{x^{(k-\nu)-1}}{\Gamma(k-\nu)} * f^{(k)}(x);\omega\right\} \\ &= \mathfrak{F}\left\{\frac{x^{(k-\nu)-1}}{\Gamma(k-\nu)};\omega\right\}\mathfrak{F}\left\{f^{(k)}(x);\omega\right\} \\ &= (-i\omega)^{-(k-\nu)}(-i\omega)^{k}F(\omega) \\ &= (-i\omega)^{\nu}F(\omega).\end{aligned} \tag{2.53}$$

The Fourier transform of the Riesz–Weyl operator may be written in a more simple form by omitting the imaginary unit [41], as

$$\mathfrak{F}\left\{{}^{\text{RW}}_{-\infty}\text{D}^{\nu}_{x}f(x);\omega\right\} \equiv -|\omega|^{\nu}F(\omega). \tag{2.54}$$

Table 2.2 summarises the most useful integral transforms of the fractional operators.

2.6 A Generalised Fourier Transform

In this final section, we analyse a possible extension of the Fourier transform in the context of the fractional calculus [67]. To do this, we start our discussion by introducing the generalised exponential

$$\exp_{\alpha}(t) = \sum_{k=0}^{\infty}\binom{\alpha}{k}\frac{t^{n}e^{t}}{\Gamma(k-\alpha+1)}, \tag{2.55}$$

Table 2.2 Integral transforms of the fractional operators of Table 2.1

Operators		**Integral Transforms**
Grünwald–Letnikov		$\mathfrak{L}\left\{{}_0^{\mathrm{GL}}\mathrm{D}_t^{\nu}f(t);s\right\} = s^p F(s)$
Riemann–Liouville	Integral	$\mathfrak{L}\left\{{}_0\mathrm{D}_t^{-\nu}f(t);s\right\} = s^{-\nu}F(s)$
	Derivative	$\mathfrak{L}\left\{{}_0^{\mathrm{RL}}\mathrm{D}_t^{\nu}f(t);s\right\} = s^{k-\nu}F(s) - \sum_{r=1}^{k} s^{r-1}\frac{d^{\nu-r}}{dt^{\nu-r}}f(t)\Big\vert_{t=0}$
Caputo		$\mathfrak{L}\left\{{}_0^{\mathrm{C}}\mathrm{D}_t^{\nu}f(t);s\right\} = s^{\nu}F(s) - \sum_{r=0}^{k-1} s^{\nu-r-1}\frac{d^{r}}{dt^{r}}f(t)\Big\vert_{t=0}$
Riesz–Weyl		$\mathfrak{F}\left\{{}_{-\infty}^{\mathrm{RW}}\mathrm{D}_x^{\nu}f(x);\omega\right\} = (-i\omega)^{\nu}F(\omega)$ or $-\vert\omega\vert^{\nu}F(\omega)$

which may be obtained from the following definition of the fractional derivative:

$$
\begin{aligned}
{}_0\mathrm{D}_t^{\alpha}\left(f(t)\right) &= \sum_{k=0}^{\infty}\binom{\alpha}{k}\,{}_0\mathrm{D}_t^{\alpha-k}(t^0)\,{}_0\mathrm{D}_t^{k}\left(f(t)\right)\\
&= \sum_{k=0}^{\infty}\binom{\alpha}{k}\frac{t^{k-\alpha}}{\Gamma\left(k-\alpha+1\right)}\mathrm{D}_t^{k}\left(f(t)\right)
\end{aligned}
\tag{2.56}
$$

by choosing $f(t) = e^t$. By using this generalised exponential, we may introduce also the generalised Fourier transform of a function, defined as

$$
f_{\alpha}(\omega) = \int_{-\infty}^{\infty} dt\,\exp_{\alpha}(i\omega t)f(t). \tag{2.57}
$$

In order to obtain the inverse of this transform, we perform the following change:

$$
g(t) = \frac{f(t)}{(i\omega)^{\alpha}} \qquad \text{or} \qquad f(t) = (i\omega)^{\alpha}g(t). \tag{2.58}
$$

Now, by replacing $f(t)$ in Eq. (2.57), we can rewrite the generalised Fourier transform as

$$
f(\omega) = \int_{-\infty}^{\infty} dt(i\omega t)^{\alpha}\exp_{\alpha}(i\omega t)g(t). \tag{2.59}
$$

In terms of the fractional derivative with respect to the frequency, the previous equation can be written as follows:

$$f(\omega) = \int_{-\infty}^{\infty} dt \mathrm{D}_{\omega}^{\alpha} \left(e^{i\omega t}\right) f(t)$$

$$= {}_0\mathrm{D}_{\omega}^{\alpha} \left[\int_{-\infty}^{\infty} dt e^{i\omega t} f(t) \right] \tag{2.60}$$

since

$${}_0\mathrm{D}_{\omega}^{\alpha} \left(e^{i\omega t}\right) = (it)^{\alpha} E_{\alpha}(i\omega t).$$

The integral in Eq. (2.60) yields the ordinary Fourier transform of $g(t)$, that is, $g(\omega)$, so that Eq. (2.60) can be formally written as

$$f(\omega) = {}_0\mathrm{D}_{\omega}^{\alpha} \left[g(\omega)\right]. \tag{2.61}$$

The previous equation can be inverted in order to obtain

$$g(\omega) = {}_0\mathrm{D}_{\omega}^{-\alpha} \left[f(\omega)\right] \tag{2.62}$$

so that the inverse Fourier transform of Eq. (2.62) yields

$$g(t) = \int_{-\infty}^{\infty} \frac{d\omega}{2\pi} e^{i\omega t} \, {}_0\mathrm{D}_{\omega}^{\alpha} \left[f(\omega)\right], \tag{2.63}$$

or in terms of the original function, using Eq. (2.58), we have

$$f(t) = (it)^{\alpha} \int_{-\infty}^{\infty} \frac{d\omega}{2\pi} e^{i\omega t} \, {}_0\mathrm{D}_{\omega}^{\alpha} \left[f(\omega)\right]. \tag{2.64}$$

This equation defines the inverse of the generalised Fourier transform defined in Eq. (2.57) in terms of a generalised exponential given in Eq. (2.55).

Besides the integral transforms of the operators, some useful integral transforms of special functions as the Mittag-Leffler and H-function of Fox have been introduced in Section 1.3 and will be helpful in solving the problems treated in the next chapters.

3
From Normal to Anomalous Diffusion

This chapter starts with a brief history of the approaches to diffusion phenomena, by emphasising the first investigations of Brownian motion, i.e., stochastic motion, the random walk problem, and its connection with the diffusion processes. Subsequently, the concepts of anomalous diffusion and continuous-time random walk are introduced. Some formal aspects of the dynamics in normal and anomalous diffusion are presented. The link between these formalisms is established by introducing memory effects in the diffusion processes. In this enlarged scenario, non-Markovian behaviour and temporal memory are incorporated into the description of the diffusive processes in the presence of external fields, thus opening the whole approach to consider the possibility of application of fractional calculus.

3.1 Historical Perspectives on Diffusion Problems

The term *diffusion* comes from the Latin *diffusio*, *diffusionem*, connected with the verb *diffundere*, meaning "to scatter", "to pour out", and is formed by *dis-* "apart, in every direction" plus *fundere* "pour". In physics, this term is applied to molecular diffusion, i.e., the random molecular motion by which matter is transported from places of higher to places of lower concentrations.

3.1.1 Pioneering Studies

The pioneering investigations of the diffusion process are usually attributed to the Scottish chemist Thomas Graham (1805–1869), who is also one of the founders of the Chemical Society of London and its first president (1841–1843). An important paper on gaseous diffusion appeared in 1829, in the *Quarterly Journal of Science*, under the title "A short account of experimental researches on the diffusion of gas through each other, and their separation by mechanical means". The first lines of the article state [68]:

> Fruitful as the miscibility of the gases has been in interesting speculations, the experimental information we possess on the subject amounts to little more than well established fact, that gases of a different nature, when brought into contact, do not arrange themselves according to their density, the heaviest undermost, the lightest uppermost, but they spontaneously diffuse, mutually and equably, through each other, and so remain in an intimate state of mixture for any length of time.

As pointed out by Graham a few lines after the previous quotation, the law of mixture was first developed by John Dalton (1766–1844), while Claude Berthollet (1748–1822) made subsequent measurements in a very careful manner. However, the true experimentalist in this field was Graham, who carried out experiments on diffusion in a systematic way. By measuring the rate at which gases diffuse through a plug of plaster of Paris, he developed the law now known as Graham's law, stating that "the diffusion rate of gases is inversely proportional to the square root of their densities".

Some of the experimental data of gases obtained by Graham [69] was used by James Clerk Maxwell (1831–1879) to determine the diffusion coefficient of two gases when one-tenth of a vertical tube was filled with the first gas (a heavy gas) and the remaining nine-tenths with a lighter gas. In equilibrium, by neglecting external forces, the diffusion equation for the pressure p is written in the form [70]:

$$\frac{dp}{dt} = D\frac{d^2p}{dx^2},$$

where D is the diffusion coefficient, t is the temporal variable, and x the spatial coordinate (Maxwell does not use a specific notation for partial derivatives). The proportion between the partial pressure, p_1, of the first gas, and the total pressure, p, is given by:

$$\frac{p_1}{p} = \frac{1}{10} - \frac{20}{\pi^2}\left[e^{-\pi^2 D/a^2 t}\sin^2\frac{\pi}{10} - e^{-2^2\pi^2 D/a^2 t}\sin^2\frac{2\pi}{10} + e^{-3^2\pi^2 D/a^2 t}\sin^2\frac{3\pi}{10} - \&c\right].$$

From this expression, Maxwell determined the diffusion coefficient as

$$D = \frac{\log 10}{10\pi^2}\frac{a^2}{T},$$

where T is the time interval and a the length of the vertical tube. The gases are CO_2 and air. From Graham's experiments, we obtain $T = 500$ and $a = 0.57$ m. Thus,

$$D = 0.235$$

in inch-grain-second ($\approx 1.5 \times 10^{-5} m^2/s$, which is accurate at 5%.). These units correspond to area/time, as Maxwell explains when he defines the coefficient of diffusion [70]:

> D is the volume of gas reduced to unit of pressure which passes in unit of time through unit area when the total pressure is uniform and equal to p, and the pressure of either gas increases or diminishes by unity in unit of distance. D may be called the coefficient of diffusion. It varies directly as the square of the absolute temperature, and inversely as the total pressure.

The dimensions of D are evidently L^2T^{-1}, where L and T are the standards of length and time.

One decade before, a phenomenological theory of diffusion had been proposed by the German physiologist Adolf Eugen Fick (1829–1901), who published in 1855 his first article in physics entitled "Über diffusion" [71]. It consists mainly in the mathematical solution of the differential equation for different boundary and initial conditions. This approach is applicable not only to molecular diffusion but also to the transfer of heat. The result comes from the analogy between diffusion and conduction of heat or electricity [71]:

> It was quite natural to suppose, that this law for the diffusion of a salt in its solvent must be identical with that, according to which the diffusion of heat in a conducting body takes place; upon this law Fourier founded his celebrated theory of heat, and it is the same which Ohm applied with such extraordinary success, to the diffusion of electricity in a conductor. According to this law, the transfer of salt and water occurring in a unit of time, between two elements of space filled with differently concentrated solutions of the same salt, must be, *ceteris paribus*, directly proportional to the difference of concentration, and inversely proportional to the distance of the elements from one another.

In 1807, Jean-Baptiste Joseph Fourier (1768–1830), French mathematician and prominent politician during the Revolution, submitted a long paper to the *Académie des Sciences* in which the problem of heat diffusion in solid bodies of different geometries (rectangle, annulus, sphere, cylinder, and prism) was formulated in terms of a differential equation. For the case of a solid cube of volume $dxdydz$, he considered that the instantaneous movement of heat may be governed by the equation

$$\frac{dv}{dt} = \frac{K}{CD}\left(\frac{d^2v}{dx^2} + \frac{d^2v}{dy^2} + \frac{d^2v}{dz^2}\right), \tag{3.1}$$

where v is the temperature at a point x after a time t is elapsed, K is the conductivity, C denotes the capacity of the substance for heat, and D is the density [72]. To express the solution of this equation, Fourier used the series solutions (as Maxwell has done some decades after) but his work did not win the prize offered by the *Académie* in view of the strong resistance of Lagrange (raised mainly because of the use of series). The other examiners, Laplace, Monge, and Lacroix, were in favour of accepting his work. According to Fourier's ideas, the differences in temperature between two points tend to disappear because a thermal flux from the higher to the low temperature occurs.

In formulating the mathematical problem of diffusion, Fick was guided by the preceding approach to heat conduction, as stated. He first considered a volume of salt solution in a vertical container whose concentration in each horizontal elementary stratum is constant and denoted by $y(x)$, where x is the height of the stratum above a reference stratum; i.e., it is the coordinate chosen perpendicular to the surface of reference. Thus dy/dx is the rate of change of concentration in the direction of the coordinate x. The function $y(x)$ has to diminish as x increases; i.e., the higher stratum must be less concentrated. In this scenario, during an element of time dt, a quantity of salt

$$-Qk\frac{dy}{dx}dt$$

will pass into the stratum, bounded by the horizontal planes $x + dx$ and $x + 2dx$, whose concentration is

$$y + \frac{dy}{dx}dx.$$

In the previous expressions, Q represents the surface of the stratum and K is a constant dependent upon the nature of the substances. It is then evident that a volume of water equal to that of the salt passes simultaneously from the upper stratum into the lower. In other words, for a given substance with nonuniform concentration, there is a flow of substance from positions of higher to places of lower concentration, resulting in a uniform distribution of the substance throughout the system. The amount of flow is proportional to the concentration difference between two points, and the constant of proportionality is the diffusion coefficient. This is the essence of the *Fick's first law*.

Exactly according to the model of Fourier's mathematical developments for a current of heat, Fick then obtained the fundamental law for the diffusion current in the form (*Fick's second law*) [71]:

$$\frac{\delta y}{\delta t} = -k\left(\frac{\delta^2 y}{\delta x^2} + \frac{1}{Q}\frac{dQ}{dx}\frac{\delta y}{\delta x}\right), \tag{3.2}$$

which reduces to

$$\frac{\delta y}{\delta t} = -k\frac{\delta^2 y}{\delta x^2}, \tag{3.3}$$

when Q is constant. Equation (3.3) was solved by Fick in the stationary regime, i.e., $\delta y/\delta t = 0$. Then the mathematical solution is a linear one, and he could check the results of a series of experiments performed in this regime.

Empirically, the Fick's first law may be expressed as

$$j = -D\frac{dy}{dx}, \tag{3.4}$$

in which j represents the diffusion flow or current per unit cross section of a substance in a mixture. It gives the amount of this substance passing perpendicularly through a reference surface of unit area during unit time. It is possible to formulate the law independent of the particular system of coordinates as

$$\mathbf{j} = -D\nabla y; \tag{3.5}$$

i.e., the vector $\mathbf{j}$ representing the diffusion current is in the opposite direction to the concentration gradient and is proportional to its absolute value. This is a general law, valid for gases and liquids, but also for solids. However, since the solids are generally anisotropic, D is a second-rank tensor such that $\mathbf{j}$ and ∇y have different directions.

The extension of the experiments on diffusion to solid bodies was carried out by William Chandler Roberts-Austen (1843–1902), an English metallurgist noted for his research on the physical properties of metal and their alloys. He began as an assistant to Thomas Graham and, referring to a statement by the chemist William Ostwald (1853–1932) that "to make accurate experiments on diffusion is one of the most difficult problem in practical physics", he affirmed [73]:

> The difficulties are obvious, but my long connection with Graham's researches made it almost a duty to attempt to extend his work on liquid diffusion to metals . . .

The starting point is also to assume that the laws established by Fick for liquids appply to metals [73]:

> It appeared probable that the law of diffusion of salts, framed by Fick, would also apply to the diffusion of one metal with another.

He thus arrives at the diffusion equation, written as governing the movement of linear diffusion as

$$\frac{dv}{dt} = k\frac{d^2v}{dx^2}, \tag{3.6}$$

where, according to him, x represents the distance in the direction in which the diffusion takes place; v is the degree of concentration of the diffusing metal; t is the time; and k is the diffusion constant [73]. In the experiments, Roberts-Austen measured the diffusivity in precious metals like gold in liquid lead and platinum in liquid lead, and had also tempted a solid interdiffusion, gold into lead. The values obtained for the diffusion coefficient are comparable to modern ones [74].

3.1.2 Brownian Motion

One of the first scientist to observe and, consequently, report on the random motion of particles was the English botanist Robert Brown (1773–1858), who observed that an aqueous suspension of pollen of the herb *Clarkia pulchella* contained microscopic particles carrying out a continuous, haphazard zigzag movement. Before him, the first observation of the phenomenon was reported by the Roman poet Titus Lucretius Carus (died about in 55 b.c.), in the poem *De rerum natura* (On the Nature of Things). In Book 2, one can find the following description of what was later named "Brownian motion" [75]:

> There's a model, you should realise,
> A paradigm of this that's dancing right before your eyes
> For look well when you let the sun peep in a shuttered room
> Pouring forth the brilliance of its beams into the gloom,
> And you'll see myriads of motes all moving in many ways
> Throughout the void and intermingling in the golden rays
> ...
> Your attention to the motes that drift and tumble in the light:
> Such turmoil means that there are secret motions, out of sight,
> That lie concealed in matter. For you'll see the motes careen
> Off course, and then bounce back again, by means of blows unseen,
> Drifting now in this direction, now that, on every side.
> You may be sure this starts with atoms; they are what provide
> The base of this unrest. For atoms are moving on their own,
> Then small formations of them, nearest them in scale, are thrown
> Into agitation by unseen atomic blows ...

The work of Brown was preceded by the observations made by John T. Needham (1713–1781) and F. Willhelm von Gleichen (1717–1783), but he was the first to carry out a detailed investigation of the phenomenon, proving that it is a characteristic of all microscopically small particles (and cannot be attributed to life in particles themselves). The quantitative theory of the translational Brownian movement was developed independently by Einstein (1905), Marian von Smoluchowski (1906), and Paul Langevin (1908).

The first paper dedicated by Albert Einstein (1879–1955) to the Brownian movement was published in his *Annus mirabilis* and the absence of the term Brownian movement in the title is explained in the second paragraph of the article [76]:

> It is possible that the movements to be discussed here are identical with the so called "Brownian molecular motion"; however, the information available to me regarding the latter is so lacking in precision, that I can form no judgement in the matter.

A first impressive result reported in the paper refers to the calculation of the coefficient of diffusion of the suspended substance as

$$D = \frac{RT}{N}\frac{1}{6\pi kP}, \tag{3.7}$$

in which P is the radius of the spherical particle; k is the coefficient of viscosity of the surrounding liquid; R is the gas constant; T is the absolute temperature; and N is the "actual number of molecules contained in a gram-molecule", i.e., Avogadro's number.

In a subsequent section of the paper, Einstein considers the relation between the irregular motion of particles and the diffusion process. In this regard, Brownian movement is described as a diffusion process governed by the equation

$$\frac{\partial f}{\partial t} = D\frac{\partial^2 f}{\partial x^2}, \tag{3.8}$$

in which $f(x, t)$ is the number of particles per unit volume around x in the time t. To obtain this equation, Einstein considered that each single particle executes a movement which is independent of the movement of all other particles. He introduced a time interval τ which is to be very small compared with the observation time, but, nevertheless, of such a magnitude that the movement executed by a particle in two consecutive intervals of time τ are to be considered as mutually independent phenomena, i.e., the movement of a particle does not depend on the story before the considered interval. In modern terms, the diffusion is a Markovian process.

If n denotes the number of suspended particles in the liquid, in an interval τ the x-coordinate of the single particles will increase by a quantity Δ. Thus, $\phi(\Delta)d\Delta$ may be the probability that a particle will be displaced between Δ and $\Delta + d\Delta$, during the interval τ. This probability is normalised and symmetric, i.e.,

$$\int_{-\infty}^{\infty} \phi(\Delta)d\Delta = 1, \quad \text{with} \quad \phi(\Delta) = \phi(-\Delta). \tag{3.9}$$

The number of particles per unit volume $f(x, t)$ which are located at the time $t + \tau$ between two planes perpendicular to the x-axis, with abscissa x and $x + dx$, will be

$$\begin{aligned} f(x, t + \tau)dx &= dx \int_{-\infty}^{\infty} f(x + \Delta, t)\phi(\Delta)d\Delta \\ &\simeq f(x, t) + \tau\frac{\partial f}{\partial x}, \end{aligned}$$

because τ is very small. Now, $f(x + \Delta)$ in the integrand in the right-hand side may be expanded to second-order terms in Δ to give

$$f(x + \Delta, t) = f(x, t) + \Delta\frac{\partial f(x, t)}{\partial x} + \frac{\Delta^2}{2!}\frac{\partial^2 f(x, t)}{\partial x^2} + \cdots .$$

Thus,

$$f(x,t+\tau)dx = f(x,t)\int_{-\infty}^{\infty}\phi(\Delta)d\Delta + \frac{\partial f}{\partial x}\int_{-\infty}^{\infty}\Delta\phi(\Delta)d\Delta$$
$$+\frac{\partial^2 f}{\partial x^2}\int_{-\infty}^{\infty}\frac{\Delta^2}{2}\phi(\Delta)d\Delta + \cdots .$$

By using (3.9), one deduces that $\int_{-\infty}^{\infty}\Delta\phi(\Delta)d\Delta = 0$, because $\phi(\Delta) = \phi(-\Delta)$ and, consequently, that

$$\tau\frac{\partial f}{\partial t} = \frac{\partial^2 f}{\partial x^2}\int_{-\infty}^{\infty}\frac{\Delta^2}{2}\phi(\Delta)d\Delta. \qquad (3.10)$$

By defining D as the second moment of the probability distribution, namely

$$D = \frac{1}{2\tau}\int_{-\infty}^{\infty}\Delta^2\phi(\Delta)d\Delta,$$

one arrives at the diffusion equation in the form (3.8). Note that this result implies that the distribution $\phi(\Delta)$ has a second moment and, consequently, that the central limit theorem is satisfied (see Section 3.1.3). If we assume that for $t = 0$ all particles are at the origin, the solution of Eq. (3.8) is

$$f(x,t) = \frac{n}{\sqrt{4\pi Dt}}e^{-x^2/4Dt}, \qquad (3.11)$$

in which $n = \int_{-\infty}^{\infty} f(x,t)dx$. The mean square displacement may be written as

$$\langle(x - \langle x\rangle)^2\rangle = \langle x^2\rangle = \frac{1}{n}\int_{-\infty}^{\infty} x^2 f(x,t)dx = 2Dt, \qquad (3.12)$$

or, by using (3.7), as

$$\langle x^2\rangle = \frac{RT}{N}\frac{1}{3\pi kP}t, \qquad (3.13)$$

which is a relation that can be used to calculate Avogadro's number, because in

$$N = \frac{RT}{\langle x^2\rangle}\frac{1}{3\pi kP}t, \qquad (3.14)$$

the quantities $\langle x^2\rangle$, t, P, and k can be measured.

The French physicist Jean Baptiste Perrin (1870–1942), who won the Nobel prize for physics in 1926 "for his work on the discontinuous structure of matter, and especially for his discovery of sedimentation equilibrium", recognises the crucial role played by the Brownian motion in the determination of the atomic structure of matter [77]:

It is due to Einstein and Smoluchowski that we have a kinetic theory for the Brownian movement which lends itself to verification.

The work of Perrin was indeed a landmark in establishing the existence of atoms and molecules from the experimental point of view [77]:

In short, if molecules and atoms exist, their relative weights are known to us, and their absolute weights would be known at the same time as Avogadro's number.

In a series of experiments, Perrin succeeded to measure the Avogadro's number according to the Einstein predictions and the "crude average" of their results amounts to $N \simeq 64 \times 10^{22}$ molecules/mole.

The work of Polish physicist Marian Smoluchowski (1872–1917) contributed with a detailed analysis of the mechanism of motion of the Brownian particle [78]. According to him, in the absence of external fields, the conditional probability that the suspended particle starting from the point x_0 reaches the point x in the time t is given by

$$W(x)dx = \frac{1}{\sqrt{4\pi Dt}} e^{-(x-x_0)^2/4Dt}\, dx. \tag{3.15}$$

The next step was to consider that

$$W(x,t|x_0,0) = \int dx' W(x,t|x',\tau) W(x',\tau|x_0,0)$$

represents the conditional probability for t ($t > \tau > 0$) and may be expressed as the probabilities for the arbitrary segments from 0 to τ and from τ to t, which are independent; i.e., as in Einstein's theory, the process is considered as a Markovian one. The conditional probability for Brownian motion in the case of an external field $f(x) = -\alpha x$ is given by

$$W(x,t|x_0,0) = \sqrt{\frac{\beta}{2\pi D(1-e^{-2\beta t})}}\, e^{-\beta/D \frac{(x-x_0 e^{-\beta t})}{1-e^{-2\beta t}}}, \tag{3.16}$$

where $\beta = \alpha D/k_B T$, and k_B is the Boltzmann constant. For this particular form of the force we have

$$\langle x^2 \rangle = \frac{D}{\beta}(1 - e^{-2\beta t}) + x_0^2 e^{-2\beta t}.$$

This result shows that for $\beta t \ll 1$ the second moment $\langle x^2 \rangle$ tends to a constant value, leading the system to a stationary state. For small β, we easily obtain

$$\langle x^2 \rangle \approx 2Dt.$$

Expressions like (3.15) and (3.16) may be obtained as solutions of the drift-diffusion equation in the form

$$\frac{\partial W}{\partial t} = D\frac{\partial^2 W}{\partial x^2} - \beta\frac{\partial}{\partial x}\left[Wf(x)\right], \tag{3.17}$$

where $f(x)$ accounts for the external forces acting on the Brownian particle. The drift-diffusion Eq. (3.17) governs the diffusion of probability and is known as Smoluchowski equation [79]. It can be obtained from the Fokker–Planck equation, which is an equation of motion for the distribution function of a fluctuating variable. Its general form, for one variable x and a distribution function $W(x,t)$, is

$$\frac{\partial W(x,t)}{\partial t} = \left[-\frac{\partial}{\partial x}D^{(1)}(x) + \frac{\partial^2}{\partial x^2}D^{(2)}(x)\right]W(x,t), \tag{3.18}$$

where $D^{(2)}(x) > 0$ is the diffusion coefficient and $D^{(1)}(x)$ is the drift coefficient. Equation (3.18) is a linear second-order partial differential equation of parabolic type and may be viewed as a diffusion equation with an additional first-order derivative with respect to x. It is also known in the mathematical literature as a forward Kolmogorov equation [80].

In 1908, Paul Langevin (1872–1946), a French physicist, proposed a very different but successful description of Brownian motion by applying Newton's second law to a representative Brownian particle [81]. It seems as if Langevin invented the "$F = ma$" of stochastic physics, which is now called the "Langevin equation" [82]. The analysis of Langevin begins by underlining the importance of the work of Louis Georges Gouy (1854–1926) in understanding that Brownian motion is connected with molecular thermal agitation. After that, he invokes the quantitative works of Einstein and Smoluchowski and expresses the following opinion [82]:

> I have been able to determine, first of all, that a correct application of the method of M. Smoluchowski leads one to recover the formula of M. Einstein PRECISELY, and, furthermore, that it is easy to give a demonstration that is infinitely more simple by means of a method that is entirely different.

The starting point of Langevin's quantitative approach is the equipartition theorem. It requires that a particle suspended in any kind of liquid possesses an average kinetic energy such that the average extended to a large number of identical particles of mass m is given by

$$m\langle\xi^2\rangle = \frac{RT}{N}, \tag{3.19}$$

where $\xi = dx/dt$ is the speed, at a given instant, of the particle in the direction that is considered. The equation of motion of the particle in the x direction may be written as

$$m\frac{d^2x}{dt^2} = -6\pi\mu a\frac{dx}{dt} + X, \tag{3.20}$$

where $-6\pi\mu a\xi$ is the viscous resistance according to Stokes's law and X is a stochastic force (according to Langevin, a "complementary force") that pushes the Brownian particle. This complementary force is such that it maintains the agitation of the particle, otherwise the viscous resistance would stop it. A more suitable form of Eq. (3.20) may be obtained by multiplying it by x, namely,

$$\frac{1}{2}m\frac{d^2x^2}{dt^2} - m\xi^2 = -3\pi\mu a\frac{dx^2}{dt} + Xx. \tag{3.21}$$

Now, by considering a large number of identical particles and taking into account the mean of terms in Eq. (3.21), we easily conclude that $\langle Xx\rangle = 0$ in view of the random character of X, and Eq. (3.21) becomes:

$$\frac{1}{2}m\frac{dz}{dt} + 3\pi\mu a z = \frac{RT}{N}, \tag{3.22}$$

where Eq. (3.19) was used and $z = \langle dx^2/dt\rangle$. The solution of Eq. (3.22) is

$$z = \frac{RT}{N}\frac{1}{3\pi\mu a} + Ce^{-\frac{6\pi\mu a}{m}t}, \tag{3.23}$$

where C is an integration constant. This solution becomes stationary when $t > m/(6\pi\mu a) \approx 10^{-8}$ for the particles for which Brownian motion is observable. Thus,

$$\left\langle\frac{dx^2}{dt}\right\rangle = \frac{RT}{N}\frac{1}{3\pi\mu a}, \tag{3.24}$$

which, for a time interval τ, may be written as

$$\langle(x - \langle x\rangle)^2\rangle = \frac{RT}{N}\frac{1}{3\pi\mu a}\tau, \tag{3.25}$$

i.e., the same result obtained by Einstein, Eq. (3.13).

Langevin introduced a method that has inspired new mathematics as well new physics, in which one deals with Gaussian white noise and a stochastic differential equation. To analyse this aspect of the approach in more detail, let us rewrite Eq. (3.20) in the form

$$\frac{d\xi}{dt} = -\gamma\xi + \Gamma(t), \tag{3.26}$$

where $\gamma = 6\pi\mu a/m$ and $\Gamma(t)$ represents the stochastic force per unit mass (also called Langevin force or random force) which is related to the properties of the system. Thus, a suitable description of the system depends on the choice of $\Gamma(t)$

which may lead to a Markovian or a non-Markovian process depending on the characteristics of the particular system under investigation. For illustrative purposes, we consider that $\Gamma(t)$ has the following simple characteristics:

$$\langle \Gamma(t) \rangle = 0$$
$$\langle \Gamma(t)\Gamma(t') \rangle = \overline{q}\delta(t - t'), \qquad (3.27)$$

with $\overline{q} = 2\gamma^2 D/m^2$, where $D = k_B T/m\gamma$ (Einstein–Smoluchowski relation). This choice for $\Gamma(t)$ is typical of a Markovian process, i.e., Brownian motion, and, in particular, it is usually called white noise.

Due to the Markovian nature of this process, it may also be described in terms of the usual form of the diffusion equation. For an anomalous process, i.e., a non-Markovian process, we have to choose other suitable expressions for $\Gamma(t)$ in order to describe different kinds of diffusive processes, as we will discuss later. Consequently, different choices for Eq. (3.27) yield different diffusion processes. A particularly simple example is represented by a random noise with a dependence

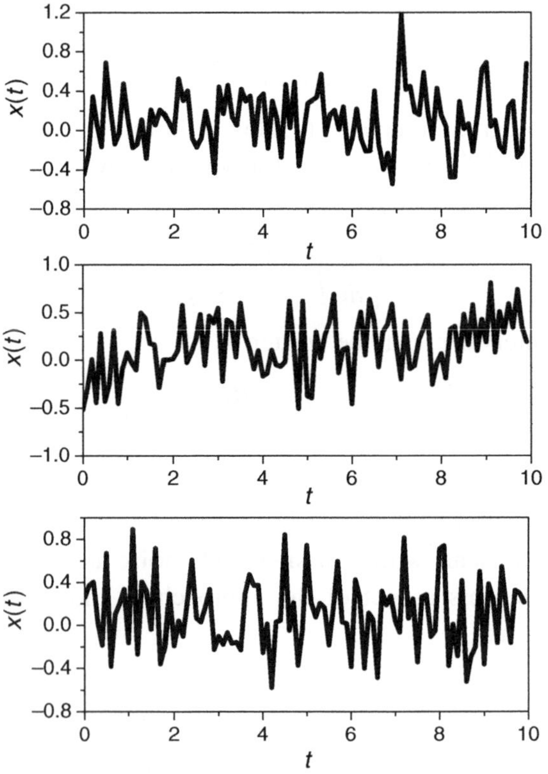

Figure 3.1 Illustrations of the behaviour predicted by Eq. (3.26) versus t for white noise.

on the distribution of the system [83], which yields the porous media equation [84, 85]. Another possibility is to consider $\langle \Gamma(t)\Gamma(t')\rangle \propto \Phi(t-t')$, with $\Phi(t-t')$ representing a dichotomous process with memory [86].

From Eqs. (3.26) to (3.27), we can obtain the dispersion relation (variance) for the system governed by Eq. (3.26), which gives information about the diffusion, i.e., what the time evolution of distribution of the system is. In this sense, it is also useful to characterise the diffusion process which, for the usual case, has a linear dependence on time, as we show below. To see this, let us solve Eq. (3.26) for the initial condition such that at $t = 0$ the velocity ξ (stochastic variable) has an initial value ξ_0. The solution may be written as

$$\xi(t) = \xi_0 e^{-\gamma t} + \int_0^t e^{-\gamma(t-t')}\Gamma(t')dt'. \tag{3.28}$$

By using (3.28), the velocity correlation function may be put in the form:

$$\langle \xi(t_1)\xi(t_2)\rangle = \xi_0^2 e^{-\gamma(t_1+t_2)} + \frac{q}{2\gamma}\left(e^{-\gamma|t_1-t_2|} - e^{-\gamma(t_1+t_2)}\right). \tag{3.29}$$

By assuming that the particle is at $x = x_0$ when $t = 0$, the mean square displacement at time t is defined as

$$\langle (x-x)^2\rangle = \left\langle \int_0^t \xi(t_1)dt_1 \int_0^t \xi(t_2)dt_2 \right\rangle = \int_0^t \int_0^t \langle \xi(t_1)\xi(t_2)\rangle dt_1 dt_2. \tag{3.30}$$

After some calculation, it is possible to show that

$$\langle (x-\langle x\rangle)^2\rangle = \left(\xi_0^2 - \frac{q}{2\gamma}\right)\frac{1}{\gamma^2}\left(1-e^{-\gamma t}\right) + \frac{q}{\gamma^2}t - \frac{q}{\gamma^3}\left(1-e^{-\gamma t}\right). \tag{3.31}$$

This equation, for very long times, i.e., $\gamma t \gg 1$, yields

$$\left\langle (x-\langle x\rangle)^2\right\rangle \approx 2Dt. \tag{3.32}$$

This behaviour is characteristic of usual diffusion and, in particular, of the Markovian nature of this process.

Anomalous diffusion instead is characterised by a different dependence on time in Eq. (3.32) which, according to the nature of the process, may be a power law, i.e.,

$$\left\langle (x-\langle x\rangle)^2\right\rangle \sim t^{\alpha} \tag{3.33}$$

in such a way that $\alpha < 1$ corresponds to subdiffusion whereas $\alpha > 1$ to superdiffusion, as illustrated in Fig. 3.2.

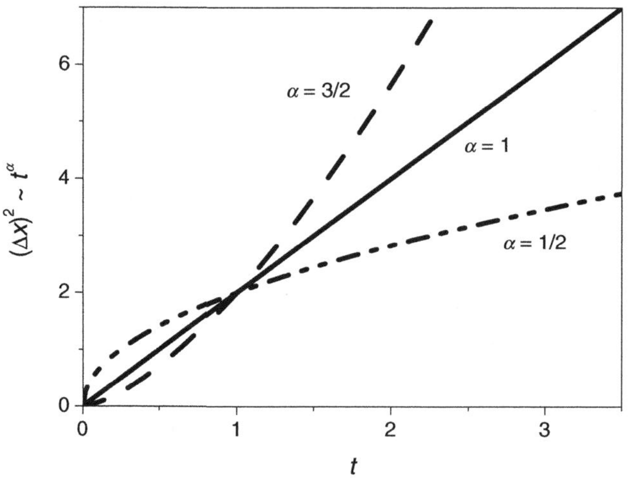

Figure 3.2 The mean square displacement in the case of sub- ($\alpha < 1$), usual ($\alpha = 1$), and superdiffusive ($\alpha > 1$) behaviour.

3.1.3 Random Walk

In 1905, Karl Pearson (1857–1936) appealed to the readers of *Nature* for a solution of the problem that is now known as the *random walk* or the *drunkard's walk*. In a letter dated July 27, he proposed the problem in the following terms [87]:

A man starts from a point O and walks l yards in a straight line; he turns through any angle whatever and walks another l yards in a second straight line. He repeats this process n times. I require the probability that after these n stretches he is at a distance between r and $r + \delta r$ from his starting point, O. The problem is one of considerable interest.

Among the respondents was Lord Rayleigh, who, on August 3, affirmed that [88]

The problem, proposed by Prof. Karl Pearson in the current number of Nature, is the same as that of the composition of n iso-periodic vibrations of unit amplitude and of phases distributed at random ... If n is very great, the probability sought is

$$\frac{2}{n} e^{-r^2/n}\, rdr.$$

Person was also interested in describing the mosquito infestation in a forest. He wanted to know the distribution of mosquitos after many steps had been taken. He answered Lord Rayleigh in the same issue. On August 10, he wrote [89]:

Lord Rayleigh's solution for n very large is most valuable, and may probably suffice for the purposes I have immediately in view ... The lesson of Lord Rayleigh's solution is that in open country the most probable place to find a drunken man who is at all capable of keeping on his feet is somehow near his starting point!

To explore the connections between the random walk problem and the diffusion process in the framework of Pearson's problem, let us consider first the random walk on an infinite one-dimensional lattice whose lattice spacing is l, along the x-axis. If we assume that the time elapsed between steps is Δt, it is possible to define a probability to find the random walker at a point $x = nl$ after N steps, $p(nl, N\Delta t)$, as

$$p\,[nl,(N+1)\Delta t] = \sum_{m=-\infty}^{m=\infty} p\,[ml, N\Delta t]\,p\,[ml, N\Delta t|nl,(N+1)\Delta t]\,, \tag{3.34}$$

where $p\,[ml, N\Delta t|nl,(N+1)\Delta t]$ is the transition probability to go from site $x = ml$ to site $x = nd$ in one step. If, for simplicity, we suppose furthermore that the random walker has an equal probability to go one lattice site to the left or right during each step, this transition probability reduces to:

$$p\,[ml, N\Delta t|nl,(N+1)\Delta t] = \frac{1}{2}\delta_{n,m+1} + \frac{1}{2}\delta_{n,m-1}. \tag{3.35}$$

Thus, Eq. (3.34) may be written as

$$p\,[nl,(N+1)\Delta t] = \frac{1}{2}p\,[(n+1)l, N\Delta t] + \frac{1}{2}p[(n-1)l, N\Delta t]. \tag{3.36}$$

To perform the continuum limit, we rewrite Eq. (3.36) as:

$$p\,[nl,(N+1)\Delta t] - p(nl, N\Delta t) = \frac{1}{2}\Big\{p\,[(n+1)l, N\Delta t] \\ +p\,[(n-1)l, N\Delta t] - 2p\,[nl, N\Delta t]\Big\}$$

or

$$\frac{p(x,t+\Delta t)-p(x,t)}{\Delta t} = \frac{l^2}{2\Delta t}\left[\frac{p(x+l,t)+p(x-l,t)-2p(x,t)}{l^2}\right], \tag{3.37}$$

if we introduce $x = nl$ and $t = N\Delta t$. By performing the limit $l \to \infty$, $\Delta t \to 0$, while keeping $D = l^2/2\Delta t$ as a finite quantity, we obtain the differential equation

$$\frac{\partial p}{\partial t} = D\frac{\partial^2 p}{\partial x^2}, \tag{3.38}$$

which is the diffusion equation for $p(x,t)$. It may be solved for the particular case in which $p(x,0) = \delta(x)$. Indeed, its Fourier transform reads:

$$\frac{\partial \overline{p}(k,t)}{\partial t} = -Dk^2\overline{p}(k,t), \tag{3.39}$$

where

$$\overline{p}(k,t) = \int_{-\infty}^{\infty} p(x,t)e^{ikx}\, dx$$

is the Fourier transform of $p(x, t)$. The solution of Eq. (3.39) is easily found as

$$\overline{p}(k,t) = e^{-Dk^2 t}. \tag{3.40}$$

Thus,

$$p(x,t) = \int_{-\infty}^{\infty} e^{-Dk^2 t} e^{-ikx}\, dk$$
$$= \sqrt{\frac{1}{4\pi Dt}} e^{-x^2/4Dt}. \tag{3.41}$$

From this probability distribution function we easily find:

$$\langle x \rangle = \int_{-\infty}^{\infty} x\, p(x,t)\, dx = 0 \tag{3.42}$$

and

$$\langle x^2 \rangle = \int_{-\infty}^{\infty} x^2\, p(x,t)\, dx = 2Dt. \tag{3.43}$$

The discrete (N steps) d-dimensional version of the probability (3.41) is

$$p_N(\mathbf{r}) \propto \frac{e^{-d\mathbf{r}^2/2\langle \mathbf{r}^2 \rangle N}}{\left(2\pi \langle \mathbf{r}^2 \rangle N/d\right)^{d/2}}. \tag{3.44}$$

For an isotropic walk, the probability distribution function of the distance $r = |\mathbf{r}|$ from the origin is

$$p(r) = A_d r^{d-1} p(\mathbf{r}),$$

where A_d is the surface area of the unit sphere in d dimensions, namely $A_1 = 1$, $A_2 = 2\pi$, $A_3 = 4\pi$, etc. In the case of Pearson problem, $d = 2$ and $\langle r^2 \rangle = a^2$. The probability becomes

$$p(r) \propto \frac{2r}{a^2 N} e^{-r^2/a^2 N}, \tag{3.45}$$

which is Lord Rayleigh's solution for unit amplitude, i.e., $a^2 = 1$ and N very large. This result is embodied by the so-called *central limit theorem*, which states that the probability density describing the distribution of outcomes of a very large number of events approaches a Gaussian form like (3.45) provided the moments of the distribution for the individual events are finite.

From the mathematical point of view one considers a sum of N independent stochastic variables x_i such that

$$y_N = \frac{1}{N}\sum_{i=1}^{N} x_i - \langle x \rangle$$

represents the deviation from the average of N statistically independent measurements of a stochastic variable x. The characteristic function of the stochastic variable $1/N(x_i - \langle x \rangle)$ is defined as [90]

$$f(k, N) = \int_{-\infty}^{\infty} e^{ik/N(x_i - \langle x \rangle)}\, P(x)\, dx_i, \tag{3.46}$$

where $P(x_i)$ is the probability associated with the stochastic variable x_i. For large values of N and finite variance, the integrand may be expanded to give

$$f(k, N) \approx 1 - \frac{1}{2}\frac{k^2}{N^2}\left(\langle x^2 \rangle - \langle x \rangle^2\right), \tag{3.47}$$

when high-order terms can be neglected. Since for independent random variables the characteristic function of their sum is the product of their characteristic functions, the characteristic function of the variable y may be written as

$$f_{y_N}(k) = \left[1 - \frac{1}{2}\frac{k^2}{N^2}\left(\langle x^2 \rangle - \langle x \rangle^2\right) + \cdots\right]^N \to e^{-k^2\sigma_x^2/2N} \quad \text{as} \quad N \to \infty, \tag{3.48}$$

where $\sigma_x^2 = \left(\langle x^2 \rangle - \langle x \rangle\right)$ and the identity $\lim_{n\to\infty}(1 + x/n)^n = e^x$ has been used. The probability distribution associated with the variable y_N is

$$P_{y_N}(y) \to \frac{1}{2\pi}\int_{\infty}^{\infty} dx e^{iky} f_y(k)\, dk = \sqrt{\frac{N}{2\pi\sigma_x^2}} e^{-Ny^2/2\sigma_x^2}, \tag{3.49}$$

when $N \to \infty$. Notice that the explicit form of $P(x)$ was not required if we assume that it has finite moments. In this general case, the average of a large number of statistically independent measurements of x will be a Gaussian whose center is in $\langle x \rangle$ and whose standard deviation is $(1/\sqrt{N})\sigma_x$.

3.1.4 Super- and Subdiffusion

In his paper "Atmospheric diffusion shown on a distance neighbour graph", published in 1926, the English mathematician (but also physicist and meteorologist) Lewis Fry Richardson (1881–1953) proposed a non-Fickian diffusion equation to treat the phenomenon of eddies in the free atmosphere [91]. Instead of considering the diffusion equation as governing the quantity $\nu(x,t)$, i.e., a continuous concentration of the substance, in the form

$$\frac{\partial \nu}{\partial t} = K\frac{\partial^2 \nu}{\partial x^2},$$

in which K is the diffusivity, Richardson considered a small dot of continuous substance, q. By treating the problem as diffusion on a line (along x), q may be viewed as the number of neighbours per unit length, which is a function of the distance between these neighbours, denoted by l, i.e., $q = q(l,t)$. Thus,

$$\frac{\partial q}{\partial t} = \frac{\partial}{\partial l}\left[F(l)\frac{\partial q}{\partial l}\right], \tag{3.50}$$

where $F(l)$ is an increasing function of l. This proposal is based on the observation that the rate of diffusion increases with the separation l between neighbouring dots. To determine $F(l)$, he constructed a curve exhibiting the logarithm of K as a function of l from the experimental data. He then found that $F(l) \approx \varepsilon\, l^{4/3}$, where ε is a constant of the order of $0.4\,\mathrm{cm}^{2/3}\mathrm{s}^{-1}$. Equation (3.50) may be rewritten as

$$\frac{\partial q}{\partial t} = \varepsilon\frac{\partial}{\partial l}\left(l^{4/3}\frac{\partial q}{\partial l}\right) \tag{3.51}$$

and has a solution in the form

$$q(l,t) = A\left(\frac{4t\varepsilon}{9}\right)^{-3/2} e^{-\frac{\alpha^2}{4t\varepsilon/9}}, \tag{3.52}$$

where $\alpha = l^{1/3}$ and A is independent of t and α. The solution (3.52) represents a process in which at $t = 0$ all neighbours are indefinitely close, and as time proceeds they spread out continuously. From his analysis, one deduces that $\langle x(t)\rangle = 0$ and

$$\langle x^2(t)\rangle = \frac{105}{16}\left(\frac{4\varepsilon t}{9}\right)^3 \propto t^3, \tag{3.53}$$

indicating that the mean square displacement is not linear in t, different from what was found in the approaches of Einstein, Smoluchowski, and Langevin. In this sense, one concludes that the phenomenon of turbulence may imply a more rapid diffusive regime, i.e., a superdiffusion process. As pointed out before, if $\langle x^2(t)\rangle \propto t^p$, with $p > 1$, the process may be called superdiffusion; likewise, if $p < 1$, it may be called as subdiffusion (see Fig. 3.2).

The presence of sub diffusive processes may be discussed briefly in connection with the work of H. Scher and E. W. Montroll (1916–1983), who studied a system involving amorphous solids for which they tried to describe the transport phenomena [92, 93]. The charge carriers in these materials tend to be trapped by local defects and then released by thermal fluctuations. To face the problem they studied random walks on a lattice with an emphasis on continuous-time walks with an asymmetric bias. These walks are characterised by random pauses between jumps, with a common pausing time distribution $\psi(t)$, which is well represented by a Pareto–Lévy-like distribution (power-law or long-tailed distribution).

In this case, the average time between jumps, $\langle t \rangle$, is a divergent quantity but $\langle x^2(t) \rangle$ is a finite one. The asymptotic behaviour of $\psi(t)$ for $t \to \infty$ is

$$\psi(t) \propto \left[A(t)t^{1+\alpha}\Gamma(1-\alpha)\right]^{-1}, \tag{3.54}$$

where $0 < \alpha < 1$ and $A(t)$ is a slowly varying function [94]. From these results, we may find for the mean square displacement that

$$\langle x^2(t) \rangle \propto t^{\alpha}, \tag{3.55}$$

i.e., with a fractional time dependence. This implies that the charge carriers diffuse in time in a nonlinear way and slower than in the normal diffusive process, which is a typical sub diffusive behaviour.

Another paradigmatic example of anomalous behaviour is the diffusion on a fractal object of dimension ν, embedded in a space of dimension d in which the probability at time t to be within a hyper-spherical shell between r and $r + dr$ centred on some origin lying on the fractal, denoted by $M(r,t)$, obeys the conservation law [95]:

$$\frac{\partial M(r,t)}{\partial t} = \frac{\partial J(r,t)}{\partial r}, \tag{3.56}$$

where $J(r,t)$ is the total radial current. If $p(r,t)$ denotes the average probability per site, then $M(r,t) \propto r^{\nu-1}p(r,t)$ and there exists a constitutive relation of the form

$$J(r,t) = K(r)r^{\nu-1}\frac{\partial p(r,t)}{\partial r}, \tag{3.57}$$

where $K(r)r^{\nu-1}$ represents the total conductivity of a shell of $r^{\nu-1}$ sites. Substitution of Eq. (3.57) into Eq. (3.56) yields a natural generalisation of the spherically symmetric diffusion equation in Euclidian space, namely

$$\frac{\partial p(r,t)}{\partial t} = \frac{1}{r^{\nu-1}}\frac{\partial}{\partial r}\left[K(r)r^{\nu-1}\frac{\partial p(r,t)}{\partial r}\right], \tag{3.58}$$

which reduces to the usual equation when $K(r) = K$. By assuming that the integrated resistance scales as $R(r) \propto r^{\alpha}$, it is possible to obtain a radius-dependent diffusion coefficient $K(r) = Kr^{-\theta}$, where $\theta = \nu + \alpha - 2$. For this case, the solution of (3.58) in terms of a propagator may be expressed as

$$p(r,t) = \int_0^{\infty} r'^{\nu-1}p(r',0)G(r,r',t)\,dr', \tag{3.59}$$

where

$$G(r,r',t) = \frac{2(rr')^{\frac{1}{2}(2-d+\theta)}}{(2+\theta)Kt} \mathrm{I}_{\frac{\nu}{2+\theta}-1}\left[\frac{2}{K(2+\theta)^2 t}(rr')^{\frac{2+\theta}{2}}\right] e^{-\frac{r^{2+\theta}+r'^{2+\theta}}{K(2+\theta)^2 t}}, \tag{3.60}$$

and $\mathrm{I}_m(x)$ are the modified Bessel function defined in Section 1.2.2. For the initial condition $p(r,0) = \delta(r)/r^{\nu-1}$, the previous solution may be simplified to

$$p(r,t) = \frac{2+\theta}{\nu\Gamma(\nu/(2+\theta))}\left[\frac{1}{K(2+\theta)^2 t}\right]^{\nu/(2+\theta)} e^{-\frac{r^{2+\theta}}{K(2+\theta)^2 t}}, \tag{3.61}$$

when the normalisation condition $\int_0^\infty \nu\, r^{\nu-1} p(r,t)dr = 1$ is imposed. The mean square displacement of this process is given by

$$\langle r^2(t)\rangle = \left[K(2+\theta)^2 t\right]^{2/(2+\theta)} \Gamma\left(\frac{\nu+2}{2+\theta}\right)\Gamma\left(\frac{\nu}{2+\theta}\right)^{-1}. \tag{3.62}$$

Equation (3.62) shows that the spatial dependence present in the diffusion coefficient as a characteristic (inhomogeneity) of the media introduces an anomalous diffusive behaviour; depending on the value of θ, a sub diffusive ($\theta > 0$) or a superdiffusive ($-2 < \theta < 0$) behaviour may be exhibited.

3.2 Continuous-Time Random Walk

It is now time to pursue a deeper understanding of the diffusion process from the point of view of random walks. To do this, we consider, for simplicity, those random walks which are space and time continuous, frequently referred to as continuous-time random walks (CTRW). These walks are based on the idea that the lengths of jumps as well as the time elapsed between these jumps are not constant and are connected by some kind of probability distribution function.

We consider that the jump probability distribution function related to the system under investigation is $\psi(x,t)$, as introduced in the previous section. From $\psi(x,t)$ we can introduce the quantity [41]

$$\omega(t) = \int_{-\infty}^{+\infty} \psi(x,t)\, dx, \tag{3.63}$$

such that $\omega(t)dt$ represents the probability for a waiting time lying in the interval between t and $t+dt$. Likewise, we can also introduce

$$\lambda(x) = \int_{0}^{+\infty} \psi(x,t)\, dt, \tag{3.64}$$

from which the quantity $\lambda(x)dx$ may be defined and represents the probability that a jump length lies in the interval between x and $x+dx$. In general, these variables involving position and time may be not independent. Indeed, if they are coupled, a jump of a certain length may involve a time cost, or, vice versa, in a given time span, the walker can only travel a maximum distance. However, in the simple case in which the jump length and the waiting time are independent random variables, we find the decoupled form $\psi(x,t) = \omega(t)\lambda(x)$ for the jump probability distribution function, $\psi(x,t)$. This situation may be illustrated by the case in which a particle is located at a given position and, after some time interval not connected with its actual position, it jumps with a probability distribution function $\lambda(x)$ with no temporal influence anymore. In this regard, the different kinds of continuous-time random walks may be characterised by the quantities

$$\langle t\rangle_\omega = \int_0^\infty dt\, t\, \omega(t), \tag{3.65}$$

which represents a characteristic time, and

$$\langle x^2\rangle_\lambda = \int_{-\infty}^\infty dx x^2 \lambda(x), \tag{3.66}$$

which represents a jump length variance. Whether these quantities are finite or not depends on the kind of distribution being considered. Since the length of the steps is a continuous quantity, it may even be zero. In this case, one can define the probability that after a given waiting time, the particle will remain at the same position, $\Phi(t)$. If we now consider that the probability that the particle will jump a length x in a given interval between t and $t + dt$ is given by $\int_0^t \omega(t')dt'$, we surely have

$$\Phi(t) + \int_0^t \omega(t')\, dt' = 1 \quad \text{or} \quad \Phi(t) = 1 - \int_0^t \omega(t')\, dt', \tag{3.67}$$

which may be interpreted as a cumulative probability.

Now, suppose we are interested in the transition probability $\eta(x,t)$ that the particle arrives at position x in the time interval between t and $t + dt$ in terms of the transition probability $\eta(x',t')$ that the particle being at x' in a given time t' will move to the position required. In the present case, the length of each step may assume continuous values in such a manner that the point x' may be located on the right or on the left of x. In addition, the waiting time at x' may assume values ranging from zero to infinity, according to the probability distribution function assumed. Thus, a continuous-time random walk process can be described by means of an appropriate generalised master equation, via a set of Langevin equations, or by [41]

$$\eta(x,t) = \int_{-\infty}^{+\infty} dx' \int_{0}^{+\infty} dt' \eta(x',t')\psi(x-x',t-t') + \delta(x)\delta(t), \tag{3.68}$$

which is obtained by summing over all the positions x' that make possible a displacement to the point x together with a summation over all the possible waiting times multiplied by the probability distribution function of a step of length $x - x'$ with a waiting time $t-t'$, represented by $\psi(x-x',t-t')$. In short, Eq. (3.68) connects the probability distribution function $\eta(x,t)$ of just having arrived at position x at time t with the event of having just arrived at x' at time t', $\eta(x',t')$. The second term of Eq. (3.68) represents the initial condition for the random walk. Consequently, the probability distribution $\rho(x,t)$ of being at point x at time t is given by

$$\rho(x,t) = \int_{0}^{t} dt' \eta(x,t')\Phi(t-t'). \tag{3.69}$$

Substitution of Eq. (3.68) into Eq. (3.69) yields

$$\rho(x,t) = \int_{0}^{\infty} dt'' \left[\int_{-\infty}^{\infty} dx' \int_{0}^{\infty} dt' \eta\left(x',t'\right) \psi\left(x-x',t-t'-t''\right) + \delta(x)\delta(t-t'') \right] \Phi(t''). \tag{3.70}$$

By changing the order of integration and introducing $\bar{t} = t + t''$, Eq. (3.70) may be rewritten as

$$\rho(x,t) = \int_{-\infty}^{\infty} dx' \int_{0}^{\infty} d\bar{t} \left[\int_{0}^{\infty} dt'' \eta(x',\bar{t}-t'')\Phi(t'') \right] \psi(x-x',t-\bar{t}) + \Phi(t)\delta(x) \tag{3.71}$$

or

$$\rho(x,t) = \int_{-\infty}^{\infty} dx' \int_{0}^{\infty} d\bar{t}\rho(x,\bar{t})\psi(x-x',t-\bar{t}) + \Phi(t)\delta(x), \tag{3.72}$$

if we recognise the term between brackets as $\rho(x',\bar{t})$. To proceed, we apply the Laplace transform to the temporal variable in Eq. (3.72) and use the convolution theorem in order to obtain

$$\begin{aligned} \rho(x,s) &= \int_{0}^{\infty} dt \rho(x,t) e^{-st} \\ &= \int_{-\infty}^{\infty} dx' \rho(x',s)\psi(x-x',s) + \Phi(s)\delta(x). \end{aligned} \tag{3.73}$$

Likewise, the Fourier transform of Eq. (3.73), together with the convolution theorem, yields

$$\rho(k,s) = \frac{1}{\sqrt{2\pi}} \int_{-\infty}^{\infty} dx \rho(x,s) e^{ikx}$$
$$= \rho(k,s)\psi(k,s) + \frac{1}{\sqrt{2\pi}} \Phi(s). \tag{3.74}$$

From the Laplace transform of Eq. (3.67), we obtain

$$\Phi(s) = \frac{1 - \omega(s)}{s}, \tag{3.75}$$

which, when used in (3.74), furnishes

$$\rho(k,s) = \frac{1}{\sqrt{2\pi}} \frac{1 - w(s)}{s} \frac{1}{1 - \psi(k,s)}. \tag{3.76}$$

By making suitable choices for $\psi(k,s)$ it is possible to relate the preceding equations to several kinds of diffusive processes. The usual diffusive process may be obtained from this formalism by considering that the probability distribution function is decoupled, i.e., $\psi(x,t) = \lambda(x)\omega(t)$. In this particular case, its Laplace transform is $\psi(k,s) = \lambda(k)\omega(s)$, in which

$$\omega(s) = \int_0^{\infty} dt e^{-st} \omega(t)$$
$$= \int_0^{\infty} dt \left[\sum_{n=0}^{\infty} \frac{(-st)^n}{n!} \right] \omega(t)$$
$$= \sum_{n=0}^{\infty} (-1)^n \frac{s^n}{n!} \langle t \rangle_{\omega} \tag{3.77}$$

is the Laplace transform of $\omega(t)$ and, likewise,

$$\lambda(k) = \int_{-\infty}^{\infty} dx e^{-ikx} \lambda(x)$$
$$= \sum_{n=0}^{\infty} (-i)^n \frac{k^n}{n!} \langle x^2 \rangle_{\lambda} \tag{3.78}$$

is the Fourier transform of $\lambda(x)$. In Eqs. (3.77) and (3.78), we have introduced the series expansion of the exponential function to rewrite them in terms of the second moment and the waiting time. The usual diffusion has a variance, i.e., the second moment is finite, and the average of the waiting time distribution is well defined.

These features allow us to approximate the jumping probability and the waiting time distribution functions by considering the limiting case $x \to \infty$ and $t \to \infty$, which corresponds, respectively, to $k \to 0$ and $s \to 0$. In this limit, Eqs. (3.77) and (3.78) may be approximated, respectively, as

$$\omega(s) \approx 1 - s\tau + \mathcal{O}(s^2) \tag{3.79}$$

and

$$\lambda(k) \approx 1 - \frac{k^2}{2}\sigma^2, \tag{3.80}$$

where $\tau = \langle t \rangle_\omega$ and $\sigma^2 = \langle x^2 \rangle_\lambda$. By using Eqs. (3.79) and (3.80) in Eq. (3.76), we obtain

$$\rho(k,s) \approx \frac{1}{s + Dk^2}, \tag{3.81}$$

where $D = \sigma^2/2\tau$, as introduced in Section 3.1.3 in connection with the discrete random walk problem, has the dimensions of a diffusion, e.g., m^2/s in SI units.

To appreciate the meaning of (3.81) in the context of usual diffusion, we consider the diffusion equation in one dimension in the space-time domain as

$$\frac{\partial}{\partial t}\rho(x,t) = D\frac{\partial^2}{\partial x^2}\rho(x,t), \tag{3.82}$$

subjected to the initial condition $\rho(x,0) = \delta(x)$. We also assume that the diffusing particle may not be found very far from the starting point even at very large time, i.e., $\rho(x \to \pm\infty, t) = 0$. The Laplace transform of Eq. (3.82) is

$$s\rho(x,s) = D\frac{\partial^2}{\partial x^2}\rho(x,s) + \rho(x,0), \tag{3.83}$$

which, when subsequently submitted to a Fourier transform, becomes

$$s\rho(k,s) = 1 - Dk^2\rho(k,s),$$

or

$$\rho(k,s) \approx \frac{1}{s + Dk^2}. \tag{3.84}$$

The diffusion equation leads to a structure that is identical to the one represented by Eq. (3.81) in the Fourier–Laplace space. The probability distribution function that leads to Eqs. (3.79) and (3.80) presents a second moment whose dependence with time is linear, as observed in the approaches of Einstein and Langevin discussed before.

3.3 Diffusion Equation

The previous approaches can be used to investigate the diffusive processes, usual or anomalous. However, when the system is subject to external fields the formalisms discussed before may require cumbersome calculations, which can be overcome if partial differential equations are considered. The diffusion equation is a typical partial differential equation that emerges from the above formalisms under certain assumptions and can be useful to deal with diffusing systems when external forces are present. For simplicity, here we show a simple way to obtain the diffusion equation from the Langevin equation, which is a stochastic equation.

To do this, we start by rewriting the Langevin equation, Eq. (3.20), in the presence of an external force $F(x)$. When the term md^2x/dt^2 may be neglected, an ordinary differential equation of first order may be obtained as

$$\frac{dx}{dt} = \zeta(t) + f(x), \tag{3.85}$$

where $\zeta(t) = \Gamma(t)/\gamma$ and $f(x) = F(x)/\gamma m$. Now, consider the problem of determining the probability distribution, $\rho(x,t,x_0)$, of a particle to be found in the interval x and $x + dx$, at the time t, when the initial position is x_0. In this framework, we may discretize the time interval such that x_n is the position of the particle at $t = n\tau$, where $\tau > 0$ is a very small quantity, and Eq. (3.85) becomes

$$x_{n+1} - x_n = \tau f(x_n) + \sqrt{\tau\Gamma}z_n, \tag{3.86}$$

where $z_n = \sqrt{\tau/\Gamma}\zeta_n$. In this way as well, we can also consider that $\delta(t-t') \to \delta_{nn'}/\tau$, with $\langle z_n\rangle = 0$ and $\langle z_n z_{n'}\rangle = \delta_{nn'}$. The next step is thus to obtain an equation governing the behaviour of $\rho(x,t)$, which will be the diffusion equation [96]. Consider $\rho_n = \rho(x_n, n\tau)$ as the probability distribution related to x_n and the corresponding characteristic function, $g_n(k)$. We may connect $\rho_n = \rho(x_n, n\tau)$ with $g_n(k)$ and use Eq. (3.86), because the distribution and the characteristic functions are linked by means of a Fourier transform as

$$g_n(k) = \langle e^{ikx_n}\rangle = \int_{-\infty}^{+\infty} e^{ikx_n}\rho_n dx_n, \tag{3.87}$$

which can be used to obtain

$$g_{n+1}(k) = \langle e^{ikx_{n+1}}\rangle = \langle e^{ik[x_n+\tau f(x_n)+\tau\zeta_n]}\rangle. \tag{3.88}$$

Since in Eq. (3.88) the quantities x_n and ζ_n are independent, it can be simplified and rewritten as follows:

$$g_{n+1}(k) = \langle e^{ik[x_n+\tau f(x_n)]}\rangle\langle e^{ik\tau\zeta_n}\rangle. \tag{3.89}$$

This result permits us to obtain $g_{n+1}(k)$ as a series expansion in τ up to the first-order terms, since τ was assumed to be very small. We have

$$\langle e^{ikx_n} e^{ik\tau f(x_n)} \rangle \approx \langle e^{ikx_n} \rangle + ik\tau \langle f(x_n) e^{ikx_n} \rangle, \tag{3.90}$$

and, using the properties $\langle \zeta_n \rangle = 0$ and $\langle \zeta_n^2 \rangle = \Gamma/\tau$, we may also write

$$\langle e^{ik\sqrt{\tau\Gamma} z_n} \rangle \approx \left\langle 1 + ik\sqrt{\tau\Gamma} z_n + \frac{(ik\sqrt{\tau\Gamma} z_n)^2}{2!} \right\rangle = 1 - \frac{1}{2} k^2 \tau \Gamma. \tag{3.91}$$

Substitution of Eqs. (3.90) and (3.91) into Eq. (3.89) yields

$$g_{n+1}(k) \approx g_n(k) + \tau \left[ik \langle e^{ikx_n} f(x_n) \rangle - \frac{1}{2} k^2 \Gamma g_n(k) \right]. \tag{3.92}$$

By using the definition of the characteristic function in Eq. (3.92) and performing integration by parts, we obtain

$$\begin{aligned} ik\langle e^{ikx_n} f(x_n) \rangle &= ik \int_{-\infty}^{+\infty} e^{ikx_n} f(x_n) \rho_n dx_n \\ &= - \int_{-\infty}^{+\infty} e^{ikx_n} \frac{d}{dx_n} [f(x_n)\rho_n] dx_n, \end{aligned} \tag{3.93}$$

where the boundary condition $\rho_n(\pm\infty) = 0$ has been used. A similar procedure, after integration by parts, yields

$$-k^2 g_n(k) = (ik)^2 \int_{-\infty}^{+\infty} e^{ikx_n} \rho_n dx_n = \int_{-\infty}^{+\infty} e^{ikx_n} \frac{d^2 \rho_n}{dx_n^2} dx_n. \tag{3.94}$$

Thus, Eq. (3.92) can be written as

$$\begin{aligned} &\int_{-\infty}^{+\infty} e^{ikx_{n+1}} \rho_{n+1} dx_{n+1} - \int_{-\infty}^{+\infty} e^{ikx_n} \rho_n dx_n \\ &= -\tau \int_{-\infty}^{+\infty} e^{ikx_n} \frac{d}{dx_n} [f(x_n)\rho_n]\, dx_n + \frac{1}{2}\tau\Gamma \int_{-\infty}^{+\infty} e^{ikx_n} \frac{d^2 \rho_n}{dx_n^2} dx_n, \end{aligned} \tag{3.95}$$

and, consequently,

$$\frac{\rho_{n+1} - \rho_n}{\tau} = -\frac{d}{dx} [f(x_n)\rho_n] + \frac{\Gamma}{2} \frac{d^2 \rho_n}{dx_n^2}. \tag{3.96}$$

Now, by taking the limit $\tau \to 0$ in Eq. (3.96), with $\rho_n \to \rho(x,t)$ and $f(x_n) \to f(x,t)$, it becomes the desired diffusion equation:

$$\frac{\partial \rho(x,t)}{\partial t} = D\frac{\partial^2 \rho(x,t)}{\partial x^2} - \frac{\partial}{\partial x}[f(x,t)\rho(x,t)], \tag{3.97}$$

with $D = \Gamma/2$. It can be formally rewritten as a Fokker–Planck equation (see Section 3.1.2)

$$\frac{\partial}{\partial t}\rho(x,t) = \frac{\partial}{\partial x}\left[D\frac{\partial}{\partial x}\rho(x,t) - F(x,t)\rho(x,t)\right], \tag{3.98}$$

in such a way that it is possible to identify the continuity equation

$$\frac{\partial}{\partial t}\rho(x,t) + \frac{\partial}{\partial x}J(x,t) = 0, \tag{3.99}$$

for the probability current density:

$$J(x,t) = -D\frac{\partial}{\partial x}\rho(x,t) + F(x,t)\rho(x,t). \tag{3.100}$$

Using Eqs. (3.99) and (3.100) it is possible to show that, for appropriate boundary conditions, $\int dx\rho(x,t) = \text{constant}$, i.e., the normalisation condition is satisfied.

To go on further, we search for solutions of the diffusion equation in (i) the absence of external forces and (ii) in the presence of a linear external force. These cases have as solution a Gaussian distribution function such that when the force is absent the solution is represented by a uniform distribution. The second case presents a stationary solution due to the external force, which is related to a harmonic potential.

In the absence of force, Eq. (3.98) becomes

$$\frac{\partial}{\partial t}\rho(x,t) = D\frac{\partial^2}{\partial x^2}\rho(x,t). \tag{3.101}$$

The Fourier and Laplace transforms with the initial condition $\rho(x,0) = \delta(x)$ applied to Eq. (3.101) yield

$$\rho(k,s) = \frac{1}{s + Dk^2}. \tag{3.102}$$

By performing the inverse integral transforms, we obtain

$$\rho(x,t) = \frac{e^{-\frac{x^2}{4Dt}}}{\sqrt{4\pi Dt}}. \tag{3.103}$$

Some examples are exhibited in Fig. 3.3 for the simplified case $D = 1$. Equation (3.101) is useful to evaluate the dispersion relation, which for this case is equal to the second moment since the initial condition is a Dirac delta function at the origin. The second moment is an important quantity to get information on the diffusion

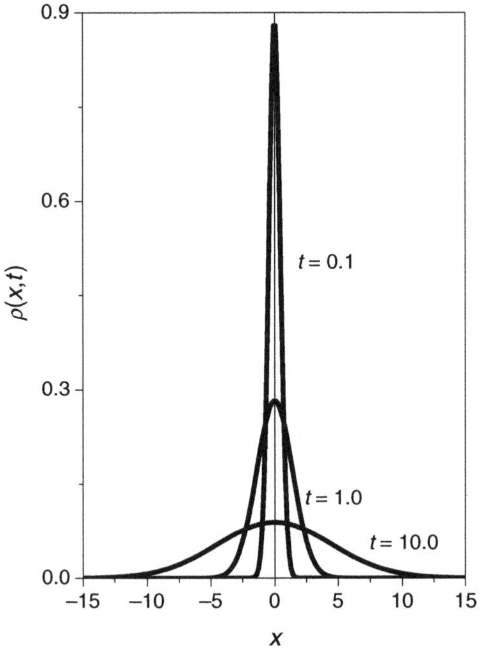

Figure 3.3 $\rho(x,t)$ as given by Eq. (3.103) versus x for typical values of t. For illustrative purposes, the curves were drawn for $D = 1$.

process; i.e., it indicates how the system is spreading. For the problem we are considering, it is given by $\langle x^2 \rangle = 2Dt$; i.e., its time dependence is linear, as expected for the typical Brownian motion occurring in the absence of an external force.

Let us incorporate in Eq. (3.100) a spatial dependence of the diffusion coefficient in the form [97]:

$$D(x) = D|x|^{-\theta}, \qquad \theta \in \mathbb{R}, \tag{3.104}$$

in which D is a constant, and consider a linear external force in the form $F = -k(t)x$. For this case, the diffusion equation which emerges from substitution of Eq. (3.100) into Eq. (3.99) is

$$\frac{\partial}{\partial t}\rho(x,t) = \frac{\partial}{\partial x}\left[D|x|^{-\theta}\frac{\partial}{\partial x}\rho(x,t)\right] - \frac{\partial}{\partial x}\left[k(t)x\rho(x,t)\right]. \tag{3.105}$$

In order to solve Eq. (3.105), we may promote the following change of variables: $x \to z = \xi(t)x$ and $t \to \beta(t)$, with $\overline{\rho}(x,t) \equiv \rho(z,\beta)$, where

$$\rho(x,t) = \exp\left[\int_0^t d\bar{t}k(\bar{t})\right]\overline{\rho}(x,t),$$

$$\beta(t) = \int_0^t dv \exp\left[(2+\theta)\int_0^v du k(u)\right],$$

and

$$\xi(t) = \exp\left[\int_0^t du k(u)\right].$$

By promoting this change of variable to Eq. (3.105), we obtain:

$$\frac{\partial}{\partial \beta}\rho(z,\beta) = \frac{\partial}{\partial z}\left[D|z|^{-\theta}\frac{\partial}{\partial z}\rho(z,\beta)\right]. \tag{3.106}$$

The solution for this equation in the framework of the Green's function approach may be obtained as:

$$\rho(z,\beta) = \int_{-\infty}^{\infty} dx' \rho(x',0) G(z,x',\beta) \tag{3.107}$$

$$\begin{aligned} G(z,x',\beta) = {} & \frac{\xi}{2(2+\theta)D\beta}|zx'|e^{-|z|^{2+\theta}/[(2+\theta)^2 D\beta]}e^{-|x'|^{2+\theta}/[(2+\theta)^2 D\beta]} \\ & \times\left\{\mathrm{I}_{-\nu}\left[\frac{2|zx'|^{\frac{1}{2}(2+\theta)}}{(2+\theta)^2 D\beta}\right] + \frac{zx'}{|zx'|}\mathrm{I}_{\nu}\left[\frac{2|zx'|^{\frac{1}{2}(2+\theta)}}{(2+\theta)^2 D\beta}\right]\right\}, \end{aligned} \tag{3.108}$$

where $\mathrm{I}_{\pm\nu}(x)$ are the modified Bessel functions (see Section 1.2.2). For the initial condition $\rho(x,0) = \delta(x)$, the previous solution can be simplified to

$$\rho(x,t) = \frac{\xi(t)}{2\Gamma\left(\frac{3+\theta}{2+\theta}\right)\left[(2+\theta)^2 D\beta(t)\right]^{\frac{1}{2+\theta}}} e^{-|\xi(t)x|^{2+\theta}/[(2+\theta)^2 D\beta]}, \tag{3.109}$$

yielding the following expression for the mean square displacement

$$\begin{aligned} (\Delta x)^2 &= \langle (x - \langle x\rangle)^2\rangle \\ &= \frac{\Gamma\left(\frac{3}{2+\theta}\right)}{2\Gamma\left(\frac{1}{2+\theta}\right)\xi^2(t)}\left[(2+\theta)^2 D\beta(t)\right]^{\frac{2}{2+\theta}}. \end{aligned} \tag{3.110}$$

Figure 3.4 illustrates the behaviour of Eq. (3.110) for different time-dependent functions. In particular, a stationary solution is found in the case $k(t) \to k =$ constant. Depending on the choice of $k(t)$ in the external force expression, the solution may present an usual or an anomalous behaviour. Equation (3.110) predicts a nonlinear time dependence for the mean square displacement. It is remarkable that we are finding an anomalous behaviour arising from Eq. (3.106) even if it

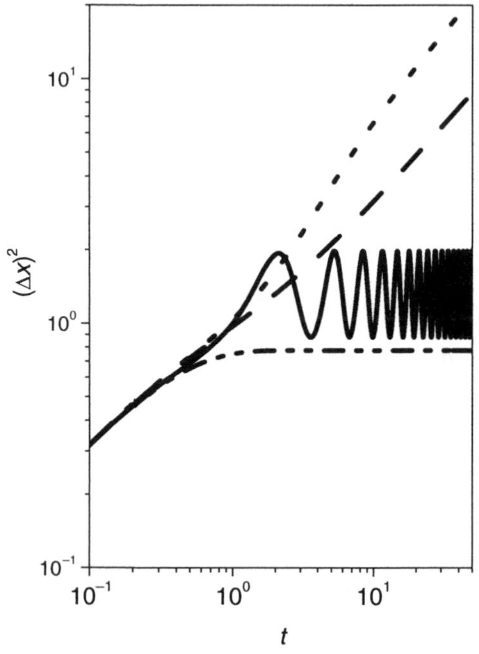

Figure 3.4 The mean square displacement given by Eq. (3.110) versus t for different time-dependent functions $k(t)$. The dashed-dotted line, which corresponds to the case $k(t) = k$, has a stationary value indicating the presence of a stationary solution for the system. The dashed and dotted lines illustrate the case characterised by $k(t) = k/(1+t)$ and $k(t) = ke^{-t}$, respectively. These cases, for long times, may have an asymptotic behaviour for the mean square displacement as a power law similar to the ones found in anomalous diffusion processes. The solid line is the case $k(t) = k\cos^2(t)$, which has a nonmonotonic behavior. For illustrative purposes, in all cases we consider $\theta = 1$, $k = 1$, and $D = 1$. Modified from A. T. Silva, E. K. Lenzi, L. R. Evangelista, M. K. Lenzi, H. V. Ribeiro, and A. A. Tateishi, Exact propagator for a Fokker–Planck equation, first passage time distribution, and anomalous diffusion, Journal of Mathematical Physics **52**, 083301 (2011), with the permission of AIP Publishing.

is a usual, i.e., nonfractional, diffusion equation with an external force term and a diffusion coefficient which is position dependent. In the following chapters, we will explore the anomalous behaviour of the solutions mainly in connection with diffusion equations expressed in terms of fractional derivatives in time, in space, or in both variables.

4

Fractional Diffusion Equations

Elementary Applications

In this chapter, we consider the one-dimensional diffusion equation written in terms of fractional operators in time and space variables – aiming to study the solutions and the consequences arising from the extension of the usual differential operators to fractional ones. In the first part of the chapter, we consider the fractional time diffusion equations, i.e., the fractional operator involving only the time variable, and, subsequently, spatial fractional diffusion equations, i.e., the fractional operator involving only the spatial variable. We also consider both – the spatial and time fractional diffusion equations – as well as the presence of linear reaction terms, the latter taking into account irreversible and reversible processes. These diffusion equations are connected with anomalous diffusion having the mean square displacement in the form $\langle (x - \langle x \rangle)^2 \rangle \propto t^{\alpha}$ or with the Lévy flights for which the mean square displacement is not defined. The continuous-time random walk formalism is also used to discuss the physical implications of the presence of the fractional derivatives. The d-dimensional case within radial symmetry as well as the nonlinear cases will be discussed later, in the following chapters.

4.1 Fractional Time Derivative: Simple Situations

Before we take up more complex situations, let us start our study about fractional diffusion equations by searching for solutions to the following equation, which is a time fractional version of Eq. (3.98):

$$\frac{\partial^{\gamma}}{\partial t^{\gamma}} \rho(x,t) = D \frac{\partial^2}{\partial x^2} \rho(x,t) - \frac{\partial}{\partial x} \left[F(x,t) \rho(x,t) \right], \tag{4.1}$$

where the fractional derivative applied to the time variable is taken in the Caputo sense, as defined in Chapter 2. In Eq. (4.1), we also have an external force $F(x,t)$ and the diffusion coefficient, D, was assumed to be constant, just to illustrate the effect only of the fractional time derivative on the solutions of the diffusion equation. The fractional diffusion equation can be obtained, for example, from the CTRW formalism by suitable choices of the waiting time, $\omega(t)$, jumping distributions $\lambda(x)$ [41, 98], master equation [99], and comb model [100]. Equation (4.1) has an equivalent formulation in terms of the Riemann–Liouville fractional derivative for $0 < \gamma \leq 1$ and it may be obtained, for example, by means of the approach employed in Ref. [98] in the absence of external forces. Finally, it is worth mentioning that Eq. (4.1) could also be formulated in terms of the fractional time derivative of distributed order [101]. The main difference in this case would be the presence of different regimes of diffusion [102].

Notice *en passant* that Eq. (4.1) reduces to the usual diffusion equation for $\gamma = 1$, as expected. As for the usual case, the distribution $\rho(x,t)$ in Eq. (4.1), subjected to the boundary condition $\rho(\pm\infty,t)=0$, may be normalised, i.e., $\int_{-\infty}^{\infty} dx\rho(x,t) = \text{constant}$. To demonstrate this, it is convenient to rewrite Eq. (4.1) as

$$\frac{\partial^\gamma}{\partial t^\gamma}\rho(x,t) - \frac{\partial}{\partial x}J(x,t) = 0, \tag{4.2}$$

with

$$J(x,t) = D\frac{\partial}{\partial x}\rho(x,t) - F(x,t)\rho(x,t). \tag{4.3}$$

By integrating Eq. (4.2) over the interval $-\infty < x < \infty$ and employing the condition $J(x = \pm\infty, t) = 0$, which embodies the boundary condition $\rho(x = \pm\infty, t) = 0$, we easily show that $\int_{-\infty}^{\infty} dx\rho(x,t) = \text{constant}$, as expected.

To proceed, we first analyse, as previously mentioned, the scenario for which D is constant and external forces are absent. We also consider the initial condition given by $\rho(x,0) = \tilde{\rho}(x)$. This case yields for the fractional diffusion equation:

$$\frac{\partial^\gamma}{\partial t^\gamma}\rho(x,t) = D\frac{\partial^2}{\partial x^2}\rho(x,t). \tag{4.4}$$

Since depending on the boundary conditions considered, Eq. (4.4) may describe different physical systems, for simplicity and to illustrate the methods to be employed in the following chapters, we start by working out the boundary condition $\rho(\pm\infty,t) = 0$. This problem is emblematic and is helpful to illustrate the influence of the fractional operators on the solution, which may be found by employing the integral transforms of Fourier (space variables) and Laplace (time variables). Application of both integral transforms to Eq. (4.4) yields

$$\rho(k,s) = \frac{\rho(k,0)}{s + Ds^{1-\gamma}k^2}, \tag{4.5}$$

with $0 < \gamma < 1$. The next step is to obtain the inverses of the Laplace and Fourier transforms. After some calculation, by using the convolution theorem, the inverse Fourier transform of Eq. (4.5) is found to be

$$\hat{\rho}(x,s) = \int_{-\infty}^{\infty} \tilde{\rho}(x-x')G(x',s)\,dx', \tag{4.6}$$

where

$$G(x,s) = \frac{1}{2s}\left(\frac{s^{\gamma}}{D}\right)^{\frac{1}{2}} \exp\left[-\left(\frac{s^{\gamma}}{D}\right)^{\frac{1}{2}}|x|\right]. \tag{4.7}$$

Note that $G(x,s)$ is the Green's function of this problem and the form of this function is related to the boundary conditions to be satisfied by the solutions of Eq. (4.4).

In order to get the inverse Laplace transform, we may use the following procedure [41]: (i) relate the Laplace transform to the Mellin transform and (ii) invert the Mellin transform and connect the result obtained to the H-function of Fox. This is a way to get the distribution $\rho(x,t)$ for this diffusing process. By following this sketched procedure, after some calculation, it is possible to show that the Laplace and Mellin transforms may be connected by the expression:

$$\rho(x,s') = \frac{1}{\Gamma(1-s')}\int_{0}^{\infty} ds\, s^{-s'}\rho(x,s), \tag{4.8}$$

where $\rho(x,s')$ represents the function after applying the Mellin transform (see Section 1.1.4):

$$\rho(x,s') = \int_{0}^{\infty} dt\, t^{s'-1}\rho(x,t) \Leftrightarrow \rho(x,t) = \frac{1}{2\pi i}\int_{L} \rho(x,s')t^{-s'}ds',$$

in which $\rho(x,s)$ represents the function in the Laplace space. Substitution of Eq. (4.8) into Eq. (4.7) yields

$$G(x,s') = \frac{1}{\gamma|x|}\left(\frac{|x|}{\sqrt{D}}\right)^{\frac{2}{\gamma}s'} \frac{\Gamma\left(1-\frac{2}{\gamma}s'\right)}{\Gamma(1-s')}. \tag{4.9}$$

The inverse Mellin transform of Eq. (4.9) is

$$G(x,t) = \frac{1}{\sqrt{4\pi D t^{\gamma}}} \mathrm{H}^{2\,0}_{1\,2}\left[\frac{x^2}{4Dt^{\gamma}}\middle|\begin{matrix}\left(1-\frac{\gamma}{2},\gamma\right) \\ \left(\frac{1}{2},1\right)\ (0,1)\end{matrix}\right], \tag{4.10}$$

where we have used the H-function of Fox, defined in Chapter 1. The behaviour of Eq. (4.10) is illustrated in Fig. 4.1, for different values of γ, and Fig. 4.2 illustrates

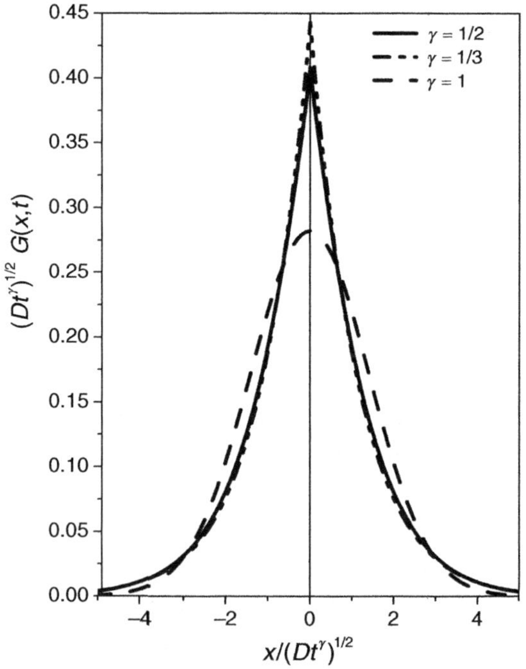

Figure 4.1 $(Dt^\gamma)^{1/2}G(x,t)$ versus $x/(Dt^\gamma)$, for typical values of γ, according to Eq. (4.10).

its asymptotic behaviour. In particular, it shows that for $\gamma < 1$ the tailed behaviour is more pronounced than for $\gamma = 1$. However, the spreading of the system for $\gamma = 1$ is faster than for $\gamma \neq 1$. This point can be checked by analysing the second moment. By using the above result, the solution, in terms of the Green's function, is

$$\rho(x,t) = \frac{1}{\sqrt{4\pi D t^\gamma}} \int_{-\infty}^{\infty} dx' \tilde{\rho}(x')\, \mathrm{H}_{1\,2}^{2\,0}\left[\frac{(x-x')^2}{4Dt^\gamma}\middle|\begin{matrix}\left(1-\frac{\gamma}{2},\gamma\right) \\ \left(\frac{1}{2},1\right)\ (0,1)\end{matrix}\right]. \tag{4.11}$$

Now, we focus our attention on the relaxation process predicted by the previous solution, which is essential to characterise the diffusion behind the fractional differential operators. In order to understand this aspect of the problem, we analyse the behaviour of the mean square displacement which, for a suitably chosen initial condition, can be directly connected with the second moment. We should also expect to obtain information about the spreading of the system, i.e., about how the system occupies the space. Consider, for simplicity, the initial condition $\rho(x,0) = \delta(x)$. It implies $\langle x^2 \rangle \propto t^\gamma$, which for $0 < \gamma < 1$ results in a subdiffusive process, since in this way the spreading of the system is slower than the spreading in the usual situation. By using the CTRW formalism, we may get some more information

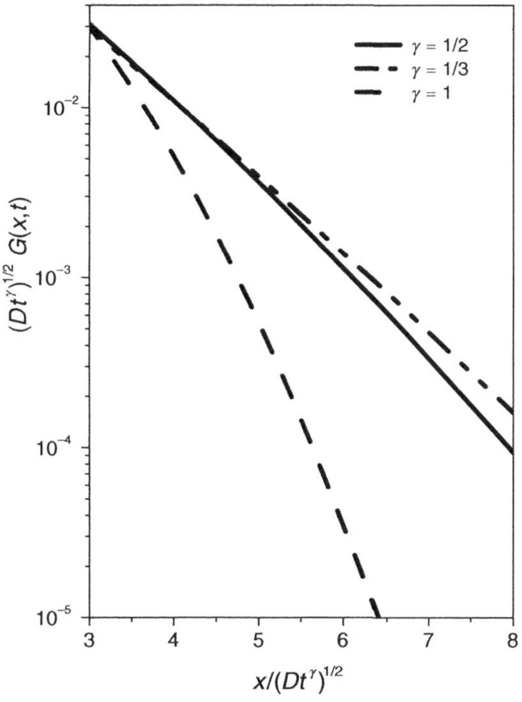

Figure 4.2 Asymptotic behaviour of the Green's function defined in Eq. (4.10).

about this kind of process, i.e., on what is happening with the system, when we analyse the waiting time distribution $\omega(t)$. It is possible to show that $\omega(t)$ related to the process analysed here is given by

$$\omega(t) = \frac{1}{\tau_0}\left(\frac{t}{\tau_0}\right)^{\gamma-1} \mathrm{E}_{\gamma,\gamma}\left(-\frac{t^\gamma}{\tau_0^\gamma}\right), \tag{4.12}$$

where τ_0 is a constant and $\mathrm{E}_{\alpha,\beta}(t)$ is the generalised Mittag-Leffler function, defined in Chapter 1. It has as asymptotic behaviour a power law, i.e., $\mathrm{E}_{\gamma,\gamma}(t) \sim 1/t^{2\gamma}$, for $t \to \infty$, in contrast to the exponential behaviour characteristic of the usual case. Thus, the presence of fractional operators applied to the time variable in the diffusion equation changes the behaviour of the waiting time distribution, $\omega(t)$, and, consequently, the resulting diffusive process is different from the usual one. Similar conclusions may be drawn if fractional spatial operators are present in the diffusion equation; however, in this case the changes occur in the jumping probability, $\lambda(x)$.

We consider now two different boundary conditions for the solution of the fractional diffusion equation (4.1): the first one is the absorbing,

$$\rho(0,t) = 0 \quad \text{and} \quad \rho(L,t) = 0, \tag{4.13}$$

whereas the second one is the reflecting,

$$\partial_x \rho(x,t)|_{x=0} = 0 \quad \text{and} \quad \partial_x \rho(x,t)|_{x=L} = 0, \tag{4.14}$$

boundary conditions. For these, we use the finite integral transforms, suitable to deal with the solutions for confined systems. A more detailed discussion about the fractional diffusion equations and the surface effects connected with them will be presented in Chapter 5. Here, we restrict our study to the changes produced on the relaxation process of the systems subjected to these boundary conditions.

To face the case with absorbing boundary, Eq. (4.13), we use a finite integral transform defined in the interval $0 < x < L$; i.e., it may be handled with the help of a Fourier series in terms of the sine functions. The solution may be written as

$$\rho(x,t) = \sum_{n=1}^{\infty} B_n(t) \sin\left(\frac{n\pi}{L}x\right), \tag{4.15}$$

with

$$B_n(t) = \frac{2}{L}\int_0^L \sin\left(\frac{n\pi}{L}x\right) \rho(x,t)\, dx. \tag{4.16}$$

In this way as well, the solution is a series expansion in terms of the eigenfunctions of the spatial operator of the fractional diffusion equation and is also a Fourier sine series. By substituting Eq. (4.15) into Eq. (4.4) and using the orthogonality of the eigenfunctions, it is possible to obtain an equation governing the Fourier coefficients, $B_n(t)$, in the form

$$\frac{d^\gamma}{dt^\gamma} B_n(t) = -\frac{n^2\pi^2}{L^2} D B_n(t). \tag{4.17}$$

To solve the above equation, we may use the Laplace transform, which simplifies the calculations and permits us to identify the solution with the Mittag-Leffler functions. After performing some standard calculations, it is possible to show that

$$B_n(t) = B_n(0)\mathrm{E}_\gamma\left(-\frac{n^2\pi^2}{L^2} D t^\gamma\right), \tag{4.18}$$

where $B_n(0)$ are determined from the initial condition. Specifically, for the initial condition $\rho(x,0) = \tilde{\rho}(x)$ the solution represented by Eq. (4.15) becomes

$$\rho(x,t) = \int_0^L dx'\; G(x,x',t)\tilde{\rho}(x')$$

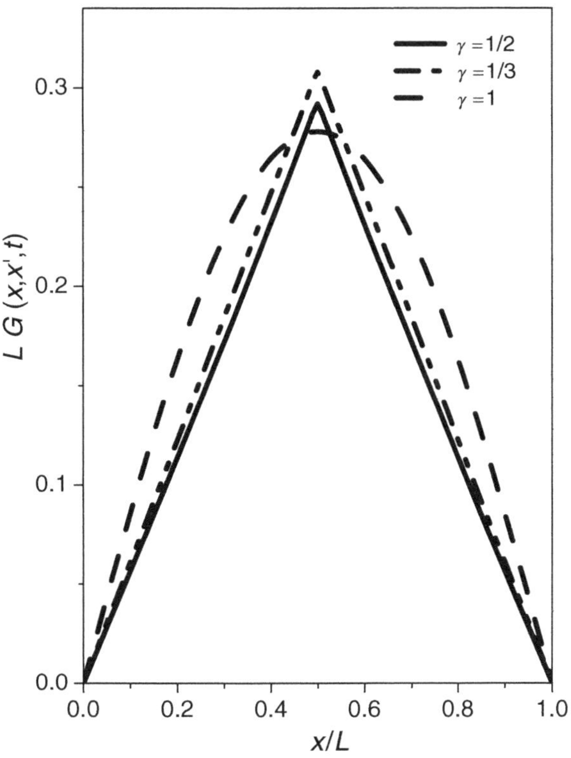

Figure 4.3 The Green's function defined in Eq. (4.19) versus x/L by considering, for illustrative purposes, $x' = L/2$ and $t = \left[L^2/(5\pi^2 D)\right]^{1/\gamma}$.

with

$$G(x,x',t) = \frac{2}{L}\sum_{n=1}^{\infty} \sin\left(\frac{n\pi}{L}x'\right) \sin\left(\frac{n\pi}{L}x\right) \mathrm{E}_{\gamma}\left(-\frac{n^2\pi^2}{L^2}Dt^{\gamma}\right), \tag{4.19}$$

whose behaviour is exhibited in Fig. 4.3 for three values of γ.

For reflecting boundary conditions, Eq. (4.14), the solution can be also expressed in terms of a Fourier series. This can be verified by analysing the eigenfunctions of the spatial operator of the fractional diffusion equation. By employing the same procedure of the previous case, we also obtain

$$\rho(x,t) = \int_0^L dx' G(x,x',t)\tilde{\rho}(x'),$$

with

$$G(x,x',t) = \frac{1}{L} + \frac{2}{L}\sum_{n=1}^{\infty} \cos\left(\frac{n\pi}{L}x'\right) \cos\left(\frac{n\pi}{L}x\right) \mathrm{E}_{\gamma}\left(-\frac{n^2\pi^2}{L^2}Dt^{\gamma}\right), \tag{4.20}$$

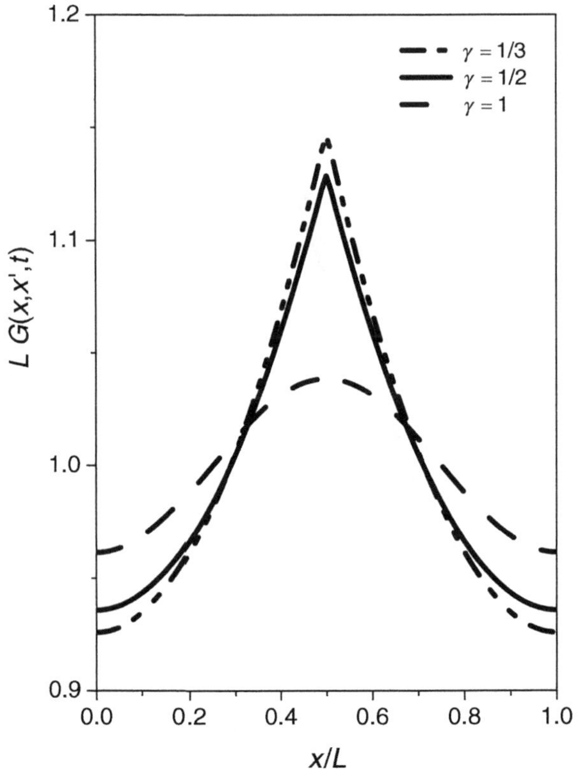

Figure 4.4 The Green's function defined in Eq. (4.20) versus x/L, for $x' = L/2$ and $t = \left[L^2/(10\pi^2 D)\right]^{1/\gamma}$.

for a given arbitrary initial condition. Figure 4.4 exhibits the general behaviour predicted by Eq. (4.20) for different values of γ. The solution is given in terms of the cosine function in contrast to the functions used for the absorbing boundary conditions. In the limit $t \to \infty$, the stationary state is reached as

$$\rho(x, t \to \infty) \to \frac{1}{L} \int_0^L dx' \tilde{\rho}(x').$$

This result can be obtained by noticing that $\lim_{t\to\infty} \mathrm{E}_\gamma(-n^2\pi^2 D t^\gamma / L^2) = 0$. This limit can be easily checked since the Mittag-Leffler function, $\mathrm{E}_\alpha(x)$, for the interval of γ considered here is a monotonically decreasing function. Using the stationary solution, it is possible to find the stationary autocorrelation function for this case [103] and to show what the connection is between the solution for an arbitrary t and the initial solution. The autocorrelation function is defined as

$$\langle x(t)x(0)\rangle_s = \int_0^L dx \int_0^L dx' x x' \rho(x,t) \rho_s(x) \, , \tag{4.21}$$

where $\rho_s(x)$ is the solution for the stationary case. By considering, for simplicity, $\rho(x,0) = \delta(x-x')$ as the initial condition, we obtain $\rho_s(x) = 1/L$. By using this result and the solution given by Eq. (4.20), after some calculation, we eventually find

$$\langle x(t)x(0)\rangle_s = \frac{L^2}{4} + \frac{8L^4}{\pi^4}\sum_{n=0}^{\infty}\frac{1}{(2n+1)^4}\mathrm{E}_\gamma\left[-\frac{(2n+1)^2}{L^2}Dt^\gamma\right]. \tag{4.22}$$

In the limit of long times ($t \to \infty$), from Eq. (4.22), we obtain $\langle x(t)x(0)\rangle \sim L^2/4$. This result is the same as the one obtained for the usual diffusion equation [103]. It shows that the diffusion equation with fractional time derivatives has an anomalous relaxation towards the equilibrium solution when suitable conditions are imposed on the system.

To go on to more general scenarios, let us now extend the previous results by incorporating a spatial dependence into the diffusion coefficient. Consider, for instance, the diffusion coefficient as given by Eq. (3.104) whose dependence has been used to investigate physical situations like the diffusion on fractals, turbulence, and the behaviour of fast electrons in hot plasma, in the presence of an electric field. For simplicity, we consider first the problem in the absence of external force, i.e., $F(x,t) = 0$. Applying the same procedure as before, it is possible to show that the solution is given by

$$\rho(x,t) = \frac{2+\theta}{2\Gamma\left(\frac{1}{2+\theta}\right)}\left[\frac{1}{(2+\theta)^2Dt^\gamma}\right]^{\frac{1}{2+\theta}}\mathrm{H}^{2\,0}_{1\,2}\left[\frac{|x|^{2+\theta}}{(2+\theta)^2Dt^\gamma}\left|\begin{matrix}\left(1-\frac{\gamma}{2+\theta},\gamma\right) \\ \left(1-\frac{1}{2+\theta},1\right)\ (0,1)\end{matrix}\right.\right], \tag{4.23}$$

satisfying the boundary conditions $\rho(\pm\infty,t) = 0$, with the initial condition $\rho(x,0) = \delta(x)$. For $\gamma = 1$ and $\theta = 0$, Eq. (4.23) is reduced to the Gaussian distribution typical of the usual diffusive process.

Let us now add to the previous problem an external force of the kind $F(x) = -kx$, which is related to the Ornstein–Uhlenbeck process [103]. The solution may be obtained by employing a number of different procedures. The simplest way is perhaps to use the method of separation of variables. This procedure consists essentially in using the eigenfunctions of the spatial operator of the diffusion equation. In this case, it is possible to show that the eigenfunctions, even (+) and odd (−), related to the spatial operator are [104, 105]:

$$\psi_+(x) = A_+(n,\theta)L_n^{-\frac{1+\theta}{\theta+2}}\left(\frac{|x|^{2+\theta}}{2+\theta}\right)e^{-\frac{|x|^{2+\theta}}{2(2+\theta)}} \tag{4.24}$$

and

$$\psi_-(x) = A_+(n,\theta)x|x|^\theta L_n^{\frac{1+\theta}{\theta+2}}\left(\frac{|x|^{2+\theta}}{2+\theta}\right)e^{-\frac{|x|^{2+\theta}}{2(2+\theta)}}, \tag{4.25}$$

where

$$A_+(n,\theta) = (2+\theta)^{\frac{1+\theta}{2\theta+4}} \sqrt{\frac{\Gamma(1+n)}{2\Gamma\left(n+\frac{1}{2+\theta}\right)}}$$

and

$$A_-(n,\theta) = (2+\theta)^{-\frac{1+\theta}{2\theta+4}} \sqrt{\frac{\Gamma(1+n)}{2\Gamma\left(n+2-\frac{1}{2+\theta}\right)}}$$

and $L_n^{(\alpha)}(x)$ denotes the generalised Laguerre polynomials. Consequently, the general solution for this case may be written as

$$\rho(x,t) = \int_0^\infty d\xi\, \tilde{\rho}(\xi) G(x,\xi,t),$$

with the Green's function defined as follows:

$$G(x,\xi,t) = \sum_{n=0}^{\infty} \left[A_+(n,\theta)\right]^2 L_n^{-\frac{1+\theta}{\theta+2}}\left(\overline{\xi}\right) L_n^{-\frac{1+\theta}{\theta+2}}\left(\overline{x}\right) e^{-\overline{x}} \mathrm{E}_\gamma\left(-\lambda_{+,n} t^\gamma\right)$$
$$+ \sum_{n=0}^{\infty} \left[A_-(n,\theta)\right]^2 x\xi |x\xi|^\theta L_n^{\frac{1+\theta}{\theta+2}}\left(\overline{\xi}\right) L_n^{\frac{1+\theta}{\theta+2}}\left(\overline{x}\right) e^{-\overline{x}} E_\gamma\left(-\lambda_{-,n} t^\gamma\right), \quad (4.26)$$

in which, for compactness, we have introduced the quantities

$$\overline{\xi} = \frac{|\xi|^{2+\theta}}{2+\theta} \quad \text{and} \quad \overline{x} = \frac{|x|^{2+\theta}}{2+\theta},$$

and the eigenvalues are defined as $\lambda_{+,n} = -(2+\theta)n$ and $\lambda_{-,n} = -(2+\theta)n - 1 - \theta$.

To obtain exact solutions for this problem is not a simple task, depending on the external force considered. For this reason, it may be useful to develop a perturbation theory to investigate these problems. To do this, we first obtain the integral equation associated with the fractional diffusion equation (4.1) by considering the simplest case represented by the boundary conditions $\rho(\pm\infty, t) = 0$. The formalism employed below may be applied to other boundary conditions, but the resulting Green's functions will be different. To build a perturbation approach by means of recursive iterations, we rewrite Eq. (4.1) in the form

$$\frac{\partial^\gamma}{\partial t^\gamma}\rho(x,t) = D\frac{\partial^2}{\partial x^2}\rho(x,t) - \alpha(x,t), \quad (4.27)$$

with

$$\alpha(x,t) = \frac{\partial}{\partial x}\left[F(x,t)\rho(x,t)\right]. \quad (4.28)$$

By applying Fourier and Laplace transforms to Eq. (4.27), we obtain

$$\rho(k,s) = \frac{\rho(k,0)}{s + Ds^{1-\gamma}k^2} - \frac{\alpha(k,s)}{s^\gamma + Dk^2}. \tag{4.29}$$

From the inverse Fourier and Laplace transforms, we get

$$\rho(x,t) = \rho^{(0)}(x,t) - \int_0^t dt' \int_{-\infty}^{\infty} G_\gamma^{(1)}(x-x', t-t')\, \alpha(x',t') dx' \tag{4.30}$$

and

$$\rho^{(0)}(x,t) = \int_{-\infty}^{\infty} dx' G_\gamma^{(0)}(x-x',t)\rho(x',0), \tag{4.31}$$

with the Green's functions defined as

$$G_\gamma^{(0)}(x,t) = \frac{1}{\sqrt{4Dt^\gamma}} \mathrm{H}_{1\,1}^{1\,0}\left[\frac{|x|}{\sqrt{Dt^\gamma}} \middle|_{(0,1)}^{(1-\frac{\gamma}{2},\frac{\gamma}{2})}\right] \tag{4.32}$$

and

$$G_\gamma^{(1)}(x,t) = \frac{1}{\sqrt{4D/t^\gamma}\,t} \mathrm{H}_{1\,1}^{1\,0}\left[\frac{|x|}{\sqrt{Dt^\gamma}} \middle|_{(0,1)}^{(\frac{\gamma}{2},\frac{\gamma}{2})}\right]. \tag{4.33}$$

Substitution of $\alpha(x,t)$ into Eq. (4.30), after performing integration by parts, yields

$$\rho(x,t) = \rho^{(0)}(x,t) + \int_0^t dt' \int_{-\infty}^{\infty} dx' G_\gamma^{(2)}(x-x',t-t') \left[F(x',t')\rho(x',t')\right], \tag{4.34}$$

in which

$$\begin{aligned} G_\gamma^{(2)}(x,t) &= \frac{d}{dx} G_\gamma^{(1)}(x,t) \\ &= -\frac{1}{2Dt\sqrt{Dt}} \mathrm{H}_{1\,0}^{1\,1}\left[\frac{|x|}{\sqrt{Dt^\gamma}} \middle|_{(0,1)}^{(0,\frac{\gamma}{2})}\right]. \end{aligned} \tag{4.35}$$

Equation (4.34) is an integral equation associated with Eq. (4.1) and can be used to investigate the influence of the external field on the solution of the problem, when it can be treated as a perturbation.

4.2 Fractional Spatial Derivative: Simple Situations

The preceding section was dedicated to investigating the influence of the fractional time derivative on solutions of diffusion equations, keeping in mind the determination of how the spreading of the system changes. The CTRW was used to understand the changes produced in the system in terms of the waiting time and jumping distributions. In this section, we consider the role of the spatial fractional derivative

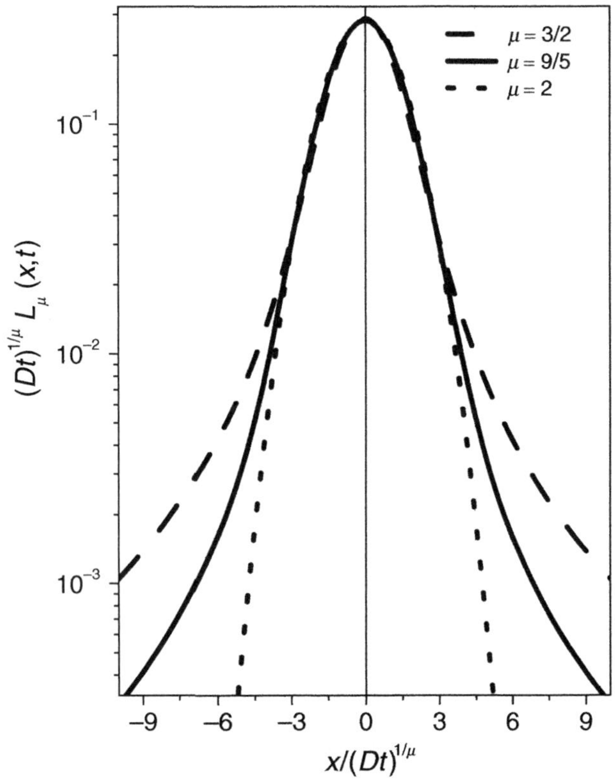

Figure 4.5 $(Dt)^{\frac{1}{\mu}} L_\mu(x,t)$ versus $x/(Dt)^{\frac{1}{\mu}}$, illustrating the behaviour predicted by Eq. (4.39), for typical values of μ.

in the diffusion equation and analyse again the consequences on the spreading of the system. An immediate consequence of incorporating the spatial fractional derivatives is that the solution may be given in terms of the Lévy distributions. These distributions predict power-law tail behaviour and not exponential behaviour, as the usual solutions, and find applications in several contexts [41, 98, 106]. Indeed, the Lévy distributions

$$L_\mu(x,t) = \int_{-\infty}^{\infty} \frac{dk}{2\pi} e^{-ikx-|k|^\mu Dt}, \tag{4.36}$$

whose behaviour is shown in Fig. 4.5 for three values of μ, satisfy

$$\frac{\partial \rho}{\partial t} = D\frac{\partial^\mu \rho}{\partial |x|^\mu} - \frac{\partial}{\partial x}\left[F(x,t)\rho(x,t)\right], \tag{4.37}$$

in the absence of external force, i.e., $F(x,t) = 0$, when the boundary condition $\rho(\pm\infty,t) = 0$ and the initial condition $\rho(x,0) = \delta(x)$ are required. In order to demonstrate this, we can use the result

$$\mathcal{F}\{\partial^{\mu}_{|x|}\rho(x,t);k\} \equiv -|k|^{\mu}\rho(k,t),$$

where $\mathcal{F}\{\ldots\}$ stands for the Fourier transform (for details, see Chapters 1 and 2). This gives the solution

$$\rho(x,t) = \mathrm{L}_{\mu}(x,t), \tag{4.38}$$

as stated before. By applying the procedure of CTRW to Eq. (4.37), as done in the previous section, we conclude that the jumping distribution changes, and is now given by $\lambda(k) = 1 - \sigma^{\mu}|k|^{\mu}$, with $\sigma^{\mu} = D\tau^{\mu}$. The usual result is reobtained for $\mu = 2$. This jumping distribution demonstrates that the distribution function associated with this fractional diffusion equation has a long-tailed behaviour, in such a way that the stochastic process described by it is non-Markovian.

An immediate extension of the above result may be achieved by incorporating a fractional time derivative in Eq. (4.37). In the context of the CTRW, this means changing the waiting time and the jumping distributions in order to associate with them a distribution with a long-tailed behaviour. By using integral transforms, it is possible to show that, in absence of external force, the solution is given by

$$\rho(x,t) = \frac{1}{\mu|x|}\mathrm{H}^{2\,1}_{3\,3}\left[\frac{|x|}{(Dt^{\gamma})^{\frac{1}{\mu}}}\middle|\begin{matrix}\left(1,\frac{1}{\mu}\right)\left(1,\frac{\gamma}{\mu}\right)\left(1,\frac{1}{2}\right)\\(1,1)\left(1,\frac{1}{\mu}\right)\left(1,\frac{1}{2}\right)\end{matrix}\right]. \tag{4.39}$$

On the other hand, if we have considered instead of fractional time derivatives the presence of the external force $F(x) = -kx$, we would obtain

$$\rho(x,t) = \frac{1}{\mu|x|}\mathrm{H}^{2\,2}_{1\,1}\left[\left(\frac{\alpha k}{D(t)}\right)^{\frac{1}{\mu}}|x|\middle|\begin{matrix}\left(1,\frac{1}{\mu}\right)\left(1,\frac{1}{2}\right)\\(1,1)\left(1,\frac{1}{2}\right)\end{matrix}\right], \tag{4.40}$$

where $D(t) = D[1 - \exp(-\alpha kt)]$. Other extensions of this approach are also possible, like incorporating a spatial or a nonlocal dependence into the diffusion coefficient [107–111].

If we consider now both the fractional spatial derivative and the action of an external force, the problem becomes difficult, as occurred with the fractional time derivative, requiring cumbersome calculations. To avoid them, a perturbation theory may be implemented as well, investigating the integral equation related to this diffusion equation, as done before. In order to perform this calculation in connection with Eq. (4.37) and to obtain the corresponding integral equation, we consider the boundary conditions, $\rho(\pm\infty,t) = 0$, together with the initial condition, $\rho(x,0) = \tilde{\rho}(x)$. Applying the procedure of the previous section, it is possible to show that

$$\rho(x,t) = \rho^{(0)}(x,t) - \int_0^t dt' \int_{-\infty}^{\infty} dx' G^{(2)}_{\mu}(x-x',t-t')F(x',t')\rho(x',t'),$$

where

$$\rho^{(0)}(x,t) = \int_{-\infty}^{\infty} dx' G_\mu(x - x', t - t')\tilde{\rho}(x'), \tag{4.41}$$

with $G_\mu(x,t) = L_\mu(x,t)$, and

$$\begin{aligned} G_\mu^{(2)}(x,t) &= -\frac{d}{dx} L_\mu(x,t) \\ &= \int_0^{\infty} \frac{dk}{\pi} k \sin(kx) e^{-tD|k|^\mu}. \end{aligned} \tag{4.42}$$

Equation (4.41) is the integral equation connected with Eq. (4.37) and may be used to obtain approximated solutions for the spatial fractional diffusion equation. It is worth mentioning that in this development the external force admits a perturbative development and no divergences are present.

4.3 Sorption and Desorption Processes

The preceding sections were dedicated to a simple class of fractional diffusion equations commonly used to discuss anomalous diffusion. Of course, there is a large class of fractional diffusion equations to be considered in connection with the anomalous diffusion phenomena. In particular, one can consider fractional diffusion equations dealing with time and spatial fractional operators in the presence of reaction terms. These reaction terms may be supposed to act at the boundaries of the system, i.e., on the surfaces that are in contact with the bulk, or directly in the bulk, being represented by appropriate terms in the diffusion equation. In the latter case, the reaction process occurs in the bulk system, where the substances diffuse, and have been extensively investigated in view of the importance of chemical reactions for scientific and industrial applications [112–115].

Let us consider in this section a different scenario which may appear during the diffusion process. The sorption or sorption–desorption process of one substance that spreads out through another has the possibility of chemically reacting [116]. To do this, we consider that the density of particles is governed by the following fractional diffusion equation:

$$\frac{\partial}{\partial t}\rho(x,t) = \mathcal{K}_\gamma \, {}_0\mathrm{D}_t^{1-\gamma}\left[\frac{\partial^\mu}{\partial |x|^\mu}\rho(x,t)\right] - \frac{\partial}{\partial t}\Gamma(x,t), \tag{4.43}$$

where $\mathcal{K}_\gamma$ is the diffusion coefficient, $0 < \gamma \le 1$ (for $\gamma = 1$ usual diffusion, $0 < \gamma < 1$ subdiffusion), and the fractional time derivative is the Riemann–Liouville, as defined in Chapter 2.[1] The spatial fractional derivative is the Riesz–Weyl one, with $1 < \mu \leq 2$, as also defined in Chapter 2. In Eq. (4.43), $\rho(x,t)$ represents the density of diffusing particles, whereas $\Gamma(x,t)$ is connected with the density of absorbed particles. The last term may represent the process of a substance being absorbed by another as well as the possibility of these substances reacting chemically. This process can thus be understood either as diffusion, in which part of the substance becomes immobilised, or as a chemical kinetics problem, in which the rate of reaction depends on the supply rate of one of the reactants by diffusion. Examples involving diffusion into living cells and microorganisms can be found in biology and biochemistry [117]. In order to keep these contexts as general as possible, we consider, for $\Gamma(x,t)$, the following kinetic equation:

$$\frac{\partial}{\partial t}\Gamma(x,t) = \int_0^t k_f(t-t')\rho(x,t')dt' - \int_0^t k_b(t-t')\Gamma(x,t')dt', \tag{4.44}$$

where $k_f(t)$ and $k_b(t)$ are, respectively, the rates of the forward and backward reactions. Thus, the immobilised solute is formed at a rate proportional to the concentration of solute free to diffuse, and disappears at a rate proportional to its own concentration. This feature produces an alternating between periods of diffusive transport and resting times, which are governed by $k_f(t)$ and $k_b(t)$. Equation (4.44) embodies, as particular cases, several problems worked out before [118, 119].

Our discussion starts by focusing the time-dependent solutions of Eq. (4.43), subjected to the boundary conditions $\rho(\pm\infty,t) = 0$ and $\Gamma(\pm\infty,t) = 0$. In order to solve this problem, we use the Laplace and Fourier transforms again. After applying the Fourier transform, Eqs. (4.43) and (4.44) can be written, respectively, as

$$-\mathcal{K}_\gamma |k|^\mu \, {}_0\mathrm{D}_t^{1-\gamma}\left[\rho(k,t)\right] - \frac{\partial}{\partial t}\Gamma(k,t) = \frac{\partial}{\partial t}\rho(k,t) \tag{4.45}$$

and

$$\frac{\partial}{\partial t}\Gamma(k,t) = \int_0^t k_f(t-t')\rho(k,t')dt' - \int_0^t k_b(t-t')\Gamma(k,t')dt'. \tag{4.46}$$

Now, by using the Laplace transform in Eqs. (4.45) and (4.46) we may write

$$\rho(k,s) = \left[\tilde{\rho}(k) + \frac{k_b(s)}{s+k_b(s)}\Gamma(k,0)\right]G(k,s), \tag{4.47}$$

where

$$G(k,s) = \frac{1}{s+\mathcal{K}_\gamma s^{1-\gamma}|k|^\mu + \Upsilon(s)}, \quad \text{with} \quad \Upsilon(s) = \frac{s k_f(s)}{s+k_b(s)}, \tag{4.48}$$

[1] We adopt here a different notation for the diffusion coefficient to avoid confusion with the symbol for the fractional derivative.

represents the Green's function in the Fourier–Laplace space of Eq. (4.43), and

$$\Gamma(k,s) = \frac{1}{s+k_b(s)}\left[k_f(s)\rho(k,s)+\Gamma(k,0)\right]. \tag{4.49}$$

The initial condition of the system depends on $\rho(k)$ and $\Gamma(k,0)$, which, for simplicity, are given by $\rho(k)=\rho_i$ and $\Gamma(k,0)=\Gamma_i$ (where $\rho_i+\Gamma_i=1$).

We notice that the processes described by Eq. (4.43), with the Green's function given by Eq. (4.48), may also be connected with a random walk. For instance, in the absence of the reaction term, i.e., when $k_b(t)=k_f(t)=0$, by performing a direct comparison between the Green's function, Eq. (4.48), and the continuous-time random walk approach [41, 98], we obtain

$$\psi(s) = \frac{1}{1+\tau^\gamma s^\gamma} \quad \text{and} \quad \lambda(k) = 1-\mathcal{K}_\gamma \tau^\gamma |k|^\mu,$$

in which τ is a characteristic time scale. The presence of the reaction term, i.e., when $k_b(t)\neq 0$ and $k_f(t)\neq 0$, implies a creation and an annihilation process in which there is addition or removal of the walkers at the starting of the waiting time between steps. Thus, for the initial conditions $\rho(x,0)=\delta(x)$ and $\Gamma(x,0)=0$, by using the random walk approach it is possible to connect Eqs. (4.43) and (4.44) with the processes described by the balance equations:

$$\begin{aligned}\rho(x,t) &= \Phi(t)\delta(x) + \int_{-\infty}^{\infty} dx' \int_0^t dt'\psi(t-t')\lambda(x-x')\rho(x',t') \\ &\quad + \int_0^t dt'\Phi_b(t-t')\Gamma(x,t') - \int_0^t dt'\Phi_f(t-t')\rho(x,t')\end{aligned} \tag{4.50}$$

and

$$\Gamma(x,t) = \int_0^t dt' \int_0^{t'} d\tilde{t} k_f(t'-\tilde{t})\rho(x,\tilde{t}) - \int_0^t dt' \int_0^{t'} d\tilde{t} k_b(t'-\tilde{t})\Gamma(x,\tilde{t}), \tag{4.51}$$

where

$$\Phi(t) = 1-\int_0^t \psi(t)dt \quad \text{and} \quad \Phi_{bf}(t) = \int_0^t dt'\Phi(t-t')k_{bf}(t').$$

After some calculation, these equations can be simplified to

$$\begin{aligned}\rho(x,t) &= \Phi(t)\delta(x) + \int_{-\infty}^{\infty} dx' \int_0^t dt'\psi(t-t')\lambda(x-x')\rho(x',t') \\ &\quad + \int_0^t dt'\Phi(t-t')R(x,t'),\end{aligned} \tag{4.52}$$

with

$$R(x,t) = -\frac{\partial}{\partial t}\int_0^t dt' M(t-t')\rho(x,t'), \tag{4.53}$$

where

$$M(t) = \mathcal{L}^{-1}\left\{\frac{k_b(s)}{[s+k_f(s)]}\right\}.$$

By considering the asymptotic limit of $t \to \infty$ ($s \to 0$), for

$$\frac{k_b(s)}{s+k_f(s)} \sim \kappa = \text{constant},$$

Eq. (4.52) can be simplified to a fractional diffusion equation and, for

$$\frac{k_b(s)}{s+k_f(s)} \sim \frac{1}{s^\gamma},$$

it can be reduced to the fractional reaction-diffusion equation (43) of Ref. [120], in which a nonnegative behaviour of $\rho(x,t)$ is exhibited. However, for $k_b(t) = 0$, nonpositive behaviours for the solutions can be manifested.

From the previous results it is possible to obtain, in the Laplace space, some quantities of interest such as the mean square displacement, i.e., $\Delta_x^2(t) = \langle (x - \langle x \rangle)^2 \rangle$, for $\mu = 2$ and the survival probability, i.e.,

$$\mathcal{S}(t) = \int_0^\infty dx \rho(x,t), \tag{4.54}$$

which is connected with the mobile particle present in the bulk. For the mean square displacement, we have

$$\overline{\Delta}_x^2(s) = \frac{2\mathcal{K}_\gamma}{s^{1+\gamma}}\left[\rho_i + \frac{k_b(s)}{s+k_b(s)}\Gamma_i\right]\left[\frac{s+k_b(s)}{s+k_b(s)+k_f(s)}\right]^2. \tag{4.55}$$

Equation (4.55) indicates that the distribution of particles spreads out depending on the parameters $k_b(s)$ and $k_f(s)$; i.e., different behaviours are found depending on the values of these parameters. The same feature can be found, for example, in the cases characterised asymptotically ($s \to 0$ or $t \to \infty$) by power laws, which imply that $\overline{k}_b(s) \sim k_b s^{\eta_b}$ and $\overline{k}_f(s) \sim k_f s^{\eta_f}$, with $0 < \eta_f < \eta_b < 1$, yielding

$$\Delta_x^2(t) \sim \begin{cases} \frac{2\mathcal{K}_\gamma}{\Gamma(1+\gamma)} t^\gamma, & t \ll t_c, \\ \frac{2\mathcal{K}_\gamma}{\Gamma(1+\xi)}\left(\frac{k_b}{k_f}\right)^2 t^\xi, & t \gg t_c, \end{cases} \tag{4.56}$$

calculated for the initial condition $\rho_i = 1$ and $\Gamma_i = 0$, with $\xi = \gamma - 2\left(\eta_b - \eta_f\right)$. In Eq. (4.56),

$$t_c = \left[\frac{\Gamma(1+\gamma)}{\Gamma(1+\xi)}\left(\frac{k_b}{k_f}\right)^2\right]^{\frac{1}{2(\eta_b-\eta_f)}}$$

is the crossover time between the two regimes that arise when these rates are considered. Note that the behaviour for $t \gg t_c$ can be subdiffusive when $0 < \gamma - 2(\eta_b - \eta_f) < 1$, superdiffusive when $1 < \gamma - 2(\eta_b - \eta_f)$, or stationary when $\gamma = 2(\eta_b - \eta_f)$.

The last condition on ξ implies that the particles present in the bulk and the kinetic process given by Eq. (4.44) reach an equilibrium situation. For the cases characterised by constant rates, we have the asymptotic limit governed by the diffusive term which depends on γ, as illustrated in Fig. 4.6. The initial condition of the system may also lead to an anomalous behaviour for short times, as shown in Fig. 4.6b. This point is also illustrated in Fig. 4.6 by considering that the system is initially immobilised, i.e., $\Gamma_i = 1$ and $\rho_i = 0$, producing an anomalous spreading of the particles for short times, i.e.,

$$t \ll (1+\gamma)\frac{k_f}{(k_f + k_b)^2}$$

as the particles are released by a physical or chemical process. In this limit, the spreading of the system is governed by

$$\Delta_x^2(t) \sim 2k_f \frac{\mathcal{K}_\gamma t^{1+\gamma}}{\Gamma(1+\gamma)}.$$

The behaviour exhibited, for example, in the solid line of Fig. 4.6a may represent the effect of molecular crowding on diffusion in the cytoplasm [121], where for short times an usual diffusion occurs, being followed by an anomalous diffusion and, then, by an usual diffusion with an effective diffusion coefficient. The other line in Fig. 4.6a represents the expected behaviour when the substance is initially bound.

For $\mu \neq 2$, $1/\rho^2(x,t)$ as a measure of the spreading of the system (see e.g. Fig. 4.7) is shown. We observe that, for the initial condition $\rho_i = 1$, with $\Gamma_i = 0$, the behaviour for short and long times is governed by $1/\rho^2(x,t) \sim t^{2/\mu}$. For $\rho_i = 0$ with $\Gamma_i = 1$, only the asymptotic behaviour is such that $1/\rho^2(x,t) \sim t^{2/\mu}$.

The survival probability, which in the above context reflects the quantity of mobile particles present in the bulk, is

$$\mathcal{S}(s) = \left[\rho_i + \frac{k_b(s)}{s + k_b(s)}\Gamma_i\right]\frac{s + k_b(s)}{s[s + k_b(s) + k_f(s)]}. \tag{4.57}$$

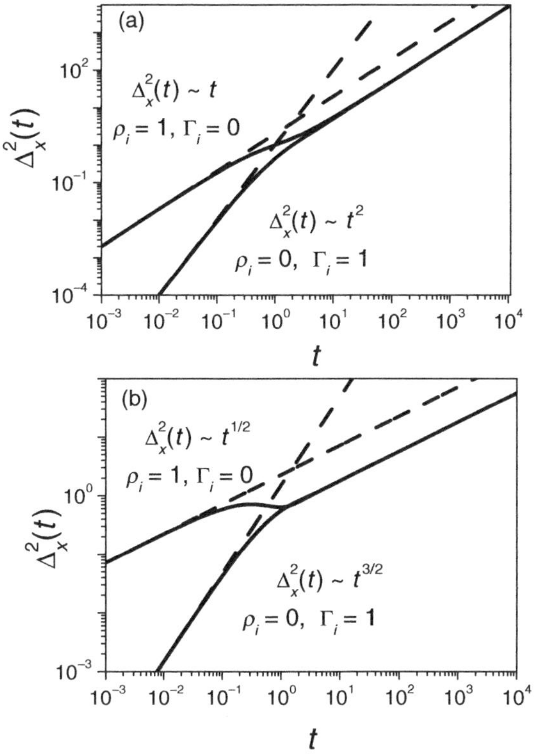

Figure 4.6 The mean square displacement for two different initial conditions and constant rates. We consider $\gamma = 1$ in (a) and $\gamma = 1/2$ in (b). For illustrative purposes, it is assumed that $k_f = 1$, $k_b = 1$, and $\mathcal{K}_\gamma = 1$ [116]. Modified from Physica A, **443/1**, E. K. Lenzi, M. A. F. dos Santos, D. S. Vieira, R. S. Zola, and H. V. Ribeiro, Solutions for a sorption process governed by a fractional diffusion equation, 32–41. Copyright (2016), with permission from Elsevier.

Equation (4.57) depends on the reaction processes in the bulk and does not show dependence on the diffusive term, because the system is not limited (or confined). Figure 4.8 shows the behaviour of Eq. (4.57) for different $k_b(t)$ and $k_f(t)$. For the case analysed before, asymptotically characterised by $k_b(s) \sim k_b s^{\eta_b}$ and $k_f(s) \sim k_f s^{\eta_f}$, however with $0 < \eta_b < \eta_f < 1$, we have:

$$\mathcal{S}(t) \sim \frac{k_b}{k_f} t^\alpha E_{\alpha,1+\alpha}\left(-\frac{k_b}{k_f} t^\alpha\right), \tag{4.58}$$

calculated for $\rho_i = 1$ and $\Gamma_i = 0$, $\alpha = \eta_f - \eta_b$, where $E_{\alpha,\beta}(x)$ is the generalised Mittag-Leffler function defined in Chapter 1. For $k_b(s) \sim k_b/s^{\eta_b}$ with $k_f(s) \sim k_f/s^{\eta_f}$, with the same initial condition, we have

$$\mathcal{S}(t) \sim E_\alpha\left(-\frac{k_f}{k_b} t^\alpha\right). \tag{4.59}$$

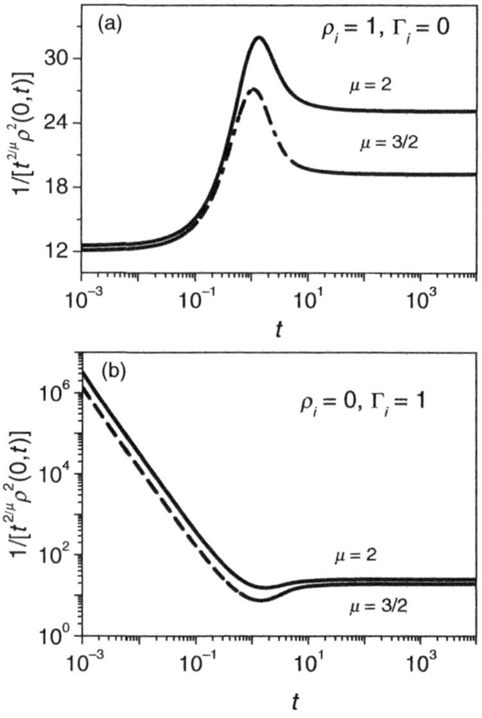

Figure 4.7 $1/[t^{2/\mu}\rho^2(0,t)]$ versus t for two different initial conditions and constant rates. For illustrative purposes, the curves were drawn for $\gamma = 1$, $k_f = 1$, $k_b = 1$, and $\mathcal{K}_\gamma = 1$ [116]. Modified from Physica A, **443**/1, E. K. Lenzi, M. A. F. dos Santos, D. S. Vieira, R. S. Zola, and H. V. Ribeiro, Solutions for a sorption process governed by a fractional diffusion equation, 32–41. Copyright (2016), with permission from Elsevier.

From the asymptotic behaviour shown by Eqs. (4.58) and (4.59), it is possible to observe two situations: the particles are absorbed and, then, desorbed (Eq. (4.58) with $\mathcal{S}(t) \to$ constant, for $t \to \infty$), or the particles are absorbed and, then, immobilised (Eq. (4.59), with $\mathcal{S}(t) \to 0$, for $t \to \infty$).

Now, let us determine the inverse Fourier and Laplace transforms of Eq. (4.48). First, we focus our attention on the Green's function, which can be written as follows:

$$G(k,s) = \frac{1}{s + \mathcal{K}_\gamma s^{1-\gamma}|k|^\mu} + \sum_{n=1}^{\infty} \frac{\left[-\overline{\Upsilon}(s)\right]^n}{\left(s + \mathcal{K}_\gamma s^{1-\gamma}|k|^\mu\right)^{n+1}}. \tag{4.60}$$

The first term of Eq. (4.60) corresponds to the Green's function of the fractional diffusion equation in the absence of the reaction term. The second term is the contribution of the reaction term or the spreading of the system. By applying the inverse Laplace transform, we obtain

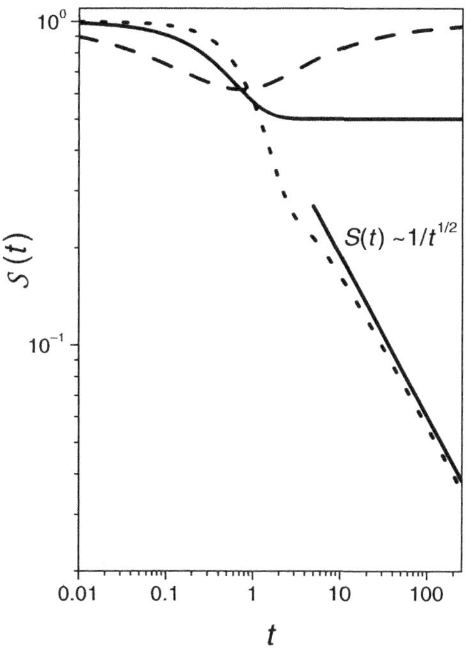

Figure 4.8 $\mathcal{S}(t)$ versus t for different $k_f(t)$ and $k_b(t)$. The dashed, solid, and dotted lines correspond to $k_f(s) \sim \overline{k}_f s^{\eta_f}$, $k_b(s) = \overline{k}_b$, $k_f(s) = \overline{k}_f$, $k_b(s) = \overline{k}_b$, and $k_f(s) = \overline{k}_f/s^{\eta_f}$, $k_b(s) = \overline{k}_b$. We observe that, depending on the rates, the survival probability may exhibit different asymptotic behaviours; i.e., we may have $\mathcal{S}(t) \to$ constant or $\mathcal{S}(t) \to 1/t^{\delta}$ $(0 < \delta < 1)$, for $t \to \infty$. The curves were drawn for $\overline{k}_f = 1$, $\eta_f = 1/2$, $\rho_i = 1$, $\Gamma_i = 0$, and $\overline{k}_b = 1$ in arbitrary units [116]. Modified from Physica A, **443**/1, E. K. Lenzi, M. A. F. dos Santos, D. S. Vieira, R. S. Zola, and H. V. Ribeiro, Solutions for a sorption process governed by a fractional diffusion equation, 32–41. Copyright (2016), with permission from Elsevier.

$$G(k,t) = E_\gamma\left(-\mathcal{K}_\gamma t^\gamma |k|^\mu\right) + \sum_{n=1}^{\infty}\frac{(-1)^n}{\Gamma(1+n)}\int_0^t dt' \Lambda_n(t-t')t'^n \mathrm{H}^{1,2}_{1,1}\left[\mathcal{K}_\gamma t'^\gamma |k|^\mu \left|^{(-n,1)}_{(0,\,1)\,(-n,\gamma)}\right.\right], \tag{4.61}$$

with $\Lambda_n(t) = \Upsilon(t)$, for $n = 1$, and

$$\Lambda_n(t) = \int_0^t dt_{n-1}\Upsilon(t-t_{n-1})\int_0^{t_{n-1}} dt_{n-2}\Upsilon(t_{n-2}-t_{n-3}) \cdots \int_0^{t_2} dt_1\Upsilon(t_2-t_1)\Upsilon(t_1), \tag{4.62}$$

for $n > 1$, with

$$\Upsilon(t) = k_f(t) - \int_0^t dt' k_f(t-t') \int_0^{t'} d\bar{t} k_b(\bar{t})$$
$$+ \int_0^t dt' k_f(t-t') \sum_{n=2}^{\infty} (-1)^n \int_0^{t'} dt_{n-1} I_{k_b}(t' - t_{n-1})$$
$$\times \int_0^{t_{n-1}} dt_{n-2} I_{k_b}(t_{n-2} - t_{n-1}) \cdots \int_0^{t_2} dt_1 I_{k_b}(t_2 - t_1) I_{k_b}(t_1), \tag{4.63}$$

where $I_{k_b}(t) = \int_0^t dt' k_b(t')$. The inverse Fourier transform of Eq. (4.61) yields

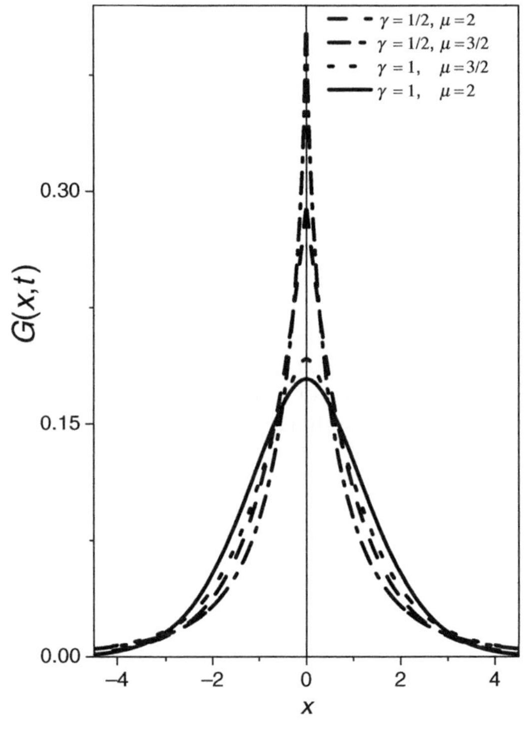

Figure 4.9 $G(x,t)$ versus x for different values of μ and γ. The curves were drawn for $k_f(t) = k'_f = 1$, $k_b(t) = k'_b = 1$, $\mathcal{K}_\gamma = 1$, and $t = 1$ in arbitrary units. We may observe a direct influence of γ and μ on the shape of the Green's function [116]. Modified from Physica A, **443**/1, E. K. Lenzi, M. A. F. dos Santos, D. S. Vieira, R. S. Zola, and H. V. Ribeiro, Solutions for a sorption process governed by a fractional diffusion equation, 32–41. Copyright (2016), with permission from Elsevier.

$$G(x,t) = \frac{1}{\mu|x|}\mathrm{H}^{2,1}_{3,3}\left[\frac{|x|}{\left(\mathcal{K}_\gamma t^\gamma\right)^{\frac{1}{\mu}}}\left|\begin{matrix}\left(1,\frac{1}{\mu}\right),\left(1,\frac{\gamma}{\mu}\right),\left(1,\frac{1}{2}\right)\\(1,1),\left(1,\frac{1}{\mu}\right),\left(1,\frac{1}{2}\right)\end{matrix}\right.\right]$$
$$+\sum_{n=1}^{\infty}\frac{(-1)^n}{\mu|x|\Gamma(1+n)}\int_0^t dt'\Lambda(t-t')t'^n$$
$$\times\,\mathrm{H}^{2,1}_{3,3}\left[\frac{|x|}{\left(\mathcal{K}_\gamma t'^\gamma\right)^{\frac{1}{\mu}}}\left|\begin{matrix}\left(1,\frac{1}{\mu}\right),\left(1+n,\frac{\gamma}{\mu}\right),\left(1,\frac{1}{2}\right)\\(1,1),\left(1+n,\frac{1}{\mu}\right),\left(1,\frac{1}{2}\right)\end{matrix}\right.\right], \tag{4.64}$$

whose behaviour is shown in Fig. 4.9. Consequently, the distribution corresponding to an arbitrary initial condition is

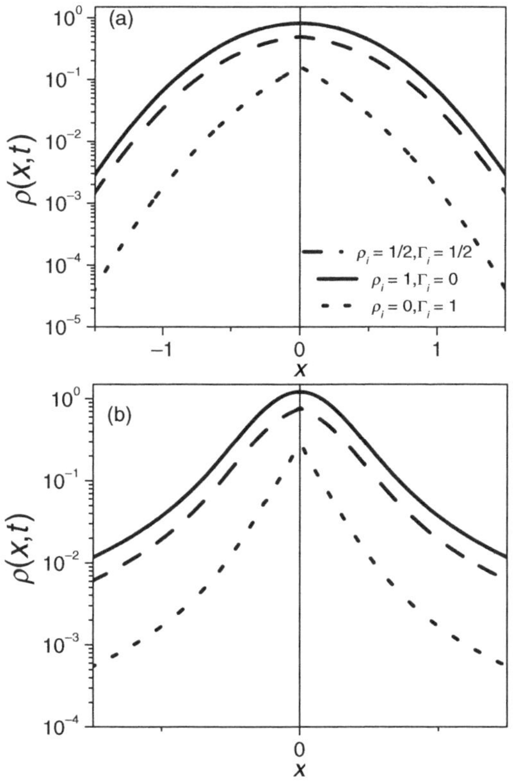

Figure 4.10 $\rho(x,t)$ versus x for different initial conditions: (a) $\mu = 2$ and (b) $\mu = 3/2$. The curves were drawn for $k_f = 1$, $k_b = 1$, $\gamma = 1$, $t = 0.1$, and $\mathcal{K}_\gamma = 1$ in arbitrary units [116]. Modified from Physica A, **443**/1, E. K. Lenzi, M. A. F. dos Santos, D. S. Vieira, R. S. Zola, and H. V. Ribeiro, Solutions for a sorption process governed by a fractional diffusion equation, 32–41. Copyright (2016), with permission from Elsevier.

$$\rho(x,t) = \int_{-\infty}^{\infty} dx' \rho(x',0) G(x-x',t) + \int_0^t dt' \int_{-\infty}^{\infty} dx' \Gamma(x',0) \Pi(t') G(x-x',t-t'), \tag{4.65}$$

with

$$\Pi(t) = I_{k_b}(t) + \sum_{n=1}^{\infty} (-1)^n \int_0^t dt_{n-1} I_{k_b}(t-t_{n-1}) \times \int_0^{t_{n-1}} dt_{n-2} I_{k_b}(t_{n-1}-t_{n-2}) \cdots \int_0^{t_2} dt_1 I_{k_b}(t_2-t_1) I_{k_b}(t_1). \tag{4.66}$$

Figure 4.10 shows the distribution when the reaction term is taken into account. We observe that it presents an anomalous spreading more pronounced when the initial condition of the system is $\rho_i = 0$ and $\Gamma_i = 1$.

4.4 Reaction Terms

In the preceding section, we have considered reaction terms to account for the possibility of describing an absorption or adsorption–desorption process in the bulk. The key point is the absence of a diffusive term in one of the fractional diffusion equations. Thus, the diffusion of the system is described by just one of the fractional diffusion equations, and, consequently, the distribution of both species is governed by this term. In this section, we shall consider a set of fractional diffusion equations which emerges from the following balance equations:

$$\begin{aligned} \rho_1(x,t) = {} & \Phi_1(t)\rho_1(x,0) + \int_{-\infty}^{\infty} dx' \int_0^t dt' \Psi_1(x-x',t-t')\rho_1(x',t') \\ & - \int_{-\infty}^{\infty} dx' \int_0^t dt' \mathcal{R}_{11}(x-x',t-t')\rho_1(x',t') \\ & + \int_{-\infty}^{\infty} dx' \int_0^t dt' \mathcal{R}_{12}(x-x',t-t')\rho_2(x',t') \end{aligned} \tag{4.67}$$

and

$$\rho_2(x,t) = \Phi_2(t)\rho_2(x,0) + \int_{-\infty}^{\infty} dx' \int_0^t dt' \Psi_2(x-x',t-t')\rho_2(x',t')$$

$$+\int_{-\infty}^{\infty} dx' \int_{0}^{t} dt' \mathcal{R}_{21}(x-x', t-t')\rho_1(x', t')$$
$$-\int_{-\infty}^{\infty} dx' \int_{0}^{t} dt' \mathcal{R}_{22}(x-x', t-t')\rho_2(x', t'), \quad (4.68)$$

where

$$\Phi_{1(2)}(t) = 1 - \int_{0}^{t}\int_{-\infty}^{\infty} \Psi_{1(2)}(x, t')dt'dx.$$

The other terms, $\mathcal{R}_{11(12)}(x,t)$ and $\mathcal{R}_{21(22)}(x,t)$, are reaction terms. $\Psi_{1(2)}(x,t)$ represent probability density functions from which the waiting time distribution and jumping probability can be obtained, i.e.,

$$\omega_{1(2)}(t) = \int_{-\infty}^{\infty} \Psi_{1(2)}(x,t)dx \quad \text{and} \quad \lambda_{1(2)}(x) = \int_{0}^{\infty} \Psi_{1(2)}(x,t)dt.$$

The fractional diffusion equations considered are [122]

$$\frac{\partial^\gamma}{\partial t^\gamma}\rho_1(x,t) = \frac{\partial^2}{\partial x^2}\int_{-\infty}^{\infty} dx' D_1(x-x')\rho_1(x',t) - r_1\rho_1(x,t) + r_2\rho_2(x,t) \quad (4.69)$$

and

$$\frac{\partial^\gamma}{\partial t^\gamma}\rho_2(x,t) = \frac{\partial^2}{\partial x^2}\int_{-\infty}^{\infty} dx' D_2(x-x')\rho_2(x',t) + r_1\rho_1(x,t) - r_2\rho_2(x,t), \quad (4.70)$$

where $0 < \gamma < 1$. In these equations, $\rho_1(x,t)$ and $\rho_2(x,t)$ represent two different diffusing systems (e.g., substances, particles, or species); $D_1(x)$ and $D_2(x)$ are the diffusion coefficients for each species and the reaction rates; and r_1 and r_2 are connected with the reaction process, which in this case can be represented by the reversible reaction $1 \rightleftharpoons 2$ or by an irreversible process, e.g., $1 \rightarrow 2$. The fractional time operator considered is that of Caputo, as defined in Chapter 2. One of the main differences between these equations and those worked out in previous section is the presence of the diffusive term in the equation which governs the processes related to species 1 and 2. For Eqs. (4.69) and (4.70), we analyse the spreading of the system and obtain exact solutions in two situations: (1) $D_1(x) \propto 1/|x|^{\mu_1-1}$, with $D_2(x) = 0$, where $1 < \mu_1 \leq 2$, and (2) $D_1(x) \propto 1/|x|^{\mu_1-1}$, $D_2(x) \propto 1/|x|^{\mu_2-1}$, where $1 < \mu_2 \leq 2$. For the first case, we can assume that one of the species remains somehow immobile in the bulk, while in the second case both species can diffuse in the bulk. Note that the kernels in the diffusive term lead to situations characterised by distributions asymptotically governed by power laws such as Lévy distributions, which may be connected with a random walk with long-tailed jump probability density. In addition, for $\mu_1 \neq 2$ and $\mu_2 \neq 2$ we have an interplay between different regimes during the time evolution of the solutions obtained for these species.

Let us analyse the behaviour of the previous set of fractional diffusion equations with linear reaction terms by considering the first case mentioned before, i.e., $D_1(x) \neq 0$ with $D_2(x) = 0$ for which Eqs. (4.69) and (4.70) can be rewritten, respectively, as

$$\frac{\partial^\gamma}{\partial t^\gamma}\rho_1(x,t) = \frac{\partial^2}{\partial x^2}\int_{-\infty}^{\infty} dx' D_1(x-x')\rho_1(x',t) - r_1\rho_1(x,t) + r_2\rho_2(x,t) \tag{4.71}$$

and

$$\frac{\partial^\gamma}{\partial t^\gamma}\rho_2(x,t) = r_1\rho_1(x,t) - r_2\rho_2(x,t), \tag{4.72}$$

with $r_1 \neq 0$ and $r_2 \neq 0$ characterising a reversible first-order reaction process in the bulk. This set of equations can be associated with an intermittent motion where the reaction terms can be related to the rate of switching the particles from the diffusive mode to the resting mode r_1 or switching them from resting to the movement r_2. As mentioned before, it can also be regarded as a problem of diffusion, in which some of the diffusing substances become immobilised as the diffusion proceeds, or a problem in chemical kinetics, in which the rate of reaction depends on the rate of supply of one of the reactants by diffusion.

To proceed, we apply the Laplace and Fourier transforms to Eqs. (4.71) and (4.72), with these equations subjected to the boundary conditions $\rho_1(\pm\infty,t) = 0$ and $\rho_2(\pm\infty,t) = 0$ and the initial conditions $\rho_1(x,0) = \varphi_1(x)$ and $\rho_2(x,0) = \varphi_2(x)$. It is possible to show that the resulting transformed distributions $\rho_1(k,s)$ and $\rho_2(k,s)$ are, respectively,

$$\rho_1(k,s) = \frac{(s^\gamma + r_2)s^{\gamma-1}}{(s^\gamma + r_2)[s^\gamma + D_1(k)k^2] + s^\gamma r_1}\left[\varphi_1(k) + \frac{r_2}{s^\gamma + r_2}\varphi_2(k)\right] \tag{4.73}$$

and

$$\rho_2(k,s) = \frac{r_1 s^{\gamma-1}}{(s^\gamma + r_2)[s^\gamma + D_1(k)k^2] + s^\gamma r_1}\left[\tilde{\varphi}_1(k) + \frac{r_2}{s^\gamma + r_2}\varphi_2(k)\right] + \frac{s^{\gamma-1}}{s^\gamma + r_2}\varphi_2(k). \tag{4.74}$$

To proceed further, we perform the inverse Fourier and Laplace transforms of $\rho_1(k,s)$ and $\rho_2(k,s)$, when $D_1(k) = \mathcal{K}_1|k|^{\mu_1-2}$ ($1 < \mu_1 \leq 2$ and $\mathcal{K}_1 = \text{constant}$), which interpolates the situations with short- ($\mu_1 = 2$) and long-tailed behaviours ($\mu_1 \neq 2$). After some calculation, it is possible to show that

$$\rho_1(x,t) = \int_{-\infty}^{\infty} dx' \varphi_1(x') \Lambda(x-x',t) + r_2 \int_{-\infty}^{\infty} dx' \varphi_2(x') \int_0^t dt' \Lambda(x-x',t-t') t'^{\gamma-1} \mathrm{E}_{\gamma,\gamma}\left(-r_2 t'^{\gamma}\right) \tag{4.75}$$

and

$$\rho_2(x,t) = r_1 \int_0^t dt' \rho_1(x,t') t'^{\gamma-1} \mathrm{E}_{\gamma,\gamma}\left(-r_2 t'^{\gamma}\right) + \mathrm{E}_{\gamma}\left(-r_2 t^{\gamma}\right) \varphi_2(x), \tag{4.76}$$

with

$$\begin{aligned} \Lambda(x,t) = G^{(1)}_{\gamma,\mu_1}(x,t) + \sum_{n=1}^{\infty} (-1)^n \int_0^t dt_n \int_{-\infty}^{\infty} dx_n \Upsilon(x-x_n, t-t_n) \\ \times \int_0^{t_n} dt_{n-1} \int_{-\infty}^{\infty} dx_{n-1} \Upsilon(x_n - x_{n-1}, t_n - t_{n-1}) \cdots \\ \times \int_0^{t_3} dt_2 \int_{-\infty}^{\infty} dx_2 \Upsilon(x_3 - x_2, t_3 - t_2) \\ \times \int_0^{t_2} dt_1 \int_{-\infty}^{\infty} dx_1 \Upsilon(x_2 - x_1, t_2 - t_1) G^{(1)}_{\gamma,\mu_1}(x_1,t_1) , \end{aligned} \tag{4.77}$$

in which

$$\Upsilon(x,t) = r_1 G^{(2)}_{\gamma,\mu_1}(x,t) - r_1 r_2 \int_0^t G^{(2)}_{\gamma,\mu_1}(x,t-t') t'^{\gamma-1} \mathrm{E}_{\gamma,\gamma}\left(-r_2 t'^{\gamma}\right) dt', \tag{4.78}$$

where the Green's functions are

$$G^{(1)}_{\gamma,\mu_1}(x,t) = \frac{1}{\mu_1 |x|} \mathrm{H}^{2,1}_{3,3}\left[\frac{|x|}{(\mathcal{K}_1 t^{\gamma})^{\frac{1}{\mu_1}}} \left| \begin{matrix} \left(1,\frac{1}{\mu_1}\right),\left(1,\frac{\gamma}{\mu_1}\right),\left(1,\frac{1}{2}\right) \\ (1,1),\left(1,\frac{1}{\mu_1}\right),\left(1,\frac{1}{2}\right) \end{matrix} \right. \right] \tag{4.79}$$

and

$$G^{(2)}_{\gamma,\mu_1}(x,t) = \frac{1}{\mu_1 |x|} t^{\gamma-1} \mathrm{H}^{2,1}_{3,3}\left[\frac{|x|}{(\mathcal{K}_1 t^{\gamma})^{\frac{1}{\mu_1}}} \left| \begin{matrix} \left(1,\frac{1}{\mu_1}\right),\left(\gamma,\frac{\gamma}{\mu_1}\right),\left(1,\frac{1}{2}\right) \\ (1,1),\left(1,\frac{1}{\mu_1}\right),\left(1,\frac{1}{2}\right) \end{matrix} \right. \right]. \tag{4.80}$$

These solutions are a particular case of the solutions found in the preceding section for $\rho(x,t)$ and $\Gamma(x,t)$, only written in a different form. They may exhibit a rich variety of behaviours which can be evidenced by analysing the asymptotic limit, i.e., $|x| \to \infty$, of the Green's functions (4.79) and (4.80). For instance, if $\gamma \neq 1$ and $\mu_1 \neq 2$, it is possible to show that

$$G^{(1)}_{\gamma,\mu_1}(x,t) \sim \frac{\mathcal{K}_1 t^{\gamma}}{\mu_1 |x|^{1+\mu_1}} \quad \text{and} \quad G^{(2)}_{\gamma,\mu_1}(x,t) \sim \frac{\mathcal{K}_1 t^{2\gamma-1}}{\mu_1 |x|^{1+\mu_1}}.$$

Others values of the parameters γ and μ_1 produce different behaviours for Eqs. (4.79) and (4.80), e.g., if $\gamma \neq 1$ with $\mu_1 = 2$, then we have

$$G^{(1)}_{\gamma,2}(x,t) = \frac{1}{\sqrt{4\mathcal{K}_1 t^{\gamma}}} \mathrm{H}^{1,0}_{1,1}\left[\frac{|x|}{\sqrt{\mathcal{K}_1 t^{\gamma}}}\middle|^{\left(1-\frac{\gamma}{2},\frac{\gamma}{2}\right)}_{(0,1)}\right] \tag{4.81}$$

and

$$G^{(2)}_{\gamma,2}(x,t) = \frac{1}{2|x|} t^{\gamma-1} \mathrm{H}^{1,0}_{1,1}\left[\frac{|x|}{\sqrt{\mathcal{K}_1 t^{\gamma}}}\middle|^{\left(\gamma,\frac{\gamma}{2}\right)}_{(1,1)}\right]. \tag{4.82}$$

Equations (4.81) and (4.82) can be approximated in the asymptotic limit, respectively, by the following expressions:

$$G^{(1)}_{\gamma,2}(x,t) \sim \frac{1}{\sqrt{4\pi\mathcal{K}_1 t^{\gamma}}}\sqrt{\frac{1}{2-\gamma}}\left(\frac{2}{\gamma}\right)^{\frac{1-\gamma}{2-\gamma}}\left(\frac{|x|}{\sqrt{\mathcal{K}_1 t^{\gamma}}}\right)^{\frac{\gamma-1}{2-\gamma}} \times \exp\left[-\frac{2-\gamma}{2}\left(\frac{\gamma}{2}\right)^{\frac{\gamma}{2-\gamma}}\left(\frac{|x|}{\sqrt{\mathcal{K}_1 t^{\gamma}}}\right)^{\frac{2}{2-\gamma}}\right] \tag{4.83}$$

and

$$G^{(2)}_{\gamma,2}(x,t) \sim \frac{1}{2|x|}\frac{t^{\gamma-1}}{\sqrt{\pi(2-\gamma)}}\left(\frac{\gamma}{2}\right)^{\frac{1-\gamma}{2-\gamma}}\left(\frac{|x|}{\sqrt{\mathcal{K}_1 t^{\gamma}}}\right)^{\frac{3-2\gamma}{2-\gamma}} \times \exp\left[-\frac{2-\gamma}{2}\left(\frac{\gamma}{2}\right)^{\frac{\gamma}{2-\gamma}}\left(\frac{|x|}{\sqrt{\mathcal{K}_1 t^{\gamma}}}\right)^{\frac{2}{2-\gamma}}\right]. \tag{4.84}$$

Another representative case is obtained when we consider $\mu_1 \neq 2$, with $\gamma = 1$, for which Eqs. (4.79) and (4.80) can be simplified to

$$G^{(1)}_{1,\mu_1}(x,t) = \frac{1}{\mu_1|x|}\mathrm{H}^{1,1}_{2,2}\left[\frac{|x|}{(\mathcal{K}_1 t)^{\frac{1}{\mu_1}}}\middle|^{\left(1,\frac{1}{\mu_1}\right),\left(1,\frac{1}{2}\right)}_{(1,1),\left(1,\frac{1}{2}\right)}\right], \tag{4.85}$$

with $G^{(1)}_{1,\mu_1}(x,t) = G^{(2)}_{1,\mu_1}(x,t)$. Equation (4.85) is a Lévy distribution, and, as expected, its asymptotic limit is characterised by a power law; i.e., it is in the form

$$G^{(1)}_{1,\mu_1}(x,t) \sim \frac{\mathcal{K}_1 t}{\mu_1 |x|^{1+\mu_1}}.$$

Let us consider now the second case mentioned before, i.e., when both species of substances are diffusing: $D_1(x) \neq 0$, with $D_2(x) \neq 0$. Similar to the first case, we may use the Fourier and Laplace transforms to find formal solutions, as

$$\rho_1(k,s) = \frac{[s^\gamma + D_2(k)k^2 + r_2]s^{\gamma-1}}{[s^\gamma + D_2(k)k^2 + r_2][s^\gamma + D_1(k)k^2] + [s^\gamma + D_2(k)k^2]r_1} \times \left[\varphi_1(k) + \frac{r_2}{s^\gamma + D_2(k)k^2 + r_2}\varphi_2(k)\right] \tag{4.86}$$

and

$$\rho_2(k,s) = \frac{r_1 s^{\gamma-1}}{[s^\gamma + D_2(k)k^2 + r_2][s^\gamma + D_1(k)k^2] + [s^\gamma + D_2(k)k^2]r_1} \times \left[\varphi_1(k) + \frac{r_2}{s^\gamma + D_2(k)k^2 + r_2}\varphi_2(k)\right] + \frac{s^{\gamma-1}}{s^\gamma + D_2(k)k^2 + r_2}\varphi_2(k). \tag{4.87}$$

In order to obtain the inverse Fourier and Laplace transforms, we consider the specific case:

$$D_1(k) = \mathcal{K}_1|k|^{\mu_1-2} \quad \text{and} \quad D_2(k) = \mathcal{K}_2|k|^{\mu_2-2},$$

for $1 < \mu_1 \leq 2$, in which $\mathcal{K}_1 =$ constant and $\mathcal{K}_2 =$ constant. These choices for $D_1(k)$ and $D_2(k)$ enable us to describe situations characterised by distributions with short- ($\mu_1 = \mu_2 = 2$) and long-tailed ($\mu_1 \neq 2$ and $\mu_2 \neq 2$) behaviours for species 1 and 2. As before, we notice that the case $\mu_1 \neq \mu_2$ corresponds to an interplay between different diffusive regimes, with the asymptotic limit governed by the distribution characterised by the lower value of the parameters μ_1 and μ_2. After performing some calculations, we obtain

$$\rho_1(x,t) = \int_{-\infty}^{\infty} dx' \varphi_1(x')\Lambda_{1,2}(x-x',t) + r_2 \int_{-\infty}^{\infty} dx' \int_{-\infty}^{\infty} dx'' \varphi_2(x-x') \times \int_0^t dt' \Lambda_{1,2}(x'-x'',t-t')G^{(4)}_{\gamma,\mu_2}(x'',t';r_2) \tag{4.88}$$

and

$$\rho_2(x,t) = r_1 \int_{-\infty}^{\infty} dx' \int_0^t dt' \rho_1(x,t-t')G^{(4)}_{\gamma,\mu_2}(x-x',t-t';r_2) + \int_{-\infty}^{\infty} dx' G^{(3)}_{\gamma,\mu_2}(x-x',t)\varphi_2(x'), \tag{4.89}$$

with

$$\Lambda_{1,2}(x,t) = G^{(1)}_{\gamma,\mu_1}(x,t) + \sum_{n=1}^{\infty}(-1)^n \int_0^t dt_n \int_{-\infty}^{\infty} dx_n I(x-x_n, t-t_n)$$
$$\times \int_0^{t_n} dt_{n-1} \int_{-\infty}^{\infty} dx_{n-1} I(x_n - x_{n-1}, t_n - t_{n-1}) \cdots$$
$$\times \int_0^{t_3} dt_2 \int_{-\infty}^{\infty} dx_2 I(x_3 - x_2, t_3 - t_2)$$
$$\times \int_0^{t_2} dt_1 \int_{-\infty}^{\infty} dx_1 I(x_2 - x_1, t_2 - t_1) G^{(1)}_{\gamma,\mu_1}(x_1,t_1) \quad (4.90)$$

where

$$I(x,t) = r_1 G^{(2)}_{\gamma,\mu_1}(x,t) - r_1 r_2 \int_0^t d\bar{t} G^{(2)}_{\gamma,\mu_1}(x,t-\bar{t}) G^{(4)}_{\gamma,\mu_2}(x,\bar{t};r_2), \quad (4.91)$$

and the Green's functions are written as

$$G^{(3)}_{\gamma,\mu_2}(x,t;r_2) = \frac{1}{\mu_2|x|}\sum_{n=0}^{\infty}\frac{(-r_2 t^{\gamma})^n}{\Gamma(1+n)} \mathrm{H}^{2,1}_{3,3}\left[\frac{|x|}{(\mathcal{K}_2 t^{\gamma})^{\frac{1}{\mu_2}}}\left|\begin{matrix}\left(1,\frac{1}{\mu_2}\right),\left(1+n\gamma,\frac{\gamma}{\mu_2}\right),\left(1,\frac{1}{2}\right)\\ (1,1),\left(1+n,\frac{1}{\mu_2}\right),\left(1,\frac{1}{2}\right)\end{matrix}\right.\right] \quad (4.92)$$

and

$$G^{(4)}_{\gamma,\mu_2}(x,t;r_2) = \frac{t^{\gamma-1}}{\mu_2|x|}\sum_{n=0}^{\infty}\frac{(-r_2 t^{\gamma})^n}{\Gamma(1+n)} \mathrm{H}^{2,1}_{3,3}\left[\frac{|x|}{(\mathcal{K}_2 t^{\gamma})^{\frac{1}{\mu_2}}}\left|\begin{matrix}\left(1,\frac{1}{\mu_2}\right),\left((1+n)\gamma,\frac{\gamma}{\mu_2}\right),\left(1,\frac{1}{2}\right)\\ (1,1),\left(1+n,\frac{1}{\mu_2}\right),\left(1,\frac{1}{2}\right)\end{matrix}\right.\right]. \quad (4.93)$$

Figure 4.11 shows the time-dependent behaviour of $1/\rho_1^2(0,t)$ and $1/\rho_2^2(0,t)$ as an indication of the spreading of species 1 and 2 in the bulk in some representative cases. It shows that the asymptotic limit is governed by the diffusive term related to the long-tailed distribution obtained from the lower value of the parameters μ_1 and μ_2.

Similar to what we have done with Eqs. (4.79) and (4.80), we can obtain approximated expressions for Eqs. (4.92) and (4.93) in the asymptotic limit of $|x| \to \infty$. When $\mu_2 \neq 2$, we have

$$G^{(3)}_{\gamma,\mu_2}(x,t;r_2) \sim \frac{1}{\mu_2|x|}\sum_{n=0}^{\infty}\sum_{m=1}^{\infty}\frac{(-r_2 t^{\gamma})^n}{\Gamma(1+n)} C_{\gamma,\mu_2}(n,m)\left(-\frac{\mathcal{K}_2 t^{\gamma}}{|x|^{\mu_2}}\right)^m \quad (4.94)$$

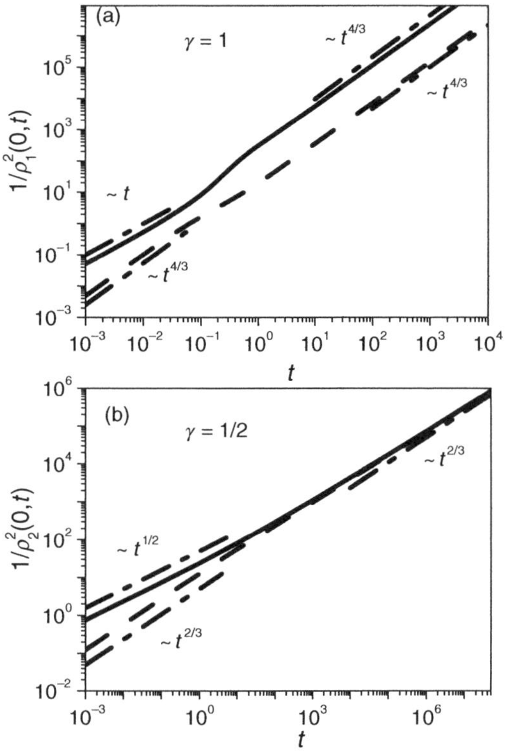

Figure 4.11 $1/\rho_1^2(0,t)$ and $1/\rho_2^2(0,t)$ obtained from Eq. (4.88) and Eq. (4.89) versus t. In (a), the dashed and solid lines refer to $\mu_1 = 3/2$ and $\mu_2 = 2$, with $r_1 = r_2/4 = 1$. In (b), the dashed and solid lines refer to $\mu_1 = 3/2$ and $\mu_2 = 2$, with $r_1 = r_2/2 = 1$. The curves were drawn for $D = 1$, $\varphi_1(x) = (1/2)\delta(x)$, and $\varphi_2(x) = (1/2)\delta(x)$. The dashed-dotted lines were included as a guide to the behaviour in the asymptotic limit [122]. Modified from Communications in Nonlinear Science and Numerical Simulation, **48**, E. K. Lenzi, M. A. F. dos Santos, M. K. Lenzi, and R. Menechini Neto, Solutions for a mass transfer process governed by fractional diffusion equations with reaction terms, 307–317. Copyright (2017), with permission from Elsevier.

and

$$G^{(4)}_{\gamma,\mu_2}(x,t;r_2) \sim \frac{1}{\mu_2|x|}t^{\gamma-1}\sum_{n=0}^{\infty}\sum_{m=1}^{\infty}\frac{(-r_2t^{\gamma})^n}{\Gamma(1+n)}C_{\gamma,\mu_2}(n,m)\left(-\frac{\mathcal{K}_2t^{\gamma}}{|x|^{\mu_2}}\right)^m, \quad (4.95)$$

where

$$C_{\gamma,\mu_2}(n,m) = \frac{\Gamma(1+\mu_2 m)\,\Gamma(1+n+m)}{\Gamma(1+m)\,\Gamma(1+(n+m)\gamma)\,\Gamma\left(1+\frac{\mu_2}{2}m\right)\Gamma\left(-\frac{\mu_2}{2}m\right)}.$$

When $\mu_2 = 2$, we obtain

$$G^{(3)}_{\gamma,2}(x,t;r_2) \sim \frac{1}{2|x|\sqrt{\pi(2-\gamma)}} X \left(\frac{2}{\gamma}\right)^{\frac{1-\gamma}{2-\gamma}} \exp\left[-\frac{r_2}{2}t^\gamma \left(\frac{\gamma}{2}\right)^{\frac{\gamma}{\gamma-2}} X^{2(1-\gamma)}\right] \times \exp\left[-\left(\frac{\gamma}{2}\right)^{\frac{\gamma}{2-\gamma}} \frac{2-\gamma}{2} X^2\right] \tag{4.96}$$

and

$$G^{(4)}_{\gamma,2}(x,t;r_2) \sim \frac{t^{\gamma-1}}{2|x|\sqrt{\pi(2-\gamma)}} X^{3-2\gamma} \exp\left[-\frac{r_2}{2}t^\gamma \left(\frac{\gamma}{2}\right)^{\frac{\gamma}{\gamma-2}} X^{2(1-\gamma)}\right] \times \exp\left[-\left(\frac{\gamma}{2}\right)^{\frac{\gamma}{2-\gamma}} \frac{2-\gamma}{2} X^2\right]. \tag{4.97}$$

where $X = [|x|/\sqrt{\mathcal{K}_2 t^\gamma}]^{1/(2-\gamma)}$. We notice that these asymptotic limits, obtained for Eqs. (4.92) and (4.93) in different conditions, show that when $\mu_2 \neq 2$, we find a long-tailed distribution characterised by power laws, in contrast to what happens when $\mu_2 = 2$, in which the distribution is essentially governed by a stretched exponential–like law.

For the particular case $D_1(k) = D_2(k) = \mathcal{K}|k|^{\mu-2}$ ($1 < \mu \leq 2$ and $\mathcal{K} =$ constant), the preceding solutions can be simplified to

$$\rho_1(x,t) = \frac{r_2}{r_1+r_2}\int_{-\infty}^{\infty} dx' G^{(1)}_{\gamma,\mu}(x-x',t)\left[\varphi_1(x') + \varphi_2(x')\right] + \frac{1}{r_1+r_2}\int_{-\infty}^{\infty} dx' G^{(3)}_{\gamma,\mu}(x-x',t;r_1+r_2)\left[r_1\varphi_1(x') + r_2\varphi_2(x')\right] \tag{4.98}$$

and

$$\rho_2(x,t) = \frac{r_1}{r_1+r_2}\int_{-\infty}^{\infty} dx' G^{(1)}_{\gamma,\mu}(x-x',t)\left[\varphi_1(x') + \varphi_2(x')\right] - \frac{1}{r_1+r_2}\int_{-\infty}^{\infty} dx' G^{(3)}_{\gamma,\mu}(x-x',t;r_1+r_2)\left[r_1\varphi_1(x') + r_2\varphi_2(x')\right]. \tag{4.99}$$

Figure 4.12 illustrates the behaviour of Eqs. (4.98) and (4.99) for different values of γ with μ. Note that for $\gamma = 1/2$ with $\mu = 2$ (solid line) the distribution may be essentially characterised by a stretched exponential-like behaviour, in contrast to the case $\gamma = 1$ with $\mu = 3/2$ (dashed-dotted line) which may be asymptotically

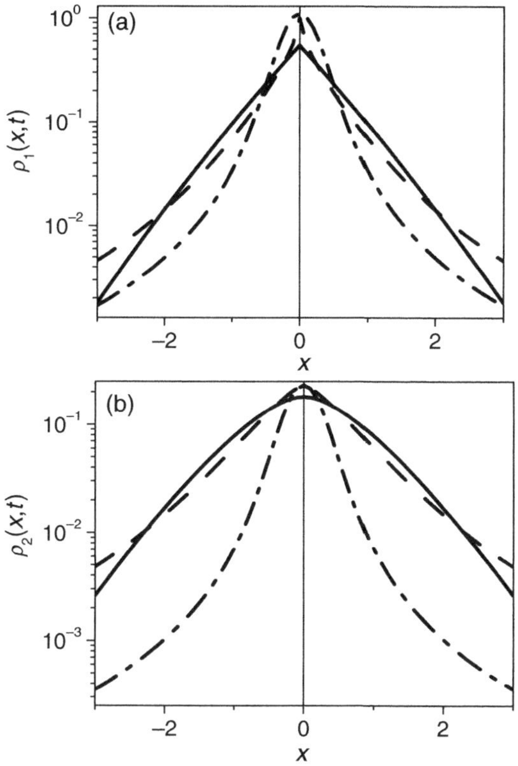

Figure 4.12 Distributions $\rho_1(x,t)$ and $\rho_2(x,t)$ obtained from Eq. (4.98) and Eq. (4.99) for $t = 0.1$ and $D = 1$. The dashed-dotted and solid lines represent the cases $\gamma = 1$ with $\mu = 3/2$ and $\gamma = 1/2$ with $\mu = 2$. The dashed line is the case $\gamma = 1/2$ with $\mu = 3/2$. The curves were drawn for $\varphi_1(x) = \delta(x)$, $\varphi_2(x) = 0$, $r_1 = 2$, and $r_2 = 1$ [122]. Modified from Communications in Nonlinear Science and Numerical Simulation, **48**, E. K. Lenzi, M. A. F. dos Santos, M. K. Lenzi, and R. Menechini Neto, Solutions for a mass transfer process governed by fractional diffusion equations with reaction terms, 307–317. Copyright (2017), with permission from Elsevier.

governed by a power law. The case $\gamma = 1/2$ with $\mu = 3/2$ is a mixing of these cases (dashed line) asymptotically characterised by stretched exponential–like and power laws.

Figure 4.13 shows

$$D_1(k) = D_2(k) = \frac{1}{1 + C_\delta |k|^\delta},$$

where C_δ is a constant, which can be related to a truncated Lévy distribution [123]. For this case, we have a transition, in the limit $t \to \infty$, from a Lévy-like to a Gaussian distribution (see Fig. 4.14). A similar situation in the absence of reaction

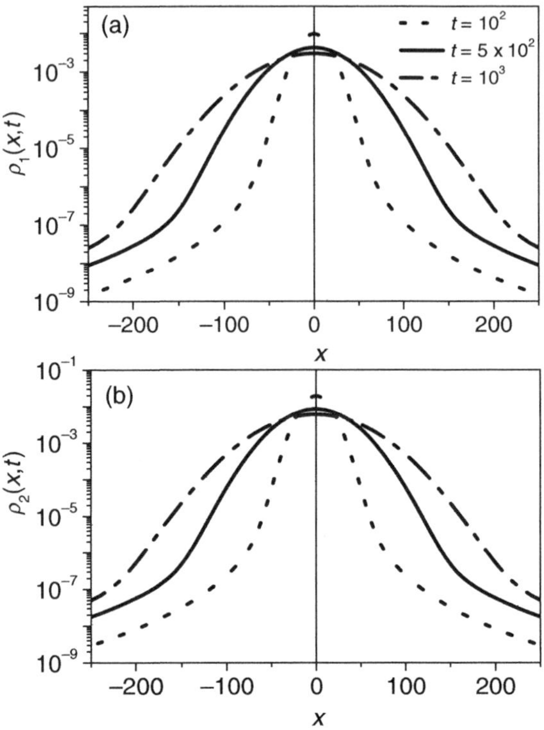

Figure 4.13 $\rho_1(x,t)$ and $\rho_2(x,t)$ obtained from Eq. (4.98) and Eq. (4.99) versus x for different times, $D_1(k) = D_2(k) = 1/(1 + \mathcal{C}_\delta |k|^\delta)$ (solid and dashed lines) with $\mathcal{C}_\delta = 1$ and $\delta = 3/2$. The curves were drawn for $\varphi_1(x) = \delta(x)$, $\varphi_2(x) = 0$, $r_1 = 2$, and $r_2 = 1$ [122]. Modified from Communications in Nonlinear Science and Numerical Simulation, **48**, E. K. Lenzi, M. A. F. dos Santos, M. K. Lenzi, and R. Menechini Neto, Solutions for a mass transfer process governed by fractional diffusion equations with reaction terms, 307–317. Copyright (2017), with permission from Elsevier.

terms has been already discussed [123]. In addition, Fig. 4.14 makes evident a normal diffusion in this limit as well as that the distributions may be approximated to the situations characterised by $D_1(k) \approx \mathcal{K}_1 =$ constant and $D_2(k) \approx \mathcal{K}_2 =$ constant.

4.5 Reaction and CTRW Formalism

From the point of view of CTRW formalism, the reaction diffusion process considered in the preceding section may essentially be related to the situation in which the walkers are included or removed from the system at an instantaneous rate. To explore this connection, in this section we discuss the consequences of considering a noninstantaneous rate of addition or removal of a walker, in order to obtain the

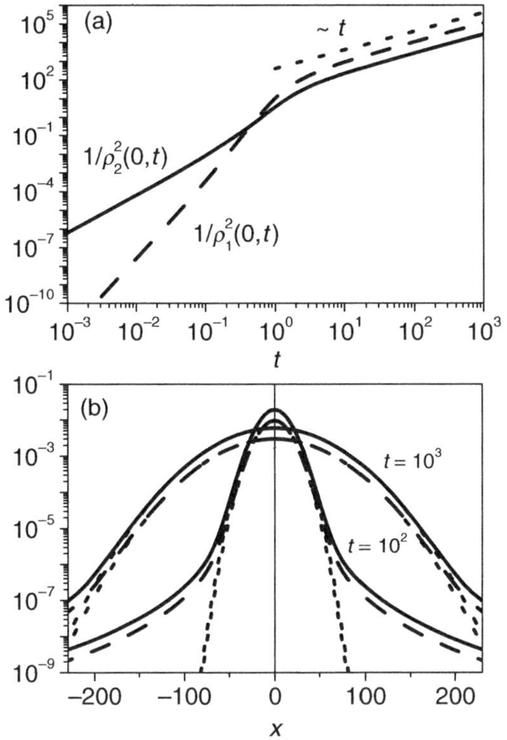

Figure 4.14 (a) $1/\rho_1^2(0,t)$ and $1/\rho_2^2(0,t)$ obtained from Eq. (4.98) and Eq. (4.99) versus t for $D_1(k) = D_2(k) = 1/(1 + \mathcal{C}_\delta|k|^\delta)$. (b) A comparison between the distributions $\rho_1(x,t)$ and $\rho_2(x,t)$, obtained for $D_1(k) = D_2(k) = 1/(1 + \mathcal{C}_\delta|k|^\delta)$ (solid and dashed lines), with the ones obtained for $D_1(k) = D_2(k) = 1$ (dotted lines). We observe a transition from a Lévy-like to a Gaussian distribution when $t \to \infty$. The curves were drawn for $\mathcal{C}_\delta = 1$, $\delta = 3/2$, $\varphi_1(x) = \delta(x)$, and $\varphi_2(x) = 0$. The dotted lines were included as a guide to the behaviour in the asymptotic limit, which in this case characterises the usual diffusion [122]. Modified from Communications in Nonlinear Science and Numerical Simulation, **48**, E. K. Lenzi, M. A. F. dos Santos, M. K. Lenzi, and R. Menechini Neto, Solutions for a mass transfer process governed by fractional diffusion equations with reaction terms, 307–317. Copyright (2017), with permission from Elsevier.

fractional diffusion-like equation to describe this process [120]. The discussion starts by considering only a single species. Subsequently, we extend the discussion to consider the problem for different species.

By considering the CTRW approach, it is possible to show, from Eq. (3.71) presented in Chapter 3, that the distribution satisfies the following equation:

$$\rho_1(x,t) = \Phi_1(t)\rho_1(x,0) + \int_{-\infty}^{\infty} dx' \int_0^t dt' \omega_1(t-t')\lambda_1(x-x')\rho_1(x',t'). \tag{4.100}$$

Now, we consider an extension of Eq. (4.100) in order to include the addition and removal of walkers at a constant per capita rate during the times that they wait before taking the next step [120]. This essentially implies considering that $\Phi_1(t) \to \Phi_1(t)e^{\pm k_1 t}$ and $\omega_1(t) \to \omega_1(t)e^{\pm k_1 t}$. After performing these changes in Eq. (4.100), it becomes

$$\rho_1(x,t) = \Phi_1(t)e^{\pm k_1 t}\rho_1(x,0) + \int_{-\infty}^{\infty} dx' \int_0^t dt' \omega_1(t-t')e^{\pm k_1(t-t')}\lambda_1(x-x')\rho_1(x',t'). \quad (4.101)$$

Let us now obtain, from Eq. (4.101), a differential equation for $\rho_1(x,t)$. By applying the Fourier–Laplace transform, it can be simplified to

$$\rho_1(k,s) = \Phi_1(s \pm k_1)\rho_1(k,0) + \omega_1(s \pm k)\lambda_1(k)\rho_1(k,s). \quad (4.102)$$

By using Eq. (3.74) in Eq. (4.102) with

$$\omega_1(s) \sim 1 - \Gamma(1-\gamma)\frac{\tau^\gamma s^\gamma}{\gamma} \quad \text{and} \quad \lambda(k) \sim 1 - \sigma^2\frac{k^2}{2},$$

we have

$$(s \pm k_1)\rho_1(k,s) = [1 \pm \omega_1(s \pm k_1)]\,\rho_1(k,0) + (s \pm k_1)\omega_1(s \pm k_1)\lambda_1(k)\rho_1(k,s) \quad (4.103)$$

and

$$s\rho_1(k,x) - \rho_1(k,0) = -\frac{\gamma}{\Gamma(1-\gamma)\tau^\gamma}(s \pm k_1)^{1-\gamma}\frac{\sigma^2 k^2}{2}\rho_1(k,s) \pm k_1\rho_1(k,s). \quad (4.104)$$

After performing the inverse Laplace and Fourier transforms, we get

$$\frac{\partial}{\partial t}\rho_1(x,t) = e^{\pm k_1 t}\mathcal{K}_\gamma\, {}_0D_t^{1-\gamma}\left[e^{\mp k_1 t}\frac{\partial^2}{\partial x^2}\rho_1(x,t)\right] \pm k_1\rho_1(x,t), \quad (4.105)$$

where $\mathcal{K}_\gamma = \sigma^2/\tau^\gamma$. The solution for this equation, by considering the conditions $\rho_1(\pm\infty,t) = 0$ and $\rho_1(x,0) = \bar{\rho}_1(x)$, is formally

$$\rho_1(x,t) = \frac{1}{\sqrt{4\pi\mathcal{K}_\gamma t^\gamma}}e^{\pm k_1 t}\int_{-\infty}^{\infty} dx' \bar{\rho}_1(x')\, \mathrm{H}_{1\,2}^{2\,0}\left[\frac{(x-x')^2}{4\mathcal{K}_\gamma t^\gamma}\middle|\begin{matrix}(1-\frac{\gamma}{2},\gamma) \\ (\frac{1}{2},1) \quad (0,1)\end{matrix}\right], \quad (4.106)$$

which is essentially the solution found before, i.e., Eq. (4.11), multiplied by the exponential $e^{\pm k_1 t}$. The extension of the previous results to the multispecies case is straightforward and yields the following equation [124]:

$$\frac{\partial}{\partial t}\Psi(x,t) = e^{\mathcal{R}t}\mathcal{K}_\gamma\, {}_0\mathrm{D}_t^{1-\gamma}\left[e^{-\mathcal{R}t}\frac{\partial^2}{\partial x^2}\Psi(x,t)\right] + \mathcal{R}\Psi(x,t), \quad (4.107)$$

where $\Psi(x,t)$ represents a vector of species concentration and $\mathcal{R}$ is a constant reaction rate matrix. In this approach, the irreversible process $1 \to 2$ can be obtained by considering that

$$\Psi(x,t) = \begin{pmatrix} \rho_1(x,t) \\ \rho_1(x,t) \end{pmatrix} \quad \text{and} \quad \mathcal{R} = \begin{pmatrix} -k_1 & 0 \\ k_1 & 0 \end{pmatrix}. \tag{4.108}$$

For this particular case, we obtain

$$\frac{\partial}{\partial t}\rho_1(x,t) = e^{-k_1 t}\mathcal{K}_\gamma \, {}_0\mathrm{D}_t^{1-\gamma}\left[e^{k_1 t}\frac{\partial^2}{\partial x^2}\rho_1(x,t)\right] - k_1\rho_1(x,t), \tag{4.109}$$

for species 1, and

$$\begin{aligned} \frac{\partial}{\partial t}\rho_2(x,t) = {} & \mathcal{K}_\gamma \, {}_0\mathrm{D}_t^{1-\gamma}\frac{\partial^2}{\partial x^2}\left(\rho_1(x,t) + \rho_2(x,t)\right) + k_1\rho_1(x,t) \\ & - e^{-k_1 t}\mathcal{K}_\gamma \, {}_0\mathrm{D}_t^{1-\gamma}\left[e^{k_1 t}\frac{\partial^2}{\partial x^2}\rho_1(x,t)\right], \end{aligned} \tag{4.110}$$

for species 2. The solution for $\rho_1(x,t)$ and $\rho_2(x,t)$, obtained from the previous set of equations for the initial conditions $\rho_1(x,0) = \bar{\rho}_1\delta(x)$ and $\rho_2(x,0) = \bar{\rho}_2\delta(x)$, are, respectively,

$$\rho_1(x,t) = \bar{\rho}_1 e^{-k_1 t}G(x,t) \tag{4.111}$$

and

$$\rho_2(x,t) = \left[\bar{\rho}_2 + \left(1 - e^{-k_1 t}\right)\bar{\rho}_1\right]G(x,t), \tag{4.112}$$

with $G(x,t)$ defined by Eq. (4.10). For the reversible reaction process $1 \rightleftharpoons 2$, for which

$$\mathcal{R} = \begin{pmatrix} -k_1 & k_2 \\ k_1 & -k_2 \end{pmatrix}, \tag{4.113}$$

it is possible to show that

$$\begin{aligned} \frac{\partial}{\partial t}\rho_1(x,t) = {} & k_2\mathcal{K}_\gamma \, {}_0\mathrm{D}_t^{1-\gamma}\left\{e^{k_1 t}\frac{\partial^2}{\partial x^2}\left[\rho_1(x,t) + \rho_2(x,t)\right]\right\} \\ & - k_1\rho_1(x,t) + k_2\rho_2(x,t) \\ & + e^{-(k_1+k_2)t}\frac{\mathcal{K}_\gamma}{k_1+k_2} \, {}_0\mathrm{D}_t^{1-\gamma}\left\{e^{(k_1+k_2)t}\frac{\partial^2}{\partial x^2}\left[k_1\rho_1(x,t) - k_2\rho_2(x,t)\right]\right\}, \end{aligned} \tag{4.114}$$

for species 1, and

$$\frac{\partial}{\partial t}\rho_2(x,t) = k_1\mathcal{K}_\gamma \, {}_0\mathrm{D}_t^{1-\gamma}\left\{e^{k_1 t}\frac{\partial^2}{\partial x^2}\left[\rho_1(x,t)+\rho_2(x,t)\right]\right\}$$
$$+ k_1\rho_1(x,t) - k_2\rho_2(x,t)$$
$$- e^{-(k_1+k_2)t}\frac{\mathcal{K}_\gamma}{k_1+k_2}\, {}_0\mathrm{D}_t^{1-\gamma}\left\{e^{(k_1+k_2)t}\frac{\partial^2}{\partial x^2}\left[k_1\rho_1(x,t) - k_2\rho_2(x,t)\right]\right\} \tag{4.115}$$

for species 2. The solution can also be found as before when a irreversible process was analysed. In particular, for the initial conditions $\rho_1(x,0) = \bar{\rho}_1\delta(x)$ and $\rho_2(x,0) = \bar{\rho}_2\delta(x)$, we have

$$\rho_1(x,t) = \frac{\bar{\rho}_1}{k_1+k_2}\left[k_2 + k_1 e^{-(k_1+k_2)t}\right]G(x,t)$$
$$+\frac{\bar{\rho}_2 k_2}{k_1+k_2}\left[1 - e^{-(k_1+k_2)t}\right]G(x,t) \tag{4.116}$$

and

$$\rho_2(x,t) = \frac{\bar{\rho}_1 k_1}{k_1+k_2}\left(1 - e^{-(k_1+k_2)t}\right)G(x,t)$$
$$+\frac{\bar{\rho}_2}{k_1+k_2}\left(k_1 + k_2 e^{-(k_1+k_2)t}\right)G(x,t). \tag{4.117}$$

These solutions are different from the ones found before, and, in the limit of long times, i.e., $t \to \infty$, they are reduced, respectively, to

$$\rho_1(x,t) \sim \frac{k_2}{k_1+k_2}\left(\bar{\rho}_1 + \bar{\rho}_2\right)G(x,t) \tag{4.118}$$

and

$$\rho_2(x,t) \sim \frac{k_1}{k_1+k_2}\left(\bar{\rho}_1 + \bar{\rho}_2\right)G(x,t), \tag{4.119}$$

with the asymptotic limit of $G(x,t)$ defined by Eq. (4.81).

The problems considered in this chapter show that reaction-diffusion models may be relevant to describe several physical phenomena in connection with anomalous diffusion behaviour [125–135]. When the fractional time derivative is incorporated into the diffusion equations, an anomalous relaxation is produced, and, if an stationary solution is present, it is the same relaxation that is obtained by means of the usual diffusion equation. When the spatial fractional derivative is incorporated, we obtain the Lévy distributions which are asymptotically characterised by a power law. Thus, the presence of fractional derivatives means changing the waiting time and the jumping distributions related to the processes. Finally, when reaction terms are considered, we observe that they can be associated with CTRWs in which the walkers are included or removed from the system with an instantaneous rate.

5

Fractional Diffusion Equations

Surface Effects

In this chapter, we consider the influence of the surfaces or membranes on diffusive processes. The main aim is to investigate how the surface may modify the diffusive process of a system governed by a fractional diffusion equation.

In the first part of the chapter, we analyse the one-dimensional problem characterised by time-dependent boundary conditions, showing how they influence the diffusive process in the system for an arbitrary initial condition, i.e., the quantities related to the diffusion process, such as the first passage time, which may have an anomalous behaviour. A similar analysis is carried out for the two-dimensional case with inhomogeneous and time-dependent boundary conditions. These results show the potential of this formalism to analyse other physical scenarios, such as describing the molecular orientation and the anchoring problem in liquid crystals confined to a cylindrical region, taking into account the adsorption phenomena at the interfaces.

The second part of the chapter is dedicated to investigating situations in which the processes occurring on the surface are coupled to the bulk dynamics by means of the boundary conditions. As a first application, we consider a surface in which, besides the adsorption–desorption process, a reaction process may occur and the system presents anomalous diffusion behaviour. Another application refers to the transport through a membrane of definite thickness, for which the processes occurring on the surface also couple with the diffusion equations governing the bulk dynamics. In all cases, the system may exhibit an anomalous diffusive behaviour for which surface effects play a remarkable role.

5.1 1D and 2D Cases: Different Diffusive Regimes

Surface effects are present in a variety of real scenarios of interest in engineering [112, 136], biological systems [137], and physics [138, 139] as a fundamental feature of several processes. For example, industrial and biochemical reactions can have the reaction rate or the sorption of reagents limited by the mass transfer between the fluid phase and the catalyst surface. In biological systems [140], the surfaces (or membranes) are responsible for the selectivity of particles by means of sorption and desorption processes, and, consequently, the particles transfer from one region to the other. Other contexts can also be found in physics such as the electrical response of water [141] or liquid crystals [142] in which the effects of the interface between electrode and fluid play an important role.

Interface (or surface) effects pose challenging problems also from the mathematical point of view. In connection with the fractional diffusion equations, these problems may arise in a variety of applications just mentioned. To explore some of these applications, in this section, we consider the fractional diffusion equation [143]:

$$\frac{\partial^\gamma}{\partial t^\gamma}\rho(r,t) = D\nabla^2\rho(r,t) + \int_0^t dt K(t-t)\nabla^2\rho(r,t), \tag{5.1}$$

where D is a dimensionless diffusion coefficient; the fractional time derivative considered here is Caputo's, as defined in Section 2.3; $K(t)$ is a time-dependent kernel, which we assume is given by

$$K(t) = D_\alpha \frac{t^{\alpha-1}}{\Gamma(\alpha)},$$

where D_α are constant coefficients; and $\Gamma(\alpha)$ is the Euler's gamma function, defined in Section 1.2.1.

The boundary conditions for Eq. (5.1) are dependent on space and time, and the initial condition is of the kind $\rho(r,0) = \tilde{\rho}(r)$. Equation (5.1) will be analysed here focusing the case $0 < \gamma \leq 1$, with $0 < \alpha + \gamma \leq 1$ ($\alpha > 0$), i.e., the subdiffusive behaviour. However, this equation could be analysed for other ranges of values of the parameters γ and α as, for instance, $0 < \gamma \leq 1$, with $\alpha > 0$, or $1 < \gamma < 2$, with $\alpha > 0$. The results found for the subdiffusive case may be extended to $1 < \gamma < 2$ by incorporating in the problem the condition $\partial_t\rho|_{t=0} = 0$. Another salient aspect associated with Eq. (5.1) is the presence of different diffusive regimes occurring when appropriate choices for $K(t)$ are done, because the same approach may be used to investigate particle diffusion in a quasi-two-dimensional bacterial bath [144–146], enhanced diffusion in active intracellular transport [147, 148], and Hamiltonian systems with long-range interactions [149, 150], among others.

Here, we search for solutions to Eq. (5.1) in two situations: (i) one-dimensional case subjected to the boundary condition $\rho(0,t) = \Phi_0(t)$ and $\rho(a,t) = \Phi_a(t)$, where $\Phi_0(t)$ and $\Phi_a(t)$ are two arbitrary time-dependent functions, and (ii) the two-dimensional case in cylindrical symmetry, with the boundary conditions

$\rho(a,\theta,t) = \widetilde{\Phi}_a(\theta,t)$ and $\rho(b,\theta,t) = \widetilde{\Phi}_b(\theta,t)$. This latter case incorporates a spatial dependence on the boundary condition and makes it possible to investigate the diffusion of a system limited by two inhomogeneous surfaces over which time-dependent phenomena occur. This model considers only the influence of the surfaces on the process characterised by different diffusive regimes. In Section 5.3, we will analyse a situation in which the surface and the bulk behaviours are coupled by the boundary conditions, in such a way that one can change the other.

The first problem to be considered here is the one-dimensional case of Eq. (5.1), subjected to the time-dependent boundary conditions $\rho(0,t) = \Phi_0(t)$ and $\rho(a,t) = \Phi_a(t)$, and the initial condition $\rho(x,0) = \widetilde{\rho}(x)$. In this case, Eq. (5.1) can be written as

$$\frac{\partial^\gamma}{\partial t^\gamma}\rho(x,t) = D\frac{\partial^2}{\partial x^2}\rho(x,t) + \frac{D_\alpha}{\Gamma(\alpha)}\int_0^t dt'(t-t')^{\alpha-1}\frac{\partial^2}{\partial x^2}\rho(x,t'), \tag{5.2}$$

with $0 < \gamma \leq 1$ and $0 \leq \alpha+\gamma \leq 1$ $(\alpha > 0)$. In order to study the surface effects on the relaxation of the system and solve Eq. (5.2), we use again the Laplace transform and the Green's function approach [12], which directly yields the contribution of the surface to the time evolution of the initial condition. This procedure yields a solution in the Laplace space as follows:

$$\widehat{\rho}(x,s) = -s^{\gamma-1}\int_0^a dx'\widehat{G}(x,x';s)\widetilde{\rho}(x') + \left[\widetilde{\Phi}_a(s)\frac{\partial}{\partial x'}\widehat{G}(x,x';s)\bigg|_{x'=a} - \widetilde{\Phi}_o(s)\frac{\partial}{\partial x'}\widehat{G}(x,x';s)\bigg|_{x'=0}\right], \tag{5.3}$$

with

$$\widetilde{\Phi}_{o,a}(s) = (D + D_\alpha s^{-\alpha})\Phi_{o,a}(s),$$

and $\widehat{G}(x,x';s)$ is governed by the equation

$$\left(D + \frac{D_\alpha}{s^\alpha}\right)\frac{d^2}{dx^2}\widehat{G}(x,x';s) - s^\gamma\widehat{G}(x;x',s) = \delta(x-x'), \tag{5.4}$$

subject to the conditions $\widehat{G}(0,x';s) = 0$ and $\widehat{G}(a,x';s) = 0$. The eigenfunctions of the Sturm–Liouville problem related to the spatial operator of Eq. (5.4) may be used to show that

$$\widehat{G}(x,x';s) = -\frac{2}{a}\sum_{n=1}^{\infty}\frac{\sin\left(n\pi x'/a\right)\sin\left(n\pi x/a\right)}{s^\gamma + (D + D_\alpha s^{-\alpha})(n\pi/a)^2}. \tag{5.5}$$

From the inverse of Laplace transform in Eq. (5.2), we obtain

$$\rho(x,t) = -\frac{1}{\Gamma(1-\gamma)}\int_0^t dt\frac{1}{(t-t)^\gamma}\int_0^a dx'\, G(x,x';t)\widetilde{\rho}(x')$$
$$+\int_0^t dt\left[\overline{\Phi}_a(t-t)\frac{\partial}{\partial x'}G(x,x';t)\bigg|_{x'=a} - \overline{\Phi}_o(t-t)\frac{\partial}{\partial x'}G(x,x';t)\bigg|_{x'=0}\right], \tag{5.6}$$

with

$$\overline{\Phi}_{o,a}(t) = D\Phi_{o,a}(t) + \frac{D_\alpha}{\Gamma(\alpha)}\int_0^t dt(t-t)^{\alpha-1}\Phi_{o,a}(t), \tag{5.7}$$

$$G(x,x';t) = -\frac{2}{a}\sum_{n=1}^{\infty}\sin\left(\frac{n\pi}{a}x'\right)\sin\left(\frac{n\pi}{a}x\right)\Upsilon_n(t,k_n), \tag{5.8}$$

and

$$\Upsilon_n(t,k_n) = \sum_{m=0}^{\infty}\frac{t^{\gamma-1}}{m!}\left(-D_\alpha t^{\gamma+\alpha}k_n^2\right)^m \mathrm{H}^{1\,1}_{1\,2}\left[Dk_n^2 t^\gamma \Big|^{(-m,1)}_{(0,1)\;(1-(\gamma+\alpha)m-\gamma,\gamma)}\right], \tag{5.9}$$

where $k_n = n\pi/a$. The quantity $\Upsilon_n(t,k_n)$ is essentially formed by a mixing of two diffusive regimes; one of them is governed by the fractional derivative and the other one is dominated by the kernel in the diffusive term. This can be demonstrated because

$$\Upsilon_n(t,k_n) = t^{\gamma-1}E_{\gamma,\gamma}(-Dk_n^2 t^\gamma),$$

when $D \neq 0$ and $D_\alpha = 0$, with $\gamma \neq 1$, and

$$\Upsilon_n(t,k_n) = t^{\gamma-1}E_{\gamma+\alpha,\gamma}(-D_\alpha k_n^2 t^{\gamma+\alpha}),$$

when $D = 0$ and $D_\alpha \neq 0$, where $E_{\gamma,\beta}(x)$ is the generalised Mittag-Leffler function. The first term of Eq. (5.6) gives the time evolution of the system for an arbitrary initial condition, whereas the second term represents surface effects. Figure 5.1a shows that Eq. (5.6) presents an unusual behaviour (e.g., oscillations and accumulation near the boundaries) which may be relevant, for example, to investigate systems with boundary condition stated in terms of a kinetic equation governing an adsorption–desorption process [151, 152]. This unexpected behaviour for the solution of Eq. (5.6) is related to the presence of the fractional derivative, the chosen kernel, and the time dependence exhibited by the boundary conditions, which change the dynamical aspects of the system. For the special case in which $\Phi_o(t) = \Phi_a(t) = 0$, a solution obtained before [153] may be reobtained.

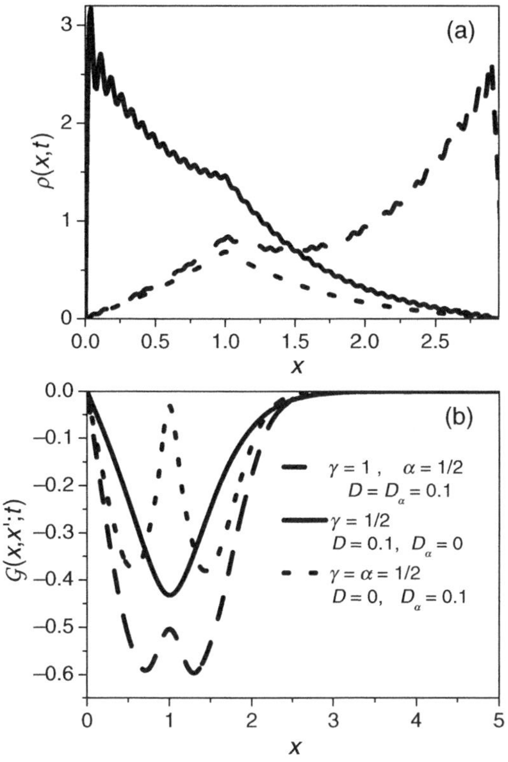

Figure 5.1 (a) $\rho(x,t)$ versus x, which illustrates Eq. (5.6) for $\gamma = 1/2$, $D = 1$, $D_\alpha = 0$, $a = 3$, $t = 0.1$, and $\rho(x,0) = \delta(x-1)$. The solid line corresponds to the solution of Eq. (5.6) subjected to the boundary conditions $\rho(0,t) = \Phi(1 + t^{-\eta}/\Gamma(1-\eta))$ ($\Phi = 1$ and $\eta = 1/2$) and $\rho(a,t) = 0$. The dotted line is the solution of Eq. (5.6) for $\rho(0,t) = \rho(a,0) = 0$, and the dashed line corresponds to the solution with the boundary conditions $\rho(0,t) = 0$ and $\rho(a,t) = \Phi(1+t^{-\eta}/\Gamma(1-\eta))$ ($\Phi = 1$ and $\eta = 1/2$). From this figure, it is possible to show that depending on the process (represented by the boundary conditions), which occurs between the system limited to the confined region and the surface, the diffusive regime can be drastically modified and presents an anomalous behaviour. (b) $G(x,x';t)$ versus x, exemplifying the behaviour of Eq. (5.8) for $a = 5$, $x' = 1$, and $t = 1$ [143]. Modified with permission from R. Rossato, M. K. Lenzi, L. R. Evangelista, and E. K. Lenzi, Physical Review E, **76**, 032102 and (2007). Copyright (2007) by the American Physical Society.

The second problem to be considered is the two-dimensional version of Eq. (5.1), in cylindrical coordinates, subjected to the boundary conditions $\rho(a,\theta,t) = \widetilde{\Phi}_a(\theta,t)$ and $\rho(b,\theta,t) = \widetilde{\Phi}_b(\theta,t)$. This formulation is more useful to treat inhomogeneous surfaces for which the boundary conditions have an angular dependence. The procedure is the same as before and yields a solution to Eq. (5.1) in the form:

$$\rho(r,\theta,t) = -\frac{1}{\Gamma(1-\gamma)}\int_0^t dt\frac{1}{(t-t)^{\gamma}}\int_0^{2\pi} d\theta' \int_a^b dr' r' G(r,\theta;r'\theta';t)\widetilde{\rho}(r',\theta')$$

$$+\int_0^t dt' \int_0^{2\pi} d\theta' \Bigg[b\widetilde{\overline{\Phi}}_b(\theta',t')\frac{\partial}{\partial r'}G(r,\theta;r',\theta';t-t')\Bigg|_{r'=b}$$

$$- a\widetilde{\overline{\Phi}}_a(\theta',t-t')\frac{\partial}{\partial r'}G(r,\theta;r',\theta';t')\Bigg|_{r'=a}\Bigg], \quad (5.10)$$

(see Fig. 5.2) with, as done in Eq. (5.7),

$$\widetilde{\overline{\Phi}}_{a,b}(t) = D\widetilde{\Phi}_{a,b}(t) + \frac{D_\alpha}{\Gamma(\alpha)}\int_0^t dt(t-t)^{\alpha-1}\widetilde{\Phi}_{a,b}(t) \quad (5.11)$$

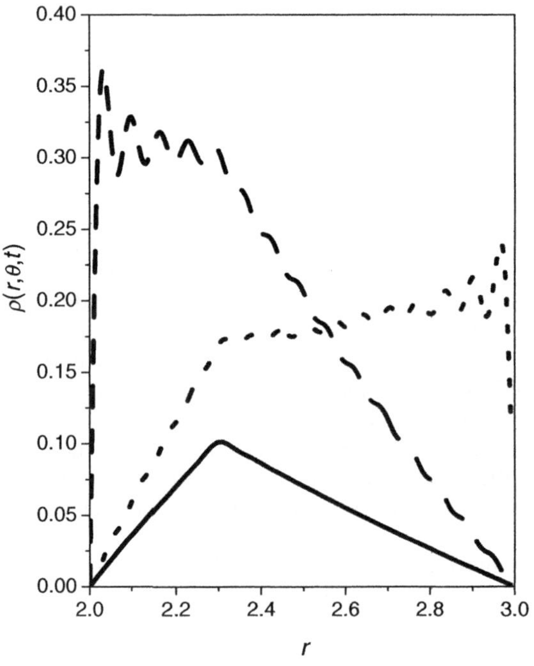

Figure 5.2 $\rho(r,\theta,t)$ versus r as given by Eq. (5.10) for $\gamma = 1/2$, $D = 1$, $D_\alpha = 0$, $\rho(r,\theta,0) = (2/r)\delta(r-\widetilde{r})$, with $\widetilde{r} = 2.3$, $a = 2$, $b = 3$, and $t = 1$. The solid line corresponds to the solution satisfying the boundary conditions $\rho(a,\theta,t) = 0$ and $\rho(b,\theta,t) = 0$. The dotted line is the solution for $\rho(a,\theta,t) = 0$ and $\rho(b,\theta,t) = 2/(3\pi)$. The dashed line corresponds to the solution with the boundary conditions $\rho(a,\theta,t) = 1/\pi$ and $\rho(b,\theta,t) = 0$ [143]. Modified with permission from R. Rossato, M. K. Lenzi, L. R. Evangelista, and E. K. Lenzi, Physical Review E, **76**, 032102 and (2007). Copyright (2007) by the American Physical Society.

and

$$
\begin{aligned}
G(r,\theta;r',\theta';t) = &-\frac{\pi}{2}\sum_{m=0}^{\infty}\sum_{n=1}^{\infty}A_{mn}\Psi_{mn}(r)\Psi_{mn}(r')\cos\left[m(\theta-\theta')\right] \\
&\times \Upsilon_{mn}(t,k_{mn}),
\end{aligned} \tag{5.12}
$$

where

$$
\Psi_{mn}(r) = \mathrm{J}_m(k_{mn}r)\mathrm{N}_m(k_{mn}a) - \mathrm{J}(k_{mn}a)\mathrm{N}_m(k_{mn}r),
$$

in which $\mathrm{J}_m(x)$ and $\mathrm{N}_m(x)$ are the Bessel functions of first and second kinds respectively (see Section 1.2.2). In the preceding equations, k_{mn} are solutions of the eigenvalue equation

$$
\mathrm{J}_m(k_{mn}b)\mathrm{N}_m(k_{mn}a) - \mathrm{J}_m(k_{mn}a)\mathrm{N}_m(k_{mn}b) = 0
$$

and the following relation holds:

$$
A_{mn} = \frac{k_{mn}^2}{\varepsilon_m\left[(\mathrm{J}_m(k_{mn}a)/\mathrm{J}_m(k_{mn}b))^2 - 1\right]},
$$

with $\varepsilon_m = 2$ for $m = 0$ and $\varepsilon_m = 1$ for $m \neq 0$. As for the case described by Eq. (5.6), the last term of Eq. (5.10) represents the surface effect on the first term, which gives the time evolution of the system for an arbitrary initial condition.

The formalism presented above can also be applied, for instance, to describe the equilibrium configuration of liquid crystal samples confined between two concentric cylindrical surfaces [154–157], and yields the analytical solution for the diffusion problem of neutral particles (e.g. dyes doping the material) in these media [158–162].

5.2 3D Case: External Force and Reaction Term

In this section, we consider the fractional diffusion equation for a three-dimensional case without the kernel term, but including a spatially dependent diffusion coefficient, an external force term, and a source (absorbent) term which may be connected with a reaction process [163]. The equation to be considered is

$$
\frac{\partial^\gamma}{\partial t^\gamma}\rho(\bar{r},t) = \nabla\cdot\left[D(\bar{r})\nabla\rho(\bar{r},t)\right] - \nabla\cdot\left[\overline{F}(\bar{r})\rho(\bar{r},t)\right] + \alpha(\bar{r},t), \tag{5.13}
$$

in spherical symmetry, with $\bar{r} = (r,\theta,\phi)$, in the subdiffusive regime $0 < \gamma \leq 1$. The diffusion coefficient will be given by

$$
D(\bar{r}) = Dr^{-\eta}, \tag{5.14}
$$

whereas the external force has a power-law form as

$$\overline{F}(\overline{r}) = \frac{\mathcal{K}}{r^{1+\eta}}\widehat{r}, \tag{5.15}$$

where D and $\mathcal{K}$ are constants. As pointed out above, $\alpha(\overline{r},t)$ is a generic source term which may be related to some kind of reaction process.

The analysis of Eq. (5.13) is first performed by considering the absence of external force with $\alpha(\overline{r},t) = 0$ and the boundary conditions

$$\rho(\overline{r},t)|_{r=a} = \widetilde{\Phi}_a(\theta,\phi,t) \quad \text{and} \quad \rho(\overline{r},t)|_{r=b} = \widetilde{\Phi}_b(\theta,\phi,t),$$

where $\widetilde{\Phi}_a(\theta,\phi,t)$ and $\widetilde{\Phi}_b(\theta,\phi,t)$ are two arbitrary time-dependent functions associated with the points $r = a$ and $r = b$ of the surfaces. This case makes it possible to investigate the diffusion in a system limited by two inhomogeneous surfaces, represented by boundary conditions that incorporate space- and time-dependent phenomena, to account for surface effects. In this first case, Eq. (5.13) is reduced to

$$\frac{\partial^\gamma}{\partial t^\gamma}\rho(\overline{r},t) = \nabla \cdot [D(\overline{r})\nabla\rho(\overline{r},t)]\,. \tag{5.16}$$

Before proceeding, we underline that depending on the boundary conditions the system may exhibit a stationary solution which is the same as the one obtained for the usual diffusion equation. This indicates that the presence of the time fractional derivative produces an anomalous relaxation to a stationary state, in contrast to what happens in the case of spatial fractional derivatives.

To obtain a solution to Eq. (5.16) and investigate the surface effects, i.e., the influence of the boundary condition on the relaxation of the system, we use again the Laplace transform and the Green's function approach [12]. Proceeding in this way, we may obtain a solution in the Laplace space as

$$\begin{aligned}\widehat{\rho}(\overline{r},s) = {} & -s^{\gamma-1}\int_a^b d\overline{r}'\overline{r}'^2\int_0^\pi d\theta'\sin\theta'\int_0^{2\pi} d\phi'\widehat{G}(\overline{r},\overline{r}';s)\widetilde{\rho}(\overline{r}') \\ & + \int_0^\pi d\theta'\sin\theta'\int_0^{2\pi} d\phi'\left[Db^{2-\eta}\widetilde{\Phi}_b(\theta,\phi,s)\frac{\partial}{\partial r'}\widehat{G}(\overline{r},\overline{r}';s)\bigg|_{r'=b}\right. \\ & \left. - Da^{2-\eta}\widetilde{\Phi}_a(\theta,\phi,s)\frac{\partial}{\partial r'}\widehat{G}(\overline{r},\overline{r}';s)\bigg|_{r'=a}\right],\end{aligned} \tag{5.17}$$

with $\widehat{G}(\overline{r},\overline{r}';s)$ governed by the equation

$$\nabla \cdot \left[D(\overline{r})\nabla\widehat{G}(\overline{r},\overline{r}';s)\right] - s^\gamma\widehat{G}(\overline{r},\overline{r}';s) = \delta(\overline{r}-\overline{r}'), \tag{5.18}$$

subjected to the conditions $\widehat{G}(\overline{r},\overline{r}';s)|_{r=b} = 0$ and $\widehat{G}(\overline{r},\overline{r}';s)|_{r=a} = 0$. Again, with the help of the eigenfunctions of the Sturm–Liouville problem related to the spatial operator of Eq. (5.18), we can write

$$\widehat{G}(\overline{r},\overline{r}';s) = -\frac{1}{2+\eta}\sum_{n=1}^{\infty}\sum_{l=0}^{\infty}\sum_{m=-l}^{l}\frac{A_{lmn}}{s^{\gamma}+Dk_{lmn}^{2}}\Psi_{lmn}(r)\Psi_{lmn}(r')$$
$$\times \mathrm{Y}_l^{m*}(\theta',\phi')\mathrm{Y}_l^m(\theta,\phi), \tag{5.19}$$

where $\mathrm{Y}_l^m(\theta,\phi)$ are the spherical harmonics, and

$$\Psi_{lmn}(r) = r^{-\frac{1}{2}(1-\eta)}\left[\mathrm{J}_\alpha\left(\frac{2k_{lmn}}{2+\theta}r^{\frac{1}{2}(2+\eta)}\right)\mathrm{N}_\alpha\left(\frac{2k_{lmn}}{2+\eta}a^{\frac{1}{2}(2+\eta)}\right)\right.$$
$$\left. -\ \mathrm{J}_\alpha\left(\frac{2k_{lmn}}{2+\eta}a^{\frac{1}{2}(2+\eta)}\right)\mathrm{N}_\alpha\left(\frac{2k_{lmn}}{2+\eta}r^{\frac{1}{2}(2+\eta)}\right)\right], \tag{5.20}$$

with

$$\alpha = \frac{2}{2+\eta}\left[\left(\frac{1-\eta}{2}\right)^2 + l(l+1)\right]^{1/2}.$$

In Eq. (5.20), $\mathrm{J}_\alpha(x)$ and $\mathrm{N}_\alpha(x)$ are, respectively, the Bessel functions of first and second kinds, k_{lmn} are solutions of the eigenvalue equation:

$$\mathrm{J}_\alpha(\varepsilon_b)\,\mathrm{N}_\alpha(\varepsilon_a) - \mathrm{J}_\alpha(\varepsilon_a)\,\mathrm{N}_\alpha(\varepsilon_b) = 0, \tag{5.21}$$

with

$$\varepsilon_a = \frac{2k_{lmn}a^{\frac{1}{2}(2+\eta)}}{2+\eta} \quad \text{and} \quad \varepsilon_b = \frac{2k_{lmn}a^{\frac{1}{2}(2+\eta)}}{2+\eta},$$

and

$$A_{lmn} = \frac{(\pi k_{mn})^2}{\left[\mathrm{J}_\alpha(\varepsilon_a)/\mathrm{J}_\alpha(\varepsilon_b)\right]^2 - 1}. \tag{5.22}$$

By taking the inverse Laplace transform of Eq. (5.17), we obtain

$$\rho(\overline{r},t) = -\frac{1}{\Gamma(1-\gamma)}\int_0^t dt'\frac{1}{(t-t')^{\gamma}}\int_a^b d\overline{r}'\overline{r}'^2\int_0^\pi d\theta'\sin\theta'\int_0^{2\pi} d\phi' G(\overline{r},\overline{r}';t')\widetilde{\rho}(\overline{r}')$$
$$+\int_0^t dt'\int_0^\pi d\theta'\sin\theta'\int_0^{2\pi} d\phi'\left[Db^{2-\eta}\widetilde{\Phi}_b(\theta,\phi,t-t')\frac{\partial}{\partial r'}G(\overline{r},\overline{r}';t')\right|_{r'=b}$$
$$\left. -\ Da^{2-\eta}\widetilde{\Phi}_a(\theta,\phi,t-t')\frac{\partial}{\partial r'}G(\overline{r},\overline{r}';t')\right|_{r'=a}\Bigg], \tag{5.23}$$

(see Fig. 5.3) with

$$G(\overline{r},\overline{r}';t) = -\frac{1}{2+\eta}\sum_{n=1}^{\infty}\sum_{l=0}^{\infty}\sum_{m=-l}^{l}A_{lmn}\Psi_{lmn}(r)\Psi_{lmn}(r')$$
$$\times \mathrm{Y}_l^{m*}(\theta',\phi')\mathrm{Y}_l^m(\theta,\phi)\Upsilon_n(t,k_n), \tag{5.24}$$

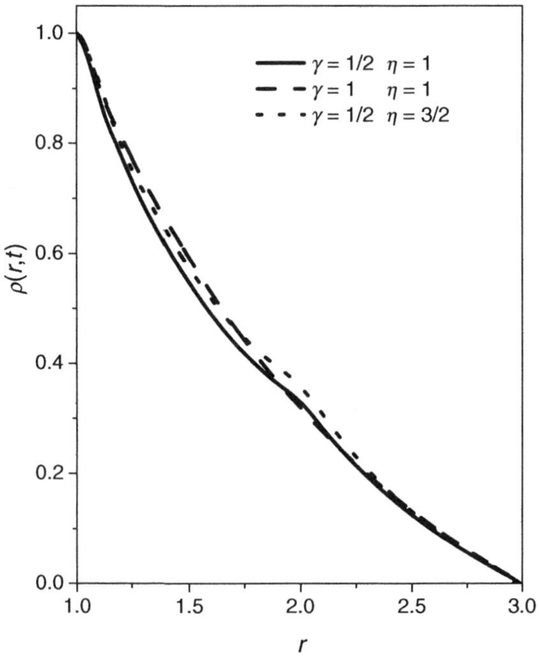

Figure 5.3 $\rho(\overline{r},t)$ versus r as predicted by Eq. (5.23) for typical values of γ and η. The boundary conditions are $\rho(a,t) = 1$ and $\rho(b,t) = 0$, $D = 1$, $t = 1$, $a = 1$, $b = 3$, and $\rho(\overline{r},0) = \delta(r-2)/r^2$ [163]. Modified from Chemical Physics, **344**/1–2, L. S. Lucena, L. R. da Silva, L. R. Evangelista, M. K. Lenzi, R. Rossato, and E. K. Lenzi, Solutions for a fractional diffusion equation with spherical symmetry using Green function approach, 90–94. Copyright (2008), with permission from Elsevier.

where $\mathrm{Y}_l^{m*}(\theta,\phi) = (-1)^m \mathrm{Y}_l^{-m}(\theta,\phi)$ and $\Upsilon_n(t,k_{lmn}) = t^{\gamma-1}\mathrm{E}_{\gamma,\gamma}\left(-k_{lmn}^2 t^\gamma\right)$. Note that $\Upsilon_n(t,k_n)$ is given in terms of the generalised Mittag-Leffler function. The presence of this function in the solution indicates an anomalous spreading of the distribution due to the fractional time derivatives in Eq. (5.13), which can be verified by analysing the behaviour of the second moment. We notice also that the surface effects represented by the second term of Eq. (5.23) may produce an anomalous behaviour in the solution. To gain some physical insights, these surface effects may be related to adsorption–desorption processes occurring at the interfaces when memory effects may also play an important role [151, 152]. Similar investigations dealing with Eq. (5.23) may be performed by considering the presence of a reactive boundary [164], and the first passage time in confined regions [165]. With regard to this latter situation, using Eq. (5.23), we obtain the survival probability $\mathcal{S}(t)$ by integrating over the spatial variables as

$$\mathcal{S}(t) = \int_0^{2\pi} d\phi \int_0^{\pi} d\theta \sin\theta \int_a^b dr r^2 \rho(\overline{r},t), \tag{5.25}$$

and determining the first passage time distribution via $\mathcal{F}(t) = -\partial S(t)/\partial t$.

In particular, if we consider the absorbing boundary condition, $\widetilde{\Phi}_a(\theta,\phi,t) = \widetilde{\Phi}_b(\theta,\phi,t) = 0$, then the survival probability obtained from Eq. (5.23), for the initial condition $\rho(r,0) = \delta(r-r')/(4\pi r^2)$, is given by

$$\mathcal{S}(t) = \frac{a^{\frac{1}{2}(1-\eta)}}{\pi}\sum_{n=1}^{\infty}\frac{\overline{\mathcal{A}}_n}{\overline{k}_n^2}\left[1-\frac{b^{\frac{1}{2}(1-\eta)}\mathrm{J}_\alpha\left(\frac{2\overline{k}_n}{2+\theta}a^{\frac{1}{2}(2+\eta)}\right)}{a^{\frac{1}{2}(1-\eta)}\mathrm{J}_\alpha\left(\frac{2\overline{k}_n}{2+\theta}b^{\frac{1}{2}(2+\eta)}\right)}\right]\overline{\Psi}_n(r')\mathrm{E}_\gamma\left(-\overline{k}_n^2 t^\gamma\right), \tag{5.26}$$

where $\overline{\mathcal{A}}_n = A_{00n}$, $\overline{k}_n = k_{00n}$, $\overline{\Psi}_n(r) = \Psi_{00n}(r)$, and $E_\gamma(x)$ is the Mittag-Leffler function (see Fig. 5.4).

We consider, now, Eq. (5.16) by incorporating a source term $\alpha(\overline{r},t) \neq 0$ and the external force given by (5.15), which represents an extension to the case of the logarithmic potential used, for instance, to establish the connection between the fractional diffusion coefficient and the generalised mobility [166]. The mathematical procedure is the same as before and the solution of Eq. (5.13) for this case may be written as

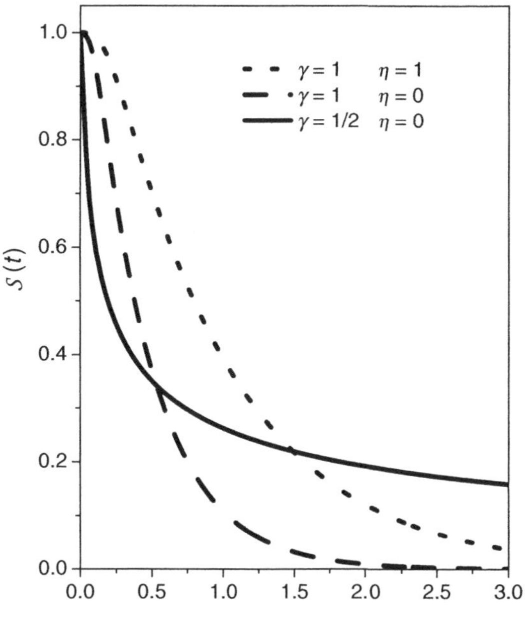

Figure 5.4 $\mathcal{S}(t)$ versus t for typical values of γ and η, with $D = 1$, $r' = 2$, $a = 1$, and $b = 3$ [163]. Modified from Chemical Physics, **344**/1–2, L. S. Lucena, L. R. da Silva, L. R. Evangelista, M. K. Lenzi, R. Rossato, and E. K. Lenzi, Solutions for a fractional diffusion equation with spherical symmetry using Green function approach, 90–94. Copyright (2008), with permission from Elsevier.

$$\rho(\bar{r},t) = -\frac{1}{\Gamma(1-\gamma)}\int_0^t dt' \frac{1}{(t-t')^\gamma}\int_a^b dr' r'^2 \int_0^\pi d\theta' \sin\theta' \int_0^{2\pi} d\phi' G(\bar{r},\bar{r}';t')\widetilde{\rho}(\bar{r}')$$
$$-\int_0^t dt' \int_a^b dr' r'^2 \int_0^\pi d\theta' \sin\theta' \int_0^{2\pi} d\phi' \alpha(\bar{r}',t') G(\bar{r},\bar{r}';t-t')$$
$$+\int_0^t dt' \int_0^\pi d\theta' \sin\theta' \int_0^{2\pi} d\phi' \left[D b^{2-\eta}\widetilde{\Phi}_b(\theta,\phi,t-t')\frac{\partial}{\partial r'}G(\bar{r},\bar{r}';t')\Big|_{r'=b} \right.$$
$$\left. - D a^{2-\eta}\widetilde{\Phi}_a(\theta,\phi,t-t')\frac{\partial}{\partial r'}G(\bar{r},\bar{r}';t')\Big|_{r'=a} \right], \tag{5.27}$$

(see Fig. 5.5) with

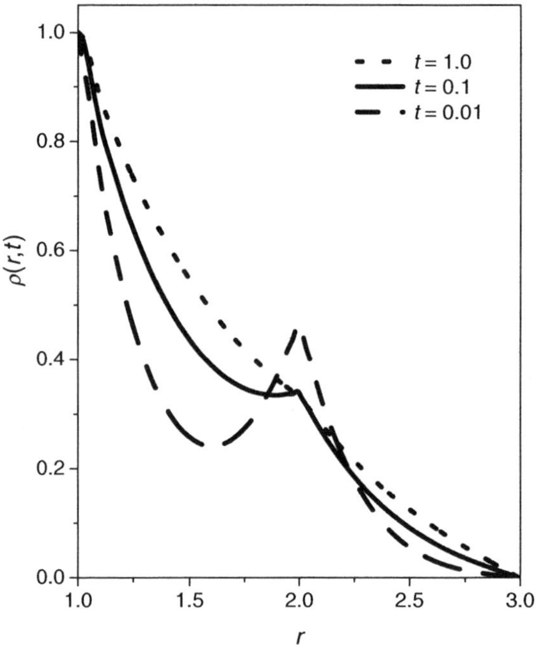

Figure 5.5 $\rho(\bar{r},t)$ versus r as given by Eq. (5.27) for some illustrative values of time. The boundary conditions are $\rho(a,t) = 1$ and $\rho(b,t) = 0$; $D = 1$, $a = 1$, $b = 3$, $\gamma = 1/2$, $\eta = 1$, and $\rho(\bar{r},0) = \delta(r-2)/r^2$ [163]. Modified from Chemical Physics, **344**/1–2, L. S. Lucena, L. R. da Silva, L. R. Evangelista, M. K. Lenzi, R. Rossato, and E. K. Lenzi, Solutions for a fractional diffusion equation with spherical symmetry using Green function approach, 90–94. Copyright (2008), with permission from Elsevier.

$$\widehat{G}(\overline{r},\overline{r}';t) = -\frac{r'^{\frac{\mathcal{K}}{D}}}{2+\eta}\sum_{n=1}^{\infty}\sum_{l=0}^{\infty}\sum_{m=-l}^{l}\widetilde{\mathcal{A}}_{lmn}\widetilde{\Psi}_{lmn}(r)\widetilde{\Psi}_{lmn}(r')\mathrm{Y}_l^{m*}(\theta',\phi')$$
$$\times \mathrm{Y}_l^m(\theta,\phi)\Upsilon_n(t,k_{lmn}), \tag{5.28}$$

where

$$\widetilde{\Psi}_{lmn}(r) = r^{-\frac{1}{2}(1-\eta+\frac{\mathcal{K}}{D})}\left[\mathrm{J}_\nu\left(\frac{2k_{lmn}}{2+\eta}r^{\frac{1}{2}(2+\eta)}\right)\mathrm{N}_\nu\left(\frac{2k_{lmn}}{2+\eta}a^{\frac{1}{2}(2+\eta)}\right)\right.$$
$$\left. - \mathrm{J}_\nu\left(\frac{2k_{lmn}}{2+\eta}a^{\frac{1}{2}(2+\eta)}\right)\mathrm{N}_\nu\left(\frac{2k_{lmn}}{2+\eta}r^{\frac{1}{2}(2+\eta)}\right)\right], \tag{5.29}$$

and

$$\nu = \frac{2}{2+\eta}\left[\left(\frac{1-\eta}{2}+\frac{\mathcal{K}}{2D}\right)^2 + l(l+1)\right]^{1/2}.$$

As before, $\mathrm{J}_\alpha(x)$ and $\mathrm{N}_\alpha(x)$ are, respectively, the Bessel functions of first and second kinds and k_{lmn} are solutions of the eigenvalue equation Eq. (5.21) in which we replace the index α of the Bessel functions $\mathrm{J}_\alpha(x)$ and $\mathrm{N}_\alpha(x)$ with ν, and

$$\widetilde{\mathcal{A}}_{lmn} = \frac{(\pi k_{mn})^2}{\left[\mathrm{J}_\nu\left(\frac{2k_{lmn}}{2+\eta}a^{\frac{1}{2}(2+\eta)}\right)/\mathrm{J}_\nu\left(\frac{2k_{lmn}}{2+\eta}b^{\frac{1}{2}(2+\eta)}\right)\right]^2 - 1}. \tag{5.30}$$

The Green's function defined in Eq. (5.28) satisfies the equation

$$\nabla\cdot\left[D(\overline{r})\nabla\widehat{G}(\overline{r},\overline{r}';s)\right] + \overline{F}(\overline{r})\cdot\nabla\widehat{G}(\overline{r},\overline{r}';s) - s^{\gamma}\widehat{G}(\overline{r},\overline{r}';s) = \delta(\overline{r}-\overline{r}') \tag{5.31}$$

subject to the conditions $\widehat{G}(\overline{r},\overline{r}';s)|_{r=b} = 0$ and $\widehat{G}(\overline{r},\overline{r}';s)|_{r=a} = 0$, whereas Eq. (5.27) was obtained by using Eqs. (5.13) and (5.31). This represents an extension of the results found in Refs. [167, 168] for the three-dimensional case, with an external force term, a space-time-dependent source term, and taking inhomogeneous boundary condition into account. In addition, they represent also an extension of results already obtained [169].

5.3 Reaction on a Solid Surface: Anomalous Mass Transfer

In this section, our goal is to investigate a system with bulk anomalous diffusion in contact with a surface in which adsorption–desorption processes and chemical reactions are present [139]. To do this, we consider that the dynamics of the particles in the bulk is governed by a fractional diffusion equation and the processes occurring on the surface are described by linear kinetic equations. These kinetic equations are coupled in order to account for reversible reaction processes. The analysis carried out here may be relevant for several systems of interest (such as heterogeneous catalysis) where surface reaction occurs when the substrate is in contact with a media supplying mass to these reactions by means of an anomalous mechanism.

Let us start our analysis by considering that the particle dynamics in the bulk is governed by the fractional diffusion equation

$$\frac{\partial}{\partial t}\rho(x,t) = D\,{}_0\mathrm{D}_t^{1-\alpha}\left[\frac{\partial^2}{\partial x^2}\rho(x,t)\right] \tag{5.32}$$

where $\rho(x,t)$ is the density of particles, D is the diffusion coefficient, and the fractional time derivative is that of Riemann–Liouville, introduced in Eq. (2.8). For the processes which may occur on the surface such as adsorption, desorption, and chemical reactions, we assume that they are governed by the kinetic equations:

$$\tau\frac{d}{dt}\Gamma_a(t) = \kappa\tau\rho(0,t) - \int_0^t dt' k_a(t-t')\Gamma_a(t') + \int_0^t dt' k_{b\to a}(t-t')\Gamma_b(t') \tag{5.33}$$

and

$$\tau\frac{d}{dt}\Gamma_b(t) = \int_0^t dt' k_{a\to b}(t-t')\Gamma_a(t') - \int_0^t dt' k_b(t-t')\Gamma_b(t'). \tag{5.34}$$

In Eqs. (5.33) and (5.34), $\Gamma_{a(b)}(t)$ is the surface density, and κ, $k_{a(b)}(t)$, and $k_{a\to b(b\to a)}(t)$ are the parameters connected to the processes on the surface. In particular, we can introduce a characteristic time

$$\tau_\kappa = \left(\frac{D}{\kappa^2}\right)^{2/(2-\alpha)}$$

which governs the diffusion and adsorption near the surface. Equation (5.33) connects the rate of adsorption of species a with the amount in the bulk, minus the desorbed amount of a plus whatever quantity is reversibly reacted from b to a. Equation (5.34) connects the rate of adsorption of species b with the amount of a that reacts to form b minus the desorbed amount of b. Typical situations governed by first-order kinetic equations can be found in several chemical contexts [170] and, in particular, in heavy metal sorption [171]. In addition, depending on $k_a(t)$ and $k_{b\to a}(t)$, Eq. (5.33) may be connected, in the steady state, to the Henry isotherm [172].

In order to solve these equations, we initially consider them subjected to the condition $\rho(x,0) = \Phi(x)$, with $\int_0^\infty dx\Phi(x) = 1$ and $\Gamma_a(0) = \Gamma_b(0) = 0$. We also consider that $\rho(\infty,t) = 0$ and

$$\Gamma_a(t)+\Gamma_b(t) + \int_0^\infty \rho(x,t)dx = \text{constant}, \tag{5.35}$$

which accounts for the conservation of the number of particles diffusing in the system. The boundary conditions may be written as

$$D\,{}_0\mathrm{D}_t^{1-\alpha}\left[\frac{\partial}{\partial x}\rho(x,t)\right]\Bigg|_{x=0} = \frac{d}{dt}\Gamma_t(t), \tag{5.36}$$

where $\Gamma_t(t) = \Gamma_a(t) + \Gamma_b(t)$. Equation (5.35) may also be extended to cover other scenarios in which the conservation of the number of particles is not satisfied [173]. In this case, suitable changes in Eq. (5.36) are required to incorporate this scenario in the approach.

Again, we use the Laplace transform and the Green's function approach to obtain the solution of Eq. (5.32) by taking Eq. (5.35) into account, in the form

$$\rho(x',s) = -\int_0^\infty G(x,x';s)\Phi(x)dx' + \rho(0,s)\, e^{-\sqrt{s^\alpha/D}x'}, \tag{5.37}$$

with the Green's function given by

$$G(x,x';s) = -\frac{1}{2s\sqrt{D/s^\alpha}}\left(e^{-\sqrt{s^\alpha/D}\,|x-x'|} - e^{-\sqrt{s^\alpha/D}\,|x+x'|}\right). \tag{5.38}$$

The last term of Eq. (5.37) represents the effect of the surface on the bulk; it indicates how the processes occurring on the surface can modify the evolution of the initial condition of the system. For the present case, the surface terms are connected with processes represented by Eqs. (5.33) and (5.34). In particular, to simulate a reversible kinetic process, e.g., a chemical reaction $A \rightleftarrows B$, we have that $k_{b\to a}(t) = k_b(t)$ and $k_{b\to a}(t) = k_a(t)$. Thus, after performing some calculations, it is possible to show that

$$\rho(0,s) = \frac{\kappa}{\kappa + s\sqrt{D/s^\alpha}}\int_0^\infty \Phi(x)e^{-\sqrt{s^\alpha/D}x}dx, \tag{5.39}$$

and that the concentrations of the species a and b on the surface are, respectively, given by

$$\Gamma_a(s) = \frac{\kappa}{\kappa + s\sqrt{D/s^\alpha}}\frac{[s\tau + k_b(s)]\,\Theta(s)}{s\tau + k_a(s) + k_b(s)} \tag{5.40}$$

and

$$\Gamma_b(s) = \frac{\kappa}{\kappa + s\sqrt{D/s^\alpha}}\frac{k_a(s)\Theta(s)}{s\tau + k_a(s) + k_b(s)}, \tag{5.41}$$

with

$$\Theta(s) = \frac{1}{s}\int_0^\infty \Phi(x)e^{-\sqrt{s^\alpha/D}x}dx. \tag{5.42}$$

These equations were obtained from Eqs. (5.37) and (5.35) by using Eqs. (5.33) and (5.34) in the Laplace space. The boundary and initial conditions for $\rho(x,t)$, $\Gamma_a(t)$, and $\Gamma_b(t)$ were also used.

Notice the presence of a stationary state when $t \to \infty$ $(s \to 0)$, if $k_a(s) \to k_a =$ constant and $k_b(s) \to k_b =$ constant, which results in

$$\Gamma_a(t) \sim \frac{k_b}{k_a + k_b} \quad \text{and} \quad \Gamma_b(t) \sim \frac{k_a}{k_a + k_b}. \tag{5.43}$$

Equations (5.35) and (5.43) imply that, after some time, the substance, initially present in the bulk, is totally adsorbed by the surface where the reversible reaction process is occurring and is not desorbed to the bulk due to the type of the kinetic process on the surface. By analysing Eq. (5.39), we also notice that this choice of kinetic process on the surface has no influence on the spreading of the species into the bulk. The influence of the surface on the bulk is manifested by κ, the adsorption rate.

By taking the inverse Laplace transform of Eqs. (5.40) and (5.41), for $k_a(s) = k_a =$ constant and $k_b(s) = k_b =$ constant, we obtain

$$\Gamma_a(t) + \Gamma_b(t) = \frac{\kappa}{\sqrt{D}} \int_0^t \frac{dt'}{\sqrt{(t-t')^\alpha}} E_{1-\alpha/2,1-\alpha/2}\left[\frac{\kappa}{\sqrt{D}}(t-t')^{1-\alpha/2}\right] h(t')$$

and

$$\Gamma_b(t) = \int_0^t dt' \Delta(t-t') h(t'), \tag{5.44}$$

with

$$\Delta(t) = \frac{k_a}{\tau} e^{-k_{a+b}t/\tau} \int_0^t \frac{\kappa}{\sqrt{Dt'^\alpha}} E_{1-\alpha/2,1-\alpha/2}\left(\frac{\kappa}{\sqrt{D}} t'^{1-\alpha/2}\right) e^{k_{a+b}t'/\tau} dt' \tag{5.45}$$

and

$$h(t') = \int_0^\infty dx \Phi(x) \mathrm{H}_{1,1}^{1,0}\left[\frac{x'}{\sqrt{Dt'^\alpha}} \middle| \begin{matrix} (1, \frac{\alpha}{2}) \\ (0, 1) \end{matrix}\right], \tag{5.46}$$

where $k_{a+b} = k_a + k_b$. Once more, we underline that the presence of the generalised Mittag-Leffler function and the H-function of Fox is connected with the anomalous spreading of the system due to the fractional time derivative in Eq. (5.32).

Figure 5.6 shows the behaviour of $\Gamma_a(t)$ and $\Gamma_b(t)$ for $\alpha = 1$ and $\alpha \neq 1$ illustrating the influence of the fractional coefficient and, thereafter, the anomalous behaviour, on the solutions of the kinetic equations. Figure 5.7 shows the

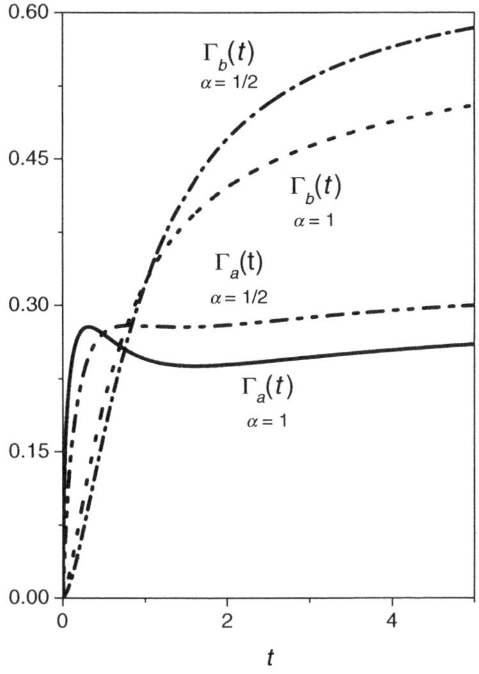

Figure 5.6 Temporal behaviour of $\Gamma_a(t)$ and $\Gamma_b(t)$ for $\alpha = 1$ and $\alpha = 1/2$. For illustrative purposes, the curves were drawn for $k_a = 2$, $k_b = 1$, $\tau = 1$, and $\tau_\kappa = 1$ in arbitrary units [139]. Modified from Physica A, **410**, E. K. Lenzi, M. K. Lenzi, R. S. Zola, H.V. Ribeiro, F.C. Zola, L.R. Evangelista, and G. Gonçalves, Reaction on a solid surface supplied by an anomalous mass transfer source, 399–406. Copyright (2014), with permission from Elsevier.

time evolution of the bulk distribution for the same cases reported in Fig. 5.6 for three different times. Clearly, the anomalous nature has great influence on both the surface and bulk densities, as evidenced from Figs. 5.6 and 5.7.

Representative cases are obtained when memory effects are introduced in Eqs. (5.33) and (5.34) due to a time dependence on $k_a(t)$ and/or $k_b(t)$, which implies that $k_a(s)$ and/or $k_b(s)$ are not constants as in the previous case. As a particular case, for $k_a(s) = k_a/(1 + s\tau_a)$ ($k_a(t) = k_a/\tau_a e^{t/\tau_a}$) [151] and for $k_b(s) = k_b =$ constant ($k_b(t) = k_b\delta(t)$), the behaviour of $\Gamma_a(t)$ and $\Gamma_b(t)$, obtained from the kinetic equations, presents a few oscillations, which can be interpreted as repeated adsorption–desorption phenomena often found in the physisorption process [174], as illustrated in Fig. 5.8.

Let us consider now the most general case, where $k_{b\to a}(t) \neq k_b(t)$ and $k_{b\to a}(t) \neq k_a(t)$, with $k_a k_b - k_{b\to a}k_{a\to b} > 0$. We obtain

$$\rho(0,s) = \frac{1}{\Delta_\kappa(s)}\kappa\left\{[s\tau + k_a(s)]\,[s\tau + k_b(s)] - k_{b\to a}(s)k_{a\to b}(s)\right\}\Theta(s), \qquad (5.47)$$

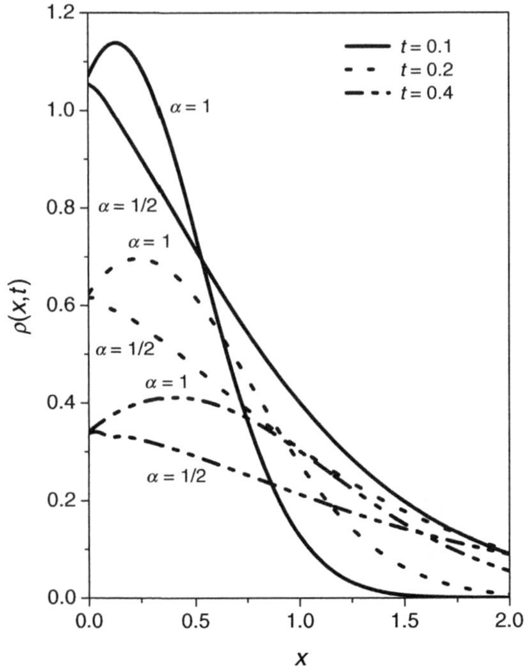

Figure 5.7 Spatial profile of Eq. (5.37), after performing the inverse Laplace transform, for $\alpha = 1$ and $\alpha = 1/2$, with $\tau_\kappa = 1$ in arbitrary units [139]. Modified from Physica A, **410**, E. K. Lenzi, M. K. Lenzi, R. S. Zola, H. V. Ribeiro, F. C. Zola, L. R. Evangelista, and G. Gonçalves, Reaction on a solid surface supplied by an anomalous mass transfer source, 399–406. Copyright (2014), with permission from Elsevier.

with

$$\Gamma_a(s) = \frac{1}{\Delta_\kappa(s)}\kappa\left[s\tau + k_b(s)\right]\Theta(s) \quad \text{and} \quad \Gamma_b(s) = \frac{1}{\Delta_\kappa(s)}\kappa k_a(s)\Theta(s), \tag{5.48}$$

in which

$$\Delta_\kappa(s) = \kappa\left[s\tau + k_{b\to a}(s) + k_b(s)\right] + \sqrt{\frac{D}{s^\alpha}}\left\{\left[s\tau + k_a(s)\right]\left[s\tau + k_b(s)\right] - k_{b\to a}(s)k_{a\to b}(s)\right\}. \tag{5.49}$$

The processes occurring on the surface have a direct influence on the diffusive behaviour of the system in the bulk; the quantities k_a, k_b, $k_{b\to a}$, and $k_{a\to b}$ in Eq. (5.47) represent the contribution given by the surface in Eq. (5.37). Another interesting feature of this case may be found in the asymptotic limit of Eqs. (5.48), for constant k_a, k_b, $k_{b\to a}$, and $k_{a\to b}$. In this limit, we obtain

$$\Gamma_a(t) \sim \frac{\kappa k_b}{k_a k_b - k_{b\to a}k_{a\to b}}\frac{1}{\sqrt{Dt^\alpha}}\int_0^\infty dx\Phi(x)\mathrm{H}^{1,0}_{1,1}\left[\frac{x}{\sqrt{Dt^\alpha}}\middle|\begin{matrix}(1-\frac{\alpha}{2},\frac{\alpha}{2})\\(0,1)\end{matrix}\right] \tag{5.50}$$

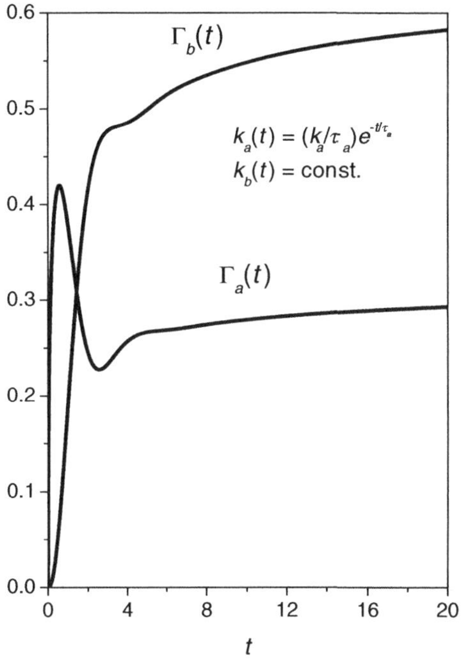

Figure 5.8 Temporal behaviour of $\Gamma_a(t)$ and $\Gamma_b(t)$ for $\alpha = 1$, when memory effects are incorporated in the kinetic equations. For illustrative purposes, we consider again $k_a = 2$, $k_b = 1$, $\tau = 1$, $\tau_a = 1$, and $\tau_\kappa = 1$ in arbitrary units [139]. Modified from Physica A, **410**, E. K. Lenzi, M. K. Lenzi, R. S. Zola, H. V. Ribeiro, F. C. Zola, L. R. Evangelista, and G. Gonçalves, Reaction on a solid surface supplied by an anomalous mass transfer source, 399–406. Copyright (2014), with permission from Elsevier.

and

$$\Gamma_b(s) \sim \frac{\kappa k_a}{k_a k_b - k_{b\to a} k_{a\to b}} \frac{1}{\sqrt{D t^\alpha}} \int_0^\infty dx \Phi(x) \mathrm{H}_{1,1}^{1,0}\left[\frac{x}{\sqrt{D t^\alpha}} \middle|_{(0,\,1)}^{(1-\frac{\alpha}{2},\frac{\alpha}{2})}\right]. \tag{5.51}$$

These equations show that, after some elapsed time, the concentration on the surface begins to saturate and the desorption process starts. The substance present in the bulk is initially adsorbed by the surface where the kinetic processes governed by Eqs. (5.33) and (5.34) occur; after some time, it is desorbed back to the bulk. This behaviour may be seen in Figs. 5.9 and 5.10 which, for short times, show that the system has an accumulation of particles on the surface and, after some time, it spreads out back to the bulk. This case is especially important in connection with heterogeneous catalysis, as in the case of the Langmuir–Hinshelwood mechanism, but now including memory effects and anomalous diffusion, which generalises the problem.

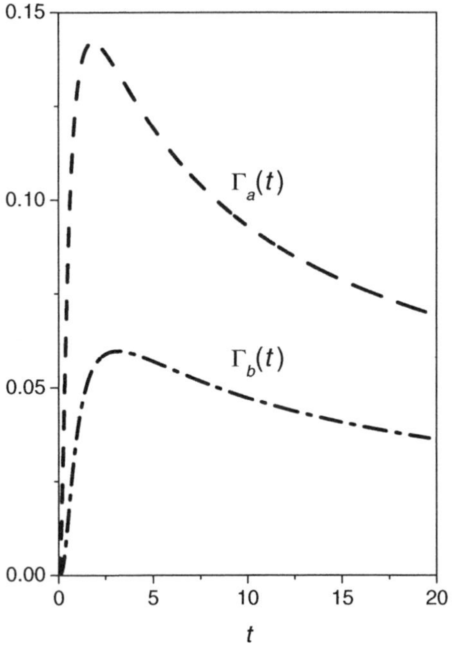

Figure 5.9 Temporal behaviour of $\Gamma_a(t)$ and $\Gamma_b(t)$ for $\alpha = 1$. The curves were drawn for $k_a = 2$, $k_b = k_{b\to a} = k_{a\to b} = 1$, $\tau = 1$, and $\tau_\kappa = 1$ in arbitrary units [139]. Modified from Physica A, **410**, E. K. Lenzi, M. K. Lenzi, R. S. Zola, H. V. Ribeiro, F. C. Zola, L. R. Evangelista, and G. Gonçalves, Reaction on a solid surface supplied by an anomalous mass transfer source, 399–406. Copyright (2014), with permission from Elsevier.

5.4 Heterogeneous Media and Transport through a Membrane

In the preceding sections, we have considered the effects of the adsorption–desorption process in a confined system. An important class of problems is represented by the case in which the surface is permeable to the passage of particles [175]. To analyse this kind of problem, we consider an infinite medium in which all the relevant quantities diffuse in one dimension (e.g., the x-direction). In practice, we shall consider that there is a membrane located at $x = 0$; thus, it divides the space in two regions, $x > 0$ (side 1) and $x < 0$ (side 2). The thickness of the membrane, ξ, is assumed to be thin enough to assure that only the kinetic processes are relevant in this region (see Fig. 5.11). We consider that the distributions of the particles in the bulk are governed by the fractional diffusion equations

$$\frac{\partial}{\partial t}\rho_1(x,t) = D_1\, {}_0\mathrm{D}_t^{1-\alpha_1}\left[\frac{\partial^2}{\partial x^2}\rho_1(x,t)\right], \qquad 0 < x < \infty, \tag{5.52}$$

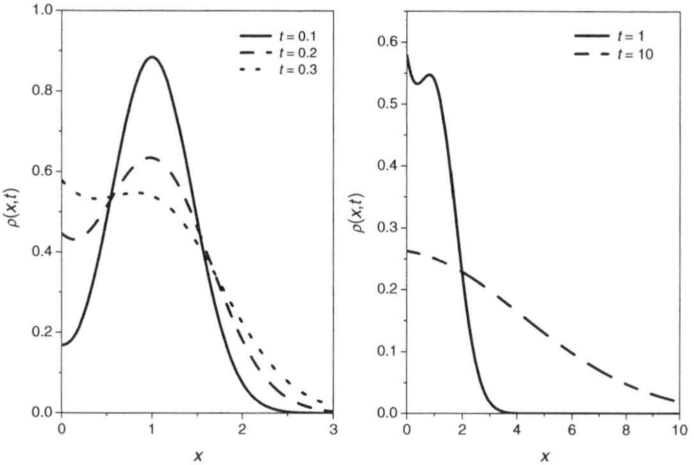

Figure 5.10 $\rho(x,t)$ versus x from Eq. (5.37) for $\alpha = 1$, when $k_{b\to a}(t) \neq k_b(t)$ and $k_{b\to a}(t) \neq k_a(t)$. For illustrative purposes, we also consider $k_a = 2$, $k_b = k_{b\to a} = k_{a\to b} = 1$, $\tau_\kappa = 1$, and $\tau = 1$ in arbitrary units. For $t = 1$, the particles are very close to the surface where the kinetic processes are occurring. For $t = 10$, the particles leave the surface by a desorption process and spread out [139]. Modified from Physica A, **410**, E. K. Lenzi, M. K. Lenzi, R. S. Zola, H. V. Ribeiro, F. C. Zola, L. R. Evangelista, and G. Gonçalves, Reaction on a solid surface supplied by an anomalous mass transfer source, 399–406. Copyright (2014), with permission from Elsevier.

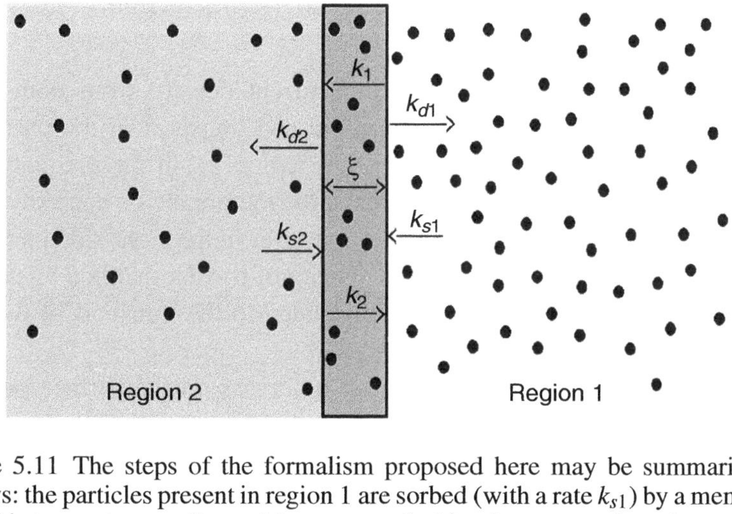

Figure 5.11 The steps of the formalism proposed here may be summarised as follows: the particles present in region 1 are sorbed (with a rate k_{s1}) by a membrane placed between two media, and, by means of a kinetic process, the substance may be transported from region 1 to region 2, with a rate k_1. Likewise, the substance in region 2 may also be sorbed (with a rate k_{s2}) by the surface and then transported to region 1, with a rate k_2. Note that the membrane thickness ξ is considered sufficiently small, i.e., $|\xi| \to 0$, to avoid diffusion in this region [175].

and

$$\frac{\partial}{\partial t}\rho_2(x,t) = D_2\, {}_0\mathrm{D}_t^{1-\alpha_2}\left[\frac{\partial^2}{\partial x^2}\rho_2(x,t)\right], \qquad -\infty < x < 0. \tag{5.53}$$

In Eqs. (5.52) and (5.53), D_1 and D_2 are the diffusion coefficients of regions 1 and 2, respectively. The quantities ρ_1 and ρ_2 represent the density of particles in each side and the fractional time derivatives in Eqs. (5.52) and (5.53) are those of Riemann-Liouville, as defined in Eq. (2.8), with $0 < \alpha_{1(2)} < 1$. The values $0 < \alpha_1 < 1$ and $0 < \alpha_2 < 1$ are connected with a subdiffusive process, whereas $\alpha_1 = \alpha_2 = 1$ corresponds to the usual diffusion. On each side of the membrane, the following equations apply:

$$D_1\, {}_0\mathrm{D}_t^{1-\alpha_1}\left[\frac{\partial}{\partial x}\rho_1(x,t)\right]\Bigg|_{x=0} = \frac{d}{dt}C_1(t) + \int_0^t k_1(t-t')\rho_1(0,t')dt' - \int_0^t k_2(t-t')\rho_2(0,t')dt' \tag{5.54}$$

and

$$D_2\, {}_0\mathrm{D}_t^{1-\alpha_2}\left[\frac{\partial}{\partial x}\rho_2(x,t)\right]\Bigg|_{x=0} = -\frac{d}{dt}C_2(t) - \int_0^t k_2(t-t')\rho_2(0,t')dt' + \int_0^t k_1(t-t')\rho_1(0,t')dt'. \tag{5.55}$$

This set of equations, which connect the current density from both sides, are intended to describe the sorption, desorption, and transport of the particles from one region (e.g., region 1) to the other (e.g., region 2). In these equations, $C_1(t)$ and $C_2(t)$ represent the density of particles at the membrane on each side, obtained from the bulk by a sorption process. The first term in the right side of Eqs. (5.54) and (5.55) gives the time variation of the density of particles sorbed by the surfaces of the membrane, and the other terms are related to the transport of the particles sorbed from one side to the other.

To describe the sorption and desorption processes on the surface, we propose that they can be modelled by the following kinetic equations [176]:

$$\frac{d}{dt}C_1(t) = k_{s1}\rho_1(0,t) - \int_0^t k_{d1}(t-t')C_1(t')dt' \tag{5.56}$$

and

$$\frac{d}{dt}C_2(t) = k_{s2}\rho_2(0,t) - \int_0^t k_{d2}(t-t')C_2(t')dt'. \tag{5.57}$$

In Eqs. (5.56) and (5.57), $\rho_1(0,t)$ and $\rho_2(0,t)$ are the bulk density just in front of the membrane, on side 1 and 2, respectively; k_{si} $(i = 1, 2)$ are parameters connected with the sorption phenomena, being related to a characteristic sorption time $\tau_i \propto 1/k_{si}$; and $k_{di}(t)$ is a kernel that governs the desorption phenomena. These equations state that the time variation of the surface density of sorbed particles at a given side depends on the bulk density of particles just in front of the membrane, and on the surface density of particles already sorbed. They extend the usual kinetic equations (Langmuir approximation) to situations characterised by unusual relaxations, i.e., non-Debye relaxations for which a nonexponential behaviour of the densities can be obtained, depending on the choice of the kernels.

The underlying physical motivation of the time-dependent rate coefficients can be related to the fractal nature, low dimensionality, or macromolecular crowding of the medium [177–180], and even to anomalous molecular diffusion [181]. Moreover, a linear response approach [182] was proposed to describe this temporal behaviour and there is a possible connection with the generalisation of the mass action law [183]. From a phenomenological point of view, the choice of the kernels for Eqs. (5.56) and (5.57) can be connected with surface irregularities [177] (which are important in the adsorption–desorption process), to diffusion, to catalytic processes, and to microscopic parameters representing the van der Waals interaction between particles and surfaces [184]. A kernel like $k_{d1(2)}(t)$ has been used in several contexts to express non-Debye relaxation, yielding a nontrivial behaviour description and allowing for different or combined effects in a single kinetic equation [176].

In a few words, Eqs. (5.56) and (5.57) are proposed to represent the sorption of the particles on each surface with the rates k_{s1} and k_{s2}. After the sorption process, the particles present on the surface can be desorbed, with the rates $k_{d1}(t)$ and $k_{d2}(t)$. They can also be transported throughout the membrane with the rates $k_1(t)$ and $k_2(t)$. Thus, the usual current balance is modified by the second and third terms on the right side of Eqs. (5.54) and (5.55) in order to represent the transport of particles from side 1 to side 2 and vice versa.

For a cell membrane, $k_1(t)$ and $k_2(t)$ may be related to a continuous-time discrete Markovian process of two stages for the transition rates, since the channel can exist either in a closed or in an open state on each side. They may be connected with the probability of opening the channel in a given side in order to make possible the transport of particles from one side to the other. Furthermore, Eqs. (5.54) and (5.55) can also be related to generalised reactive boundary conditions, and the presence of fractional time derivatives is connected to the possibility of unusual relaxation in the bulk, as we discuss below.

Formal derivations and possible applications in biological contexts of the fractional diffusion equation with reactive boundary conditions have been already considered [185]. In particular, due to the non-Markovian nature of subdiffusion, these conditions seem to be more suitable than the usual reaction-diffusion equation with a reaction term independent of the transport one [130]. Finally, we mention that Eqs. (5.54) and (5.55) are coupled with Eqs. (5.56) and (5.57) in such a way that the processes occurring in one side modify the dynamics of the other side. The searched solutions are also subjected to the homogeneous boundary conditions $\partial_x \rho_1(\infty, t) = 0$ and $\partial_x \rho_2(-\infty, t) = 0$ to guarantee the conservation of the number of particles present in the system. For a nonconservative situation, it is possible to consider inhomogeneous boundary conditions dealing with a source or a sink.

The boundary conditions incorporated in the diffusion equations imply that

$$C_1(t) + C_2(t) + \int_{-\infty}^{0} \rho_2(x,t)dx + \int_{0}^{\infty} \rho_1(x,t)dx = \text{constant}, \tag{5.58}$$

which is a direct consequence of the conservation of the total number of particles populating the system. Equation (5.58) can be obtained by performing an integration in Eqs. (5.52) and (5.53), which, after using Eqs. (5.54) and (5.55), become, respectively,

$$\frac{d}{dt}\left[\int_{0}^{\infty} \rho_1(x,t)dx + C_1(t)\right] = \int_{0}^{t} k_2(t-t')\rho_2(0,t')dt' - \int_{0}^{t} k_1(t-t')\rho_1(0,t')dt' \tag{5.59}$$

and

$$\frac{d}{dt}\left[\int_{-\infty}^{0} \rho_2(x,t)dx + C_2(t)\right] = \int_{0}^{t} k_1(t-t')\rho_1(0,t')dt' - \int_{0}^{t} k_2(t-t')\rho_2(0,t')dt'. \tag{5.60}$$

By adding Eqs. (5.59) and (5.60), we find

$$\frac{d}{dt}\left[\int_{0}^{\infty} \rho_1(x,t)dx + C_1(t)\right] = -\frac{d}{dt}\left[\int_{-\infty}^{0} \rho_2(x,t)dx + C_2(t)\right], \tag{5.61}$$

from which Eq. (5.58) can be obtained. Equation (5.61) also shows that the mass (number of particles) variation on side 1 is connected with the variations on side 2. The negative sign shows that the variation of the number of particles on one side (gain or loss) produces an opposite variation on the other side.

Let us consider some concrete, specific cases covered by Eqs. (5.56) to (5.58) in order to illustrate how the previous formalism works. To do this, we initially suppose that all the particles are located in the region $x > 0$ (region 1), such that the initial conditions are, in a general form, $\rho_1(x,0) = \varphi(x)$ (with $\int_0^\infty dx \rho_1(x,0) = 1$), $\rho_2(x,0) = 0$, and $C_1(0) = C_2(0) = 0$. As the time evolves, region 2 gets occupied by particles from region 1 as a product of the membrane transport, and, consequently, the density of particles in region 2 is totally dependent on the role developed by the membrane.

The general solutions may be established by again using the Laplace transform and the Green's function approach. The solution of Eq. (5.53), in the Laplace space, is given by

$$\rho_2(x,s) = \omega_2(s) e^{-\sqrt{\frac{s^{\alpha_2}}{D_2}}|x|}, \tag{5.62}$$

with

$$\omega_2(s) = \frac{[s + k_{d2}(s)]k_1(s)}{\left[s\sqrt{D_2/s^{\alpha_2}} + k_2(s)\right][s + k_{d2}(s)] + sk_{s2}} \rho_1(0,s), \tag{5.63}$$

from which it is possible to show that

$$C_2(s) = \frac{[s + k_{d1}(s)]k_{s2}k_1(s)}{\left[s\sqrt{D_2/s^{\alpha_2}} + k_2(s)\right][s + k_{d2}(s)]k_{s1} + sk_{s2}k_{s1}} C_1(s), \tag{5.64}$$

which enables us to reduce Eq. (5.54) to

$$D_1 s^{1-\alpha_1} \left[\frac{\partial}{\partial x}\rho_1(x,s)\right]\bigg|_{x=0} = \omega_1(s)\rho_1(0,s), \tag{5.65}$$

with

$$\begin{aligned} \omega_1(s) = {} & \frac{s}{s + k_{d1}(s)} k_{s1} + k_1(s) \\ & - \frac{[s + k_{d2}(s)]k_2(s)k_1(s)}{\left[(s\sqrt{D_2/s^{\alpha_2}} + k_2(s)\right][s + k_{d2}(s)] + sk_{s2}}. \end{aligned} \tag{5.66}$$

The Green's function approach yields the solution for $\rho_1(x,s)$ in the form

$$\rho_1(x,s) = -\int_0^\infty G_1(x,x';s)\varphi(x')dx', \tag{5.67}$$

in which the Green's function is defined as

$$\begin{aligned} G_1(x,x';s) = {} & -\frac{1}{2s\sqrt{D_1/s^{\alpha_1}}}\left(e^{-\sqrt{s^{\alpha_1}/D_1}|x-x'|} + e^{-\sqrt{s^{\alpha_1}/D_1}|x+x'|}\right) \\ & + \frac{2\omega_1(s)}{s\sqrt{D_1/s^{\alpha_1}} + \omega_1(s)} \frac{1}{2s\sqrt{D_1/s^{\alpha_1}}} e^{-\sqrt{s^{\alpha_1}/D_1}|x+x'|}. \end{aligned} \tag{5.68}$$

Equation (5.62) can be rewritten, by using the previous results, as

$$\begin{aligned} \rho_2(x,s) = {} & \frac{[s + k_{d2}(s)]\,k_1(s)}{\left[s\sqrt{D_2/s^{\alpha_2}} + k_2(s)\right][s + k_{d2}(s)] + sk_{s2}} \\ & \times \frac{1}{s\sqrt{D_1/s^{\alpha_1}} + \omega_1(s)} e^{-\sqrt{s^{\alpha_1}/D_1}|x'|} e^{-\sqrt{s^{\alpha_2}/D_2}|x|}. \end{aligned} \tag{5.69}$$

By analysing Eq. (5.68), we note that the processes occurring at the membrane have direct influence on the dynamics of the particles in region 1. The presence of $k_2(s)$, $k_{d2}(s)$, and k_{s2} in $\omega(s)$ implies that the sorption and desorption processes in region 2 and, consequently, the particle transport (from region 2 to region 1) modify the dynamic of the particles in region 1. After taking the inverse Laplace transform with $k_1(s) = k_1 =$ constant, $k_{d1}(s) = k_{d1} =$ constant, $k_2(s) = k_2 =$ constant, and $k_{d2}(s) = k_{d2} =$ constant, Eqs. (5.68) and (5.62) can be rewritten, respectively, as

$$G_1(x,x';t) = -\frac{1}{\sqrt{4D_1 t^{\alpha_1}}}\left\{\mathrm{H}^{1,0}_{1,1}\left[\frac{|x-x'|}{\sqrt{D_1 t^{\alpha_1}}}\left|\begin{matrix}(1-\frac{\alpha_1}{2},\frac{\alpha_1}{2})\\(0,1)\end{matrix}\right.\right] + \mathrm{H}^{1,0}_{1,1}\left[\frac{|x+x'|}{\sqrt{D_1 t^{\alpha_1}}}\left|\begin{matrix}(1-\frac{\alpha_1}{2},\frac{\alpha_1}{2})\\(0,1)\end{matrix}\right.\right]\right\} + \int_0^t \frac{dt'}{\sqrt{D_1 t'^{\alpha_1}}}\Pi(t-t')\mathrm{H}^{1,0}_{1,1}\left[\frac{|x+x'|}{\sqrt{D_1 t'^{\alpha_1}}}\left|\begin{matrix}(1-\frac{\alpha_1}{2},\frac{\alpha_1}{2})\\(0,1)\end{matrix}\right.\right] \tag{5.70}$$

and

$$\rho_2(x,t) = \int_0^t \frac{dt'}{\sqrt{4D_1 t'^{\alpha_2}}}\Upsilon(t-t')\mathrm{H}^{1,0}_{1,1}\left[\frac{|x|}{\sqrt{D_2 t'^{\alpha_2}}}\left|\begin{matrix}(1-\frac{\alpha_2}{2},\frac{\alpha_2}{2})\\(0,1)\end{matrix}\right.\right], \tag{5.71}$$

with

$$\Upsilon(t) = \Phi(t) + \sum_{n=1}^{\infty}(-k_{s2})^n \int_0^t dt_n \Xi(t_n - t_{n-1}) \times \int_0^{t_n} dt_{n-1}\Xi(t_{n-1}-t_{n-2})\cdots\int_0^{t_2} dt_1 \Xi(t_2-t_1)\Phi(t_1), \tag{5.72}$$

where

$$\Xi(t) = \frac{1}{\sqrt{D_2 t^{\alpha_2}}}\mathrm{E}_{\alpha',\beta'}\left(-\frac{k_2}{\sqrt{D_2}}t^{\alpha'}\right) - k_{d2}e^{-k_{d2}t'}\int_0^t dt' \frac{e^{-k_{d2}t'}}{\sqrt{D_2 t'^{\alpha_2}}}\mathrm{E}_{\alpha',\beta'}\left(-\frac{k_2}{\sqrt{D_2}}t^{\alpha'}\right)$$

and

$$\Phi(t) = k_1\rho_1(0,t) - k_1\int_0^t dt' \frac{k_2\rho_1(0,t')}{\sqrt{D_2(t-t')^{\alpha_2}}}\mathrm{E}_{\alpha',\beta'}\left[-\frac{k_2}{\sqrt{D_2}}\left(t-t'\right)^{\alpha'}\right], \tag{5.73}$$

for $\alpha' = \beta' = 1 - \alpha_2/2$. The quantity $\Pi(t)$ introduced above is defined as

$$\Pi(t) = I_{\omega_1}(t) + \sum_{n=1}^{\infty}(-1)^n \int_0^t dt_n I_{\omega_1}(t) \cdots \int_0^{t_2} dt_1 I_{\omega_1}(t_2 - t_1) I_{\omega_1}(t_1), \tag{5.74}$$

where

$$I_{\omega_1}(t) = \frac{1}{\sqrt{D_1}\Gamma\left(1 + \frac{\alpha_1}{2}\right)} \int_0^t dt' \frac{\omega_1(t')}{(t - t')^{\frac{\alpha_1}{2}}}. \tag{5.75}$$

Let us apply now the preceding results to the situation characterised by the particles being transported (e.g., osmotic process) from one side to the other, i.e., $k_2 = 0$ with $k_1 \neq 0$. In connection with a cell membrane, the constant k_1 may also be related to the opening channels rate, which are able to conduct the substance from one side to the other, as mentioned before. In this case, Eqs. (5.59) and (5.60) imply that

$$\frac{d}{dt}\left[\int_0^{\infty} \rho_1(x,t)dx + C_1(t)\right] = -k_1\rho_1(0,t) \tag{5.76}$$

and

$$\frac{d}{dt}\left[\int_{-\infty}^{0} \rho_2(x,t)dx + C_2(t)\right] = k_1\rho_1(0,t). \tag{5.77}$$

These equations show that in region 1 the amount of substance (number of particles) is decreasing and in region 2 it is increasing due to the transport across the membrane with the rate k_1.

Figure 5.12 shows the behaviour of the distribution for two different values of α_2 and the mean square displacement in region 2. In particular, we consider that $\alpha_1 = 1$ in region 1, and in region 2 two values for α_2 are allowed: $\alpha_2 = 1$ and $\alpha_2 = 1/2$. In Fig. 5.12b, we observe that the behaviour of the mean square displacement is asymptotically governed by the bulk equations for $k_{2s} = 0$ and may exhibit an anomalous behaviour depending on the α_2 value. For this case, in the asymptotic limit of long times ($t \to \infty$), we have

$$\langle (x - \langle x \rangle)^2 \rangle \to t^{\alpha_2}$$

showing that bulk effects have a pronounced role on this diffusive regime. The initial behaviour exhibited in Fig. 5.12b by the particles in region 2 is evidence of the confinement of particles by the membrane during the process of passage from region 1 to 2. It is transitory until the movement of the system reaches the regime governed by the bulk equation for long times. Indeed, for this case, in the asymptotic limit of long times ($t \to \infty$), we have $\langle (\Delta_2 x)^2 \rangle \to t^{\alpha_2}$ showing that bulk effects have a pronounced role on this diffusive regime.

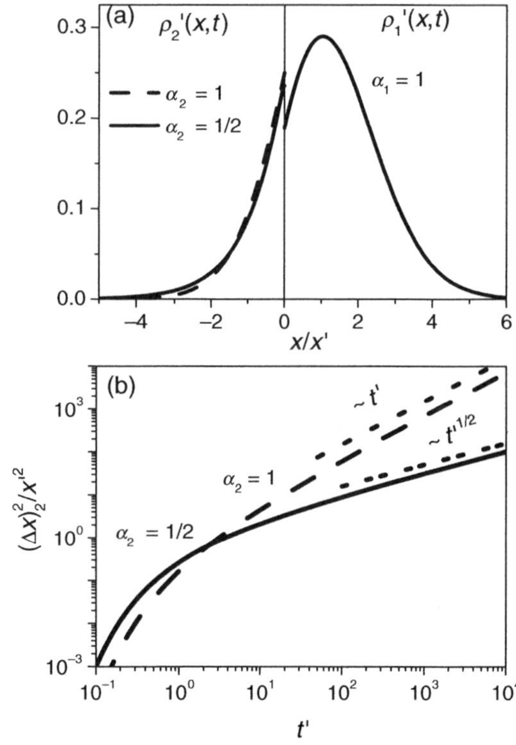

Figure 5.12 (a) Spatial profiles of the distributions $\rho_1'(x,t)$ and $\rho_2'(x,t)$ (where $\rho_1'(x,t) = x'\rho_1(x,t)$ and $\rho_2'(x,t) = x'\rho_2(x,t)$) for two different values of α_2, when $t = \tau_1$ (where $\tau_1 = \left(x'^2/\mathcal{K}_1\right)^{1/\alpha_1}$). (b) Behavior of the mean square displacement ($(\Delta_2 x)^2 = \langle (x - \langle x \rangle)^2 \rangle$ where the subindex 2 denotes the region 2) for the particles present in region 2 with $t' = t/\tau_1$. In panel (b), the dotted lines are a guide to the eyes representing the diffusive regimes. We consider, for illustrative purposes, $k_1(t) = \bar{k}_1\delta(t)$ (where $\bar{k}_1 = x'/\tau_1$), $k_{1d}(t) = k_{2d}(t) = 0$, $\tau_1 = \tau_2$ (where $\tau_2 = \left(x'^2/\mathcal{K}_2\right)^{1/\alpha_2}$), $k_{1s} = k_{2s} = 0$, and $k_2 = 0$, in arbitrary units [175].

The spreading of the system also depends on the processes at the surface in region 2, which may again involve different diffusive regimes. This is illustrated in Fig. 5.13a by comparing the cases $k_{s2} = 0$ (solid line) and $k_{s2} \neq 0$ (dashed line) with $k_2 = 0$. We verify that initially the behaviour of the particles transported from region 1 to 2, after leaving the membrane, may be transient or superdiffusive (dotted lines) and is followed by a usual behaviour for $k_{s2} = 0$ or by a subdiffusive one for $k_{s2} \neq 0$. In Fig. 5.13b, we show that in region 1, which initially presents usual diffusion, the process evolves towards an anomalous diffusion for long times. This fact can be attributed to the surface, which is sorbing, desorbing, and transporting particles out of region 1.

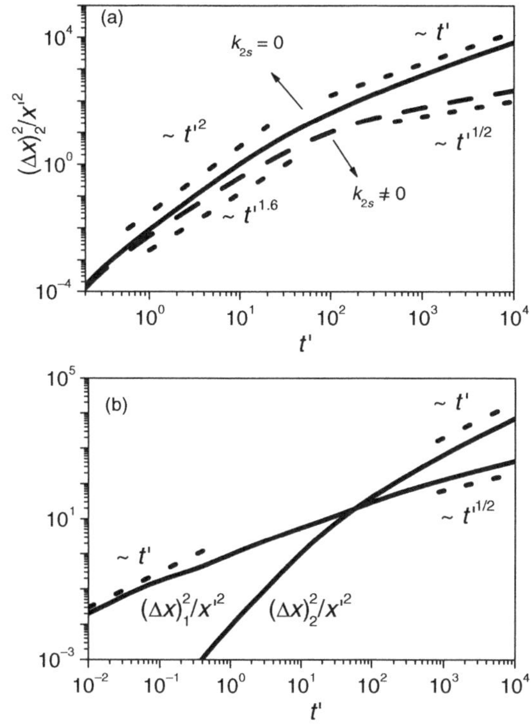

Figure 5.13 (a) Mean square displacement in region 2 for the cases $k_{2s} = 0$ and $k_{2s} \neq 0$. These cases pass through a transient characterized by different regimes until they reach a critical concentration (region 2) and the behaviour becomes usual for $k_{2s} = 0$ or subdiffusive for $k_{2s} \neq 0$. (b) Behaviour of the mean square displacement for region 1 and in region 2, with $k_{2s} = 0$. Note that, in region 1, the regime is initially usual but, after some time has elapsed, it becomes subdiffusive since the membrane promotes adsorption, desorption, and transport of particles, thus inducing particle crowding in region 1. The dotted lines are a guide to the eyes representing the diffusive regimes. We consider, for illustrative purposes, $k_{1s} = 10^2 x'/\tau_1$, $k_1(t) = \overline{k}_1\delta(t)$ (where $\overline{k}_1 = x'/\tau_1$), $t' = t/\tau_1$, $k_{1d}(t) = \overline{k}_{1d}\delta(t)$ (where $\overline{k}_{1d} = 2/\tau_1$), $k_{2d}(t) = 0$, $\alpha_1 = \alpha_2 = 1$, $\tau_1 = \tau_2$ (where $\tau_1 = \left(x'^2/\mathcal{K}_1\right)^{1/\alpha_1}$ and $\tau_2 = \left(x'^2/\mathcal{K}_2\right)^{1/\alpha_2}$), $k_{2s} = x'/\tau_2$, and $k_2(t) = 0$, in arbitrary units [175].

Figure 5.14a shows the behaviour of the mean square displacement for both regions when the case $k_2 \neq 0$ is considered, i.e., the membrane also permits a reverse (e.g., osmotic) process. In Fig. 5.14b, the cases $k_2 = 0$ and $k_2 \neq 0$ are compared for region 1. We observe that $k_2 \neq 0$ leads to the recovering of the usual diffusion in the asymptotic limit, while the case $k_2 = 0$ results in a subdiffusive regime. Again, we verify that the particles present in region 1 exhibit different diffusive regimes due to the membrane. In particular, for $k_2 \neq 0$, the asymptotic limit leads us to a regime governed by the bulk equation with $\langle(\Delta_1 x)^2\rangle \to t^{\alpha_1}$

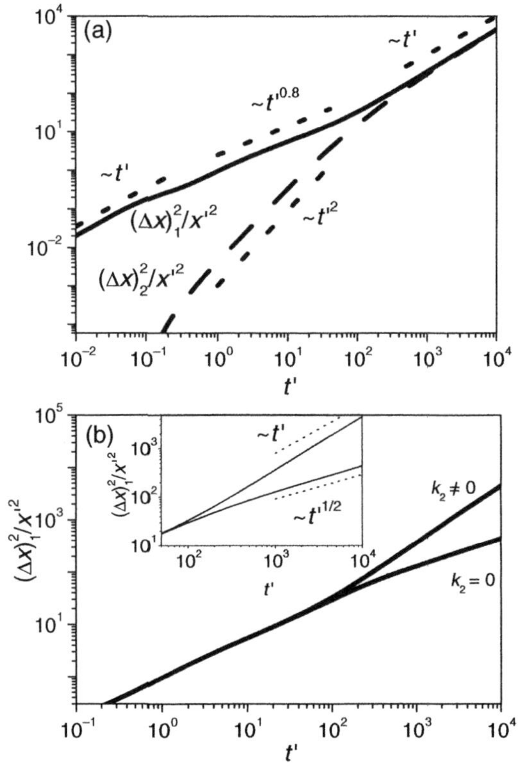

Figure 5.14 (a) Mean square displacement in regions 1 and 2 with $k_2 \neq 0$ and $k_{2s} \neq 0$. Region 1 starts with the usual diffusion and passes through different regimes, and for long times recovers the usual behaviour. In particular, this asymptotic behaviour in region 1 is due to the reverse transport of particles, i.e., $k_2 \neq 0$, from region 2 to region 1, as shown in (b) when compared with the case $k_2 = 0$. In the inset of Fig. 5.14b, we show that $k_2 \neq 0$ leads us asymptotically to a usual diffusion and $k_2 = 0$ a subdiffusive regime. Region 2, similar to the previous case, manifests a different behaviour followed by a usual diffusion. We consider, for illustrative purposes, $k_{1s} = 10^2 x'/\tau_1$, $k_1 = \overline{k}_1 \delta(t)$ (where $\overline{k}_1 = x'/\tau_1$), $k_{1d} = (2/\tau_1)\,\delta(t)$, $t' = t/\tau_1$, $\tau_1 = \tau_2$ (where $\tau_1 = \left(x'^2/\mathcal{K}_1\right)^{1/\alpha_1}$ and $\tau_2 = \left(x'^2/\mathcal{K}_2\right)^{1/\alpha_2}$), $\alpha_1 = \alpha_2 = 1$, $k_{2d}(t) = \left[1/(2\tau_2)\right]\delta(t)$, $k_{2s} = x'/\tau_2$, and $k_2(t) = \overline{k}_2 \delta(t)$ (where $\overline{k}_2 = x'/\tau_2$), in arbitrary units [175].

for region 1. For region 2, the spreading of the system is asymptotically governed by $\langle(\Delta_2 x)^2\rangle \rightarrow t^{(\alpha_2+\alpha_1)/2}$. This result depends on α_2 and α_1, in contrast to the cases discussed for $k_2 = 0$ in Figs. 5.12 and 5.13, which only depend on α_2 in the asymptotic limit and on the sorption–desorption processes exhibited by the membrane.

6

Fractional Nonlinear Diffusion Equations

In this chapter, we investigate situations more general than the ones treated in the preceding chapters, by considering nonlinear terms in the diffusion equation, and d-dimensional scenarios, by taking into account the linear and nonlinear cases for the fractional diffusion equations. We start by analysing the one-dimensional problems related to the nonlinear diffusion equation $\partial_t \rho = \partial_x [D(x,t)\partial_x] \rho^\nu$ to illustrate the differences with respect to the usual diffusion equation for what concerns the solutions which, for the nonlinear case, may be asymptotically characterised by power laws. We also consider an intermittent process and how this process can be described by starting from a random walk approach because, in this framework, different diffusive regimes may be found. Afterwards, we focus our attention on the solutions of the fractional diffusion equation:

$$\frac{\partial}{\partial t}\rho(x,t) = \frac{\partial}{\partial |x|}\left\{D(x,t)\frac{\partial^{\mu-1}}{\partial |x|^{\mu-1}}[\rho(x,t)]^\nu - F(x,t)\rho(x,t)\right\}, \tag{6.1}$$

where μ and $\nu \in \mathbb{R}$. It has as particular cases the usual diffusion equation, the fractional diffusion equation, and the porous media equation. We essentially analyse Eq. (6.1) by considering two situations: (i) the diffusion coefficient, given as in Eq. (3.104) by $D(x) = D\,|x|^{-\theta}$ ($\theta \in \mathbb{R}$) in the absence of external forces, and after that by incorporating an arbitrary time dependence in the diffusion coefficient, in the form $D(x,t) = D(t)|x|^{-\theta}$ and (ii) the action of an external force $F(x) \propto x|x|^{\alpha-1}$ with a spatial dependence on the diffusion coefficient, where $\alpha \in \mathbb{R}$.

Subsequently, we consider d-dimensional scenarios for linear and nonlinear fractional diffusion equations by taking into account a diffusion coefficient with spatial dependences and external forces. For both cases we show that a rich variety of behaviours can be found in connection with anomalous diffusion.

6.1 Nonlinear Diffusion Equations

Finding solutions for nonlinear diffusion equations is a challenge since the principle of superposition is not applicable as in the linear case. This leads us to consider alternative approaches to analyse them, in particular, the ones for which the x (position) and t (time) variables appear in a scaled form [186–190], i.e.,

$$\rho\left(x,t\right)=\frac{1}{\Phi\left(t\right)}\tilde{\rho}\left[\frac{x}{\Phi\left(t\right)}\right], \tag{6.2}$$

where $\Phi(t)$ is a generic time-dependent function. This form of solution is very close to the ones obtained by means of the method of similarity [191]. By using the transformation (6.2), it will be possible to simplify the fractional nonlinear partial differential equation to an ordinary differential equation, whose solution will depend on the boundary conditions and will be subjected here to normalisation, i.e.,

$$\int_{-\infty}^{\infty} dx\rho(x,t)=\text{constant}. \tag{6.3}$$

Before applying this procedure to general cases of Eq. (6.1), we first study the specific case $\mu = 2$, $F(x) = 0$, and $D(x) = D =$ constant, which represents the porous media equation. The resulting equation plays an important role in several scenarios such as percolation of gases through porous media [192], thin saturated regions in porous media [193], a standard solid-on-solid model for surface growth [194], thin liquid films spreading under gravity [195], etc. Thus, by substituting Eq. (6.2) into Eq. (6.1) and taking into account the previous assumptions about the porous media equation, we obtain

$$-\frac{\dot{\Phi}\left(t\right)}{\Phi\left(t\right)^{2}}\frac{d}{dz}\left[z\tilde{\rho}\left(z\right)\right]=\frac{D}{\Phi\left(t\right)^{2+\nu}}\frac{d^{2}}{dz^{2}}\left[\tilde{\rho}\left(z\right)\right]^{\nu}, \tag{6.4}$$

where $z = x/\Phi(t)$. By introducing a separation constant k in Eq. (6.4), it can be simplified to

$$\Phi\left(t\right)^{\nu}\dot{\Phi}\left(t\right)=k \tag{6.5}$$

and

$$-k\frac{d}{dz}\left[z\tilde{\rho}\left(z\right)\right]=D\frac{d^{2}}{dz^{2}}\left[\tilde{\rho}\left(z\right)\right]^{\nu}. \tag{6.6}$$

The solution for $\Phi(t)$ is

$$\Phi(t)=\left[(1+\nu)kt\right]^{1/(1+\nu)}, \tag{6.7}$$

obtained by simple integration. Note that we are considering $\Phi(0) = 0$ in order to obtain a solution whose initial condition is $\rho(x, 0) = \delta(x)$. For the ordinary differential equation governing the behaviour of $\rho(z)$, we obtain

$$\tilde{\rho}(z) = \left[1 - \frac{(\nu - 1)}{2D\nu} kz^2\right]^{\frac{1}{\nu-1}}, \tag{6.8}$$

whose behaviour is depicted in Fig. 6.1. This solution may be identified with the q-exponential, defined as [196]:

$$\exp_q(x) \equiv \begin{cases} [1 + (1-q)x]^{\frac{1}{1-q}}, & \text{if} \quad (1-q)x > -1, \\ 0, & \text{if} \quad (1-q)x < -1. \end{cases} \tag{6.9}$$

Thus, for $q = 2 - \nu$, we can write the solution for Eq. (6.2) as

$$\rho(x, t) = \frac{1}{\Phi(t)} \exp_q \left[-\frac{k}{2D\nu} z^2\right]. \tag{6.10}$$

The presence of the q-exponential function may suggest a connection between this diffusion equation and a generalised formalism, intended to give a thermostatistical

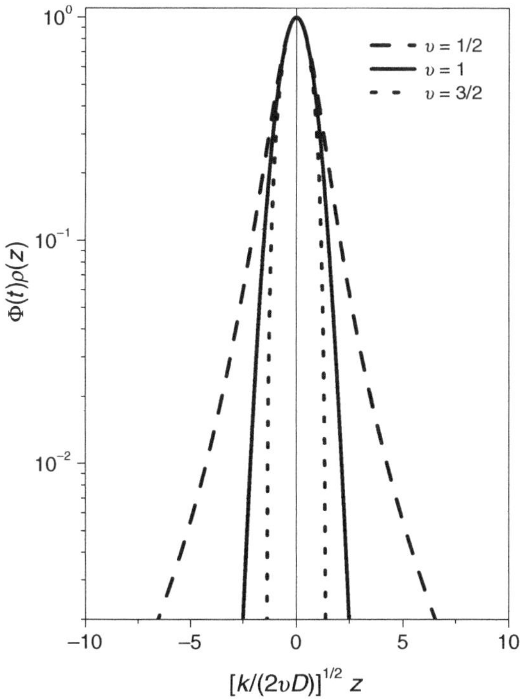

Figure 6.1 The relation between $\Phi(t)\rho(z)$ and $\left[k/(2\nu D)\right]^{1/2} z$, as predicted by Eq. (6.10), for typical values of ν. It shows for $\nu > 1$ a compact behaviour and a long-tailed behaviour for $\nu < 1$.

basis for this equation and, consequently, to the anomalous diffusion processes. This identification suggests also the possibility of obtaining Eq. (6.10) from a maximum entropy principle, where the entropic form (known as Tsallis entropy)

$$S_q = \frac{1 - \int_{-\infty}^{\infty} dx \rho(x,t)^q}{q-1} \tag{6.11}$$

is maximised under suitable constraints [186]. Another remarkable aspect of this connection concerns the H-theorem, which is satisfied in this context [197]. The curves exhibited in Fig. 6.1 show that, depending on the value of ν, the distribution may exhibit a compact or a long-tailed behaviour characterised by a power-law progress. Indeed, depending on the value of ν, the last case may be related to the Lévy distributions.

To proceed further, we may incorporate in the previous discussion a spatial dependence on the diffusion coefficient. A typical choice is to consider the diffusion coefficient as we have done in Eq. (3.104), where $\theta = 0$ corresponds to the case $D(x) = D = \text{const}$. As pointed out before, this spatial dependence of the diffusion coefficient may be meaningful in several contexts such as diffusion on fractals [95, 198], deposition and diffusion of platinum nanoparticles in porous carbon [199], ergodicity breaking, ageing, and confinement [200], etc. For the specific case represented by Eq. (3.104), employing the previous procedure, the solution may also be found, in the form

$$\rho(x,t) = \frac{1}{\Phi(t)} \exp_q \left[-\frac{k}{\nu(2+\theta)D} z^{2+\theta} \right], \tag{6.12}$$

with

$$\Phi(t) = [(\nu + \theta + 1)kt]^{1/(1+\nu+\theta)} . \tag{6.13}$$

The physical solutions, i.e., the solutions which satisfy the normalisation condition and the boundary conditions, are verified for $\theta > -2$. It is also possible to verify that the cases $\theta + \nu > 1$, $\theta + \nu = 1$, and $\theta + \nu < 1$ correspond to the subdiffusive, normal and superdiffusive regimes, respectively.

The solution can be generalised by incorporating an arbitrary time dependence in the diffusion coefficient given by Eq. (3.104), as, for instance,

$$D(x,t) = \overline{D}(t)|x|^{-\theta} . \tag{6.14}$$

The main change is verified in the time-dependent function $\Phi(t)$, which is now given as follows:

$$\Phi(t) = \left\{ [\Phi(0)]^{1+\theta+\nu} + (1+\nu+\theta) \int_0^t dt \overline{D}(t) \right\}^{\frac{1}{1+\theta+\nu}} . \tag{6.15}$$

Notice that in Eq. (6.15) we consider $\Phi(0) \neq 0$, which implies an initial condition characterised by a distribution with an initial variance, in contrast to the previous case in which D was defined by (3.104).

Let us generalise further the results found until now for the nonlinear diffusion equation with a spatial dependent diffusion coefficient in the form of Eq. (3.104), by incorporating also the external force $F(x) = kx|x|^{\alpha-1}$, which has as particular case the logarithmic potential [166]. To find a solution for arbitrary α is a hard task; however, for a suitable choice of α, for which Eq. (6.2) is verified, it is possible to find a solution. Consider $\alpha = q - \theta - 2$, i.e., $\alpha + \theta + \nu = 0$. For these conditions, we obtain

$$\rho(x,t) = \frac{1}{\Phi(t)} \exp_q \left\{ -\frac{1}{D\nu} \left[\frac{k}{2+\theta} \left(\frac{|x|}{\Phi(t)} \right)^{2+\theta} - k \ln_{2-q} \left(\frac{|x|}{\Phi(t)} \right) \right] \right\}, \tag{6.16}$$

where

$$\ln_q x \equiv \frac{x^{1-q} - 1}{1-q}$$

is the q-logarithmic function (inverse function of the q-exponential function) and $\Phi(t)$ is given by Eq. (6.13), where the separation constant k may be determined with the help of Eq. (6.3), i.e., the normalisation condition.

In the next section, we investigate an intermittent process obtained from the combination of a usual (i.e., nonfractional) nonlinear diffusion equation and pauses. We will consider the porous media equation with reaction terms related to the rate of switching the particles from the diffusive mode to the resting mode or switching them from resting to movement. Furthermore, we will also analyse a physical situation, which emerges when the diffusive term is a mixing of linear and nonlinear terms, characterised by the existence of different diffusive regimes [201] even in the framework of the usual nonlinear diffusion equation.

6.2 Nonlinear Diffusion Equations: Intermittent Motion

Let us start our discussion regarding the processes characterised by diffusion with pauses by first considering the standard case, i.e., the usual diffusion of the particles in motion [201]. In this scenario, we may assume that the diffusion is essentially governed by the Einstein equation (see Section 3.1.2) [202]:

$$\rho(x, t+\tau) = \int_{-\infty}^{\infty} \rho(x-z, t)\Phi(z)dz, \tag{6.17}$$

in the absence of pauses. Following the development reported in Ref. [202], in order to introduce the reaction terms related to the motion and pauses, we may consider that, during the period τ, the probability of not switching from the diffusive to the resting regime is $e^{-k_1\tau}$ and for the reverse it is $e^{-k_2\tau}$. The transition between these states (diffusion $\rightleftharpoons$ rest) is typical of a two-level continuous-time discrete Markovian process. Furthermore, we introduce $\rho_1(x,t)$ and $\rho_2(x,t)$ for describing the density of particle while moving and while resting, respectively. Thus, Einstein's equation turns into the system of equations

$$\rho_1(x,t+\tau) = \int_{-\infty}^{\infty} e^{-k_1\tau}\rho(x-z,t)\Phi(z)dz + (1-e^{-k_2\tau})\rho_2(x,t) \tag{6.18}$$

and

$$\rho_2(x,t+\tau) = (1-e^{-k_1\tau})\rho_1(x,t) - e^{-k_2\tau}\rho_2(x,t). \tag{6.19}$$

Equations (6.18) and (6.19), in the limit of $\tau \to 0$ and $z \to 0$, with

$$\frac{\langle z^2\rangle}{\tau} \sim \text{constant}, \quad \langle z^2\rangle = \int_{-\infty}^{\infty} z^2\Phi(z)dz,$$

become, respectively,

$$\frac{\partial}{\partial t}\rho_1(x,t) = D\frac{\partial^2}{\partial x^2}\rho_1(x,t) - k_1\rho_1(x,t) + k_2\rho_2(x,t) \tag{6.20}$$

and

$$\frac{\partial}{\partial t}\rho_2(x,t) = k_1\rho_1(x,t) - k_2\rho_2(x,t), \tag{6.21}$$

with $D = \langle z^2\rangle/(2\tau)$. In Eqs. (6.20) and (6.21), ρ_1 and ρ_2 refer to two different states. The first one represents the particles (species, or substance) diffusing, i.e., in motion, and the second one corresponds to the particles (species, or substance) that are immobilised. We notice that Eq. (6.20) has the diffusive term, i.e., $D \neq 0$, which promotes the spreading of the system, while Eq. (6.21) has no diffusive term, which leads the particles to rest. Thus, this system of equations can be associated with the diffusion $\rightleftharpoons$ pauses process, as expected. We notice also that Eq. (6.20) may directly be extended to the context of the fractional diffusion equations by making suitable choices of $\Phi(z)$ appearing in Eq. (6.18) in order to incorporate a long-tailed behaviour. The extension to the nonlinear case will be discussed later on.

It is possible to obtain exact solutions for these equations by using the standard calculus techniques. In particular, for the boundary conditions $\rho_1(\pm\infty,t) = 0$ and $\rho_2(\pm\infty,t) = 0$ and the initial conditions $\rho_1(x,0) = \delta(x)$ and $\rho_2(x,0) = 0$, they are

$$\rho_1(x,t) = \int\limits_{-\infty}^{\infty} dx' \mathcal{G}(x-x',t)\rho_1(x',0) \tag{6.22}$$

and

$$\rho_2(x,t) = k_1 \int\limits_0^t dt' \int\limits_{-\infty}^{\infty} dx' \mathcal{G}(x-x',t-t')e^{-k_2 t'}\rho_1(x',0), \tag{6.23}$$

with

$$\begin{aligned}\mathcal{G}(x,t) = \mathcal{G}^{(1)}(x,t) + \sum_{n=1}^{\infty}(-1)^n \int\limits_0^t dt_n \int\limits_{-\infty}^{\infty} dx_n \mathcal{G}^{(2)}(x-x_n,t-t_n)\\ \times \int\limits_0^{t_n} dt_{n-1} \int\limits_{-\infty}^{\infty} dx_{n-1}\mathcal{G}^{(2)}(x_n - x_{n-1}, t_n - t_{n-1}) \cdots\\ \times \int\limits_0^{t_2} dt_1 \int\limits_{-\infty}^{\infty} dx_1 \mathcal{G}^{(2)}(x_2 - x_1, t_2 - t_1)\mathcal{G}^{(1)}(x_1,t_1),\end{aligned} \tag{6.24}$$

where

$$\mathcal{G}^{(1)}(x,t) = \frac{1}{\sqrt{4\pi Dt}} e^{-x^2/(4Dt)} \tag{6.25}$$

and

$$\mathcal{G}^{(2)}(x,t) = k_1\mathcal{G}^{(1)}(x,t) + k_1 k_2 \int\limits_0^t dt' \mathcal{G}^{(1)}(x,t-t')e^{k_2 t'} dt'. \tag{6.26}$$

Figure 6.2 shows, for this case, the behaviour of the mean square displacements:

$$(\Delta x)_1^2 = \left\langle (x - \langle x \rangle_1)^2 \right\rangle_1 \quad \text{and} \quad (\Delta x)_2^2 = \left\langle (x - \langle x \rangle_2)^2 \right\rangle_2 .$$

We observe that, for short times, i.e., $t \ll \max\{1/k_1, 1/k_2\}$, the influence of the reaction terms is not pronounced and the spreading of $\rho_1(x,t)$ essentially behaves as in the usual diffusion. A similar feature is verified for long times, i.e., $t \gg \max\{1/k_1, 1/k_2\}$, as shown in Fig. 6.2; however, this case is characterised by an effective diffusion coefficient. For intermediate times, $t \sim \max\{1/k_1, 1/k_2\}$, the reaction terms play an important role, and we have a pronounced effect on an intermittent process characterised by an interchange between motion and pauses. This is made evident by the mean square displacement, which, in this time interval, exhibits a subdiffusion instead of the usual one.

The nonlinear case related to the anomalous diffusion and the Tsallis statistics may be obtained from the previous approach by incorporating, in the dispersal

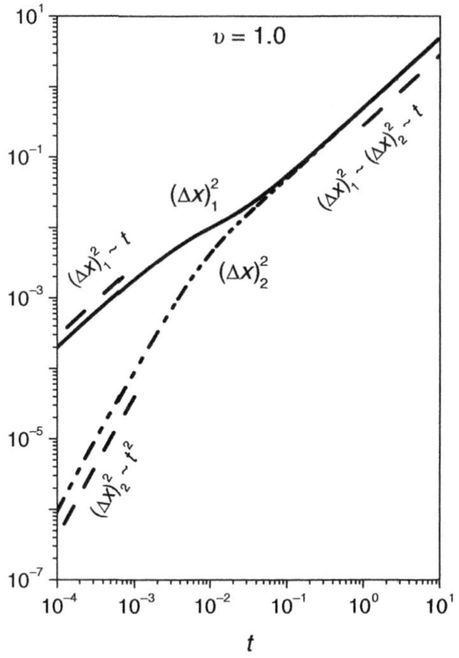

Figure 6.2 The mean square displacement obtained from Eqs. (6.22) and (6.23), for $k_1 = k_2 = 10^2\ [T]^{-1}$ and $D = 1\ [L]^2[T]^{-1}$, where $[L]$ and $[T]$, here and in subsequent figures, represent arbitrary units of length and time, respectively [201].

term, a nonlinear dependence on the distribution, e.g., $\Psi(\rho_1)$, and modifying the additional terms related to the probability of diffusing or resting in order to preserve the linearity of the reaction terms. These reaction terms act on the particles promoting the transition between the states of motion and resting. Thus, the nonlinearity present in the dispersal term will only appear in the diffusive term, which promotes the spreading of the particles. This nonlinear dependence on the dispersal term is connected with the feature that the jumping probability depends explicitly on the distribution $\rho_1(x,t)$. In this case, it may describe situations characterised by distributions with a compact form or a long-tailed distribution, which asymptotically may be connected with the Lévy distribution [203]. This implies that the diffusion coefficient in any element of the system depends on the history of the element. We have

$$\rho_1(x,t+\tau) = \int_{-\infty}^{\infty} e^{-k_1\tau}\Psi[\rho_1(x-z,t)]\rho_1(x-z,t)\Phi(z)dz + (1-e^{-k_2\tau})\rho_2(x,t) + e^{-k_1\tau}(\rho_1(x,t) - \Psi[\rho_1(x,t)]\rho_1(x,t)) \tag{6.27}$$

and

$$\rho_2(x,t+\tau) = -e^{-k_2\tau}\rho_2(x,t) + (1-e^{-k_1\tau})\rho_1(x,t). \tag{6.28}$$

By taking into account the limit $\tau \to 0$ and $z \to 0$ in the previous set of equations, we obtain, respectively,

$$\frac{\partial}{\partial t}\rho_1(x,t) = \frac{\partial}{\partial x}\left[D(\rho_1)\frac{\partial}{\partial x}\rho_1(x,t)\right] - k_1\rho_1(x,t) + k_2\rho_2(x,t) \tag{6.29}$$

and

$$\frac{\partial}{\partial t}\rho_2(t) = k_1\rho_1(x,t) + k_2\rho_2(x,t), \tag{6.30}$$

with

$$D(\rho_1) = \frac{\langle z^2\rangle}{2\tau}\frac{d}{d\rho_1}\left(\Psi(\rho_1)\rho_1\right). \tag{6.31}$$

It is also possible to assume that the kernel $\Phi(z)$ may depend explicitly on $\rho_1(x,t)$; however, for this case, the definition of diffusion coefficient will be different from the previous one. Equation (6.29) is equal to Eq. (6.20) for $\Psi(\rho_1) = 1$, and, consequently, the diffusion process is usual. For $\Psi(\rho_1) = \rho_1^{\nu-1}$, we obtain the cases described by the following differential equations:

$$\frac{\partial}{\partial t}\rho_1(x,t) = D\frac{\partial^2}{\partial x^2}\rho_1^{\nu}(x,t) - k_1\rho_1(x,t) + k_2\rho_2(x,t) \tag{6.32}$$

and

$$\frac{\partial}{\partial t}\rho_2(x,t) = k_1\rho_1(x,t) - k_2\rho_2(x,t). \tag{6.33}$$

The diffusive term in Eq. (6.32) has a nonlinear dependence on the distribution $\rho_1(x,t)$, characteristic of an anomalous correlated diffusion and suggests a connection with the Tsallis statistics [197] that is based on the entropy defined in Eq. (6.11). The solutions for Eq. (6.32) may have a compact or a long-tailed behaviour depending on the choice of ν and, similar to the linear case, the additional terms promote the transition between the motion and pauses during the diffusive process.

For the particular case $k_2(t) = 0$ – i.e., the particles are only switched from the diffusive to the resting mode – we may obtain a formal solution for the previous set of equations by considering the boundary conditions $\rho_1(\pm\infty,t) = 0$ and $\rho_2(\pm\infty,t) = 0$, and the initial conditions $\rho_1(x,0) = \delta(x)$ and $\rho_2(x,0) = 0$. It is possible to show that

$$\rho_1(x,t) = \frac{e^{-k_1 t}}{\mathcal{Z}(t)}\exp_q\left[-\beta(t)x^2\right] \tag{6.34}$$

and

$$\rho_2(x,t) = k_1 \int_0^t e^{-k_1 t'} \frac{\exp_q\left[-\beta(t')x^2\right]}{\mathcal{Z}(t')} dt', \tag{6.35}$$

with $\nu = 2 - q$, $\mathcal{Z}(t)\sqrt{\beta(t)} = I^{(0)}(q)$, where

$$\beta(t) = \left\{2\nu \frac{1+\nu}{(1-\nu)k_1} D\left[I^{(0)}(q)\right]^{q-1}\left[e^{k_1(1-\nu)t} - 1\right]\right\}^{-\frac{2}{3-q}}, \tag{6.36}$$

in which

$$I^{(n)}(q) = \int_{-\infty}^{\infty} x^n \exp_q\left[-x^2\right] dx.$$

The presence of the q-exponential in Eqs. (6.34) and (6.35) permits us to obtain a short- ($q < 1$) or a long- ($q > 1$) tailed behaviour for the solution depending on the value of the parameter q. Equation (6.34) has a compact behaviour for $q < 1$ due to the "cut-off" exhibited by the q-exponential to retain the probabilistic interpretation associated with $\rho_1(x,t)$. Consequently, $\rho_2(x,t)$ exhibits a similar behaviour for $q < 1$. On the other hand, for $q > 1$, Eq. (6.34) has the asymptotic limit governed by a power-law behaviour, which may be also related to a Lévy distribution [203]. In this case, the solutions obtained for the previous set of equations may be asymptotically related to some solutions, obtained for fractional diffusion equations, which are asymptotically governed by power laws [204]. The mean square displacement obtained with the distribution $\rho_1(x,t)$ is

$$(\Delta x)_1^2 = \langle (x - \langle x \rangle)^2 \rangle_1 = e^{-k_1 t} \frac{I^{(2)}(q)}{I^{(0)}(q)\beta(t)}, \tag{6.37}$$

which, for short times, i.e., $t \ll k_1$, becomes $(\Delta x)_1^2 \sim t^{2/(3-q)}$ and, for long times, i.e., $t \to \infty$, $\langle (\Delta x)^2 \rangle \sim e^{-\frac{3\nu-1}{1+\nu} k_1 t}$ for $\nu < 1$. Similarly, for $\rho_2(x,t)$, we have

$$(\Delta x)_2^2 = \langle (x - \langle x \rangle)^2 \rangle_2 = \frac{I^{(2)}(q)}{I^{(0)}(q)} = k_1 \int_0^t \frac{e^{-k_1 t'}}{\beta(t')} dt', \tag{6.38}$$

with $(\Delta x)_2^2 \sim t^{(5-q)/(3-q)}$, for short times, and $(\Delta x)_2^2 \sim$ constant, for long times. We may conclude that, for long times, all the particles are switched from the diffusive to the resting mode.

When $k_2 \neq 0$, there is an interplay between diffusion and pauses. In this case, it is possible to show that

$$\int_{-\infty}^{\infty} dx \left(\rho_1(x,t) + \rho_2(x,t)\right) = \text{constant}$$

and that the survival probabilities ($\mathcal{S}_{1(2)}(t) = \int_{-\infty}^{\infty} \rho_{1(2)}(x,t)dx$) corresponding to $\rho_1(x,t)$ and $\rho_2(x,t)$ satisfy the following set of equations:

$$\frac{d}{dt}\mathcal{S}_1(t) = k_2\mathcal{S}_2(t) - k_1\mathcal{S}_1(t) \tag{6.39a}$$

and

$$\frac{d}{dt}\mathcal{S}_2(t) = k_1\mathcal{S}_1(t) - k_2\mathcal{S}_2(t). \tag{6.39b}$$

The solutions are

$$\mathcal{S}_1(t) = \frac{k_2}{k_t} + [k_1\mathcal{S}_1(0) - k_2\mathcal{S}_2(0)]\frac{e^{-k_t t}}{k_t} \tag{6.40a}$$

and

$$\mathcal{S}_2(t) = \frac{k_1}{k_t} + [k_2\mathcal{S}_2(0) - k_1\mathcal{S}_1(0)]\frac{e^{-k_t t}}{k_t}, \tag{6.40b}$$

with $k_t = k_1 + k_2$. The previous equations show that, for long times, the equilibrium between particles switched from motion to rest (or from rest to motion) is reached for $\mathcal{S}_1 \rightarrow k_2/k_t$ and $\mathcal{S}_2 \rightarrow k_1/k_t$. An important feature of these solutions is the absence of the parameter ν and the presence of only the rates k_1 and k_2. This implies that the interchange between the states of motion and pause is independent of the diffusion process.

Performing some numerical calculations, it is possible to get more information about the behaviour of the mean square displacement and the distributions $\rho_1(x,t)$ and $\rho_2(x,t)$. Using the numerical algorithms based on central differences [205], it is possible to numerically solve Eqs. (6.32) and (6.33), which are coupled by the reaction terms.

Figures 6.3 and 6.4 show the behaviour of the mean square displacement for $\rho_1(x,t)$ and $\rho_2(x,t)$ obtained numerically with the requirement that $Ddt/\left(dx^2\right) < 1/2$ to assure the stability of the solutions during the time evolution of the initial condition [201]. The initial behaviour, i.e., for $t \ll \max\{1/k_1, 1/k_2\}$, of $\rho_1(x,t)$ is characterised by the porous media equation in the absence of reaction terms. It may be subdiffusive or superdiffusive depending on the values of ν, i.e., $\nu > 1$ or $\nu < 1$. In Fig. 6.3, we have a subdiffusive ($\nu = 1.5$) behaviour for short times, in contrast to Fig. 6.4, in which the superdiffusive behaviour ($\nu = 0.8$) is obtained. For long times, we also observe that the system is essentially governed by the porous media equation with an effective diffusion coefficient, as a result of the intermittent motion of the particles, which are constantly switching from the diffusive mode to the resting mode or switching from resting to movement. For intermediate times, we have a different behaviour from the one obtained with the porous media equation due to the interchanges between motion and pauses.

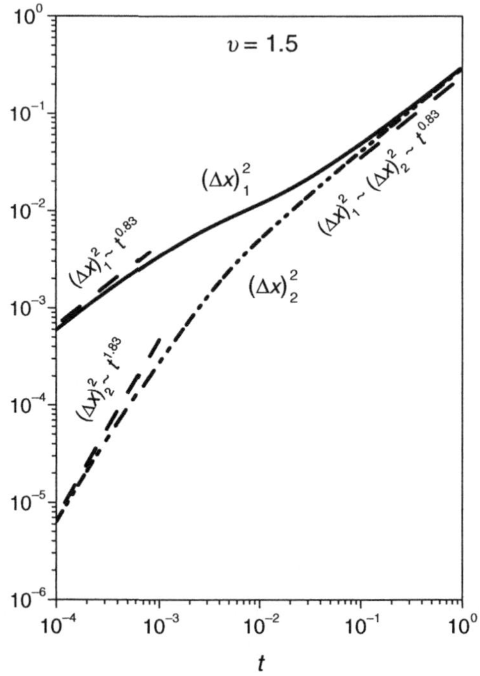

Figure 6.3 The mean square displacement obtained from Eqs. (6.32) and (6.33) for $\nu = 1.5$, $\rho_1(x,0) = \delta(x)$, $\rho_2(x,0) = 0$, $k_1 = k_2 = 10^2\ [T]^{-1}$, and $D = 1\ [L]^{1+\nu}[T]^{-1}$ [201].

In order to investigate the behaviour of the solutions, in the asymptotic limit of long times, we may analyse the following equation:

$$\frac{\partial}{\partial t}\left[\rho_1(x,t) + \rho_2(x,t)\right] = D\frac{\partial^2}{\partial x^2}\rho_1^{\nu}(x,t), \tag{6.41}$$

which can obtained by adding the set of equations (6.32) and (6.33). For $t \to \infty$, by taking into account $\rho_2(x,t) \approx (k_1/k_2)\,\rho_1(x,t)$, Eq. (6.41) can be approximated to

$$\frac{\partial}{\partial t}\rho_1(x,t) \approx \frac{k_2 D}{k_1 + k_2}\frac{\partial^2}{\partial x^2}\rho_1^{\nu}(x,t), \tag{6.42}$$

which also has the solutions expressed in terms of the q-exponential functions. After performing some calculations, it is possible to show that

$$\rho_1(x,t) \approx \frac{1}{\bar{\mathcal{Z}}(t)}\,\exp_q\left[-\bar{\beta}(t)x^2\right], \tag{6.43}$$

with $\nu = 2 - q$ and

$$\bar{\mathcal{Z}}(t)\sqrt{\bar{\beta}(t)} = \bar{I}^{(0)}(q) = \frac{k_2 + k_1}{k_2}\,I^{(0)}(q),$$

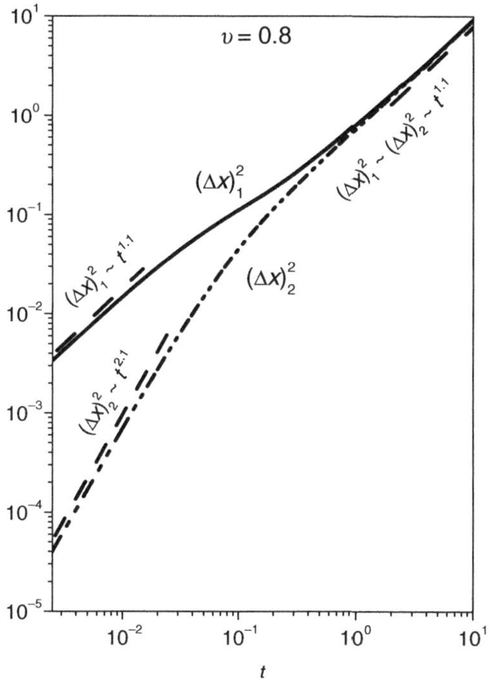

Figure 6.4 The same as in Fig. 6.3 for $\nu = 0.8$ and $k_1 = k_2 = 10^2\ [T]^{-1}$ and $D = 1\ [L]^2[T]^{-1}$ [201].

where, now,

$$\overline{\beta}(t) = \left\{ 2\nu(1+\nu)\frac{k_2 D}{k_1 + k_2}\left[\overline{I}^{(0)}(q)\right]^{q-1} t \right\}^{-\frac{2}{3-q}}. \tag{6.44}$$

Figure 6.5 permits us to compare the solution obtained numerically with the approximated solution, obtained above for long times. Indeed, in this limit we notice a complete agreement between them.

Other expressions for $\Psi(\rho_1)$ are also possible – in particular,

$$\Psi(\rho_1) = 1 + \frac{\overline{D}}{D}\rho_1^{\nu-1}, \tag{6.45}$$

where $\overline{D}$ is a constant diffusion coefficient, which may be related to different diffusive behaviours – one usual and the other one anomalous. A particular application of Eq. (6.45) may be found in the spatial distribution of dispersing animals [206]. Figures 6.6 and 6.7 exhibit the trend of the mean square displacement for the boundary conditions $\rho_1(\pm\infty, t) = 0$ and $\rho_2(\pm\infty, t) = 0$ and the initial conditions $\rho_1(x, 0) = \delta(x)$ and $\rho_2(x, 0) = 0$, when $\Psi(\rho_1)$ is of the form (6.45). We observe that the behaviour for short times is subdiffusive, if $\nu > 1$, and usual, if $\nu < 1$. On the other hand, for long times, we have an usual diffusion, if $\nu > 1$, and a

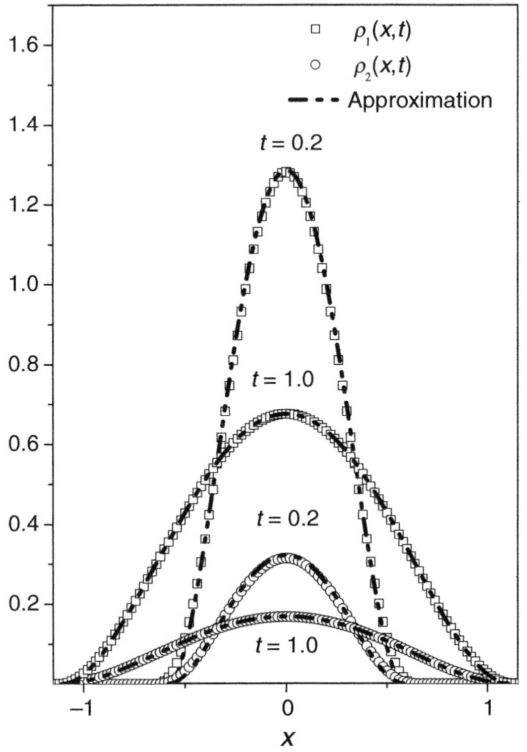

Figure 6.5 Comparison between the trends of $\rho_1(x,t)$ and $\rho_2(x,t)$, obtained numerically from Eqs. (6.20) and (6.21), with the approximated ones, obtained from Eq. (6.43) and $\rho_2(x,t) \approx (k_1/k_2)\,\rho_1(x,t)$ for $\nu = 1.5$. For illustrative purposes, we consider $\rho_1(x,0) = \delta(x)$, $\rho_2(x,0) = 0$, $k_1 = 4 \times 10^2\ [T]^{-1}$, $k_2 = 10^2\ [T]^{-1}$, and $D = 1\ [L]^2[T]^{-1}$ [201].

superdiffusion, if $\nu < 1$. The different diffusive regimes exhibited by these systems are due to the mixing between the linear and nonlinear diffusive terms, and, as pointed out in [102], they are also characterised by crossover times.

The results show that in the asymptotic limit of short and long times, the spreading of the system is essentially governed by the diffusive term. The behaviour exhibited for intermediate times depends on the rates characterising the reaction terms. Thus, we may conclude that, in the asymptotic limits, the distributions for this process are expressed in terms of power laws which, in turn, may be related to the q-exponential present in the Tsallis statistics.

6.3 Fractional Spatial Derivatives

In Section 6.1, we solved Eq. (6.1) for $\mu = 2$ by taking into account a spatial and time diffusion coefficient and the presence of external forces. We have also

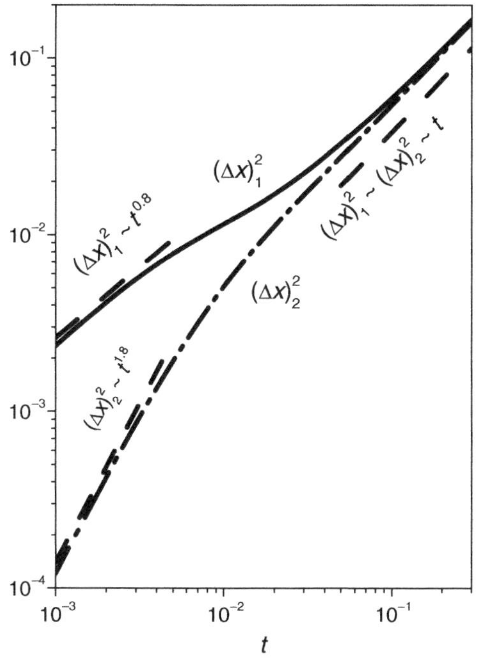

Figure 6.6 The mean square displacement obtained from Eqs. (6.29) and (6.30), for $\Psi(\rho_1) = 1 + (\overline{D}/D)\,\rho_1^{\nu-1}$, with $\nu = 1.5$. For illustrative purposes, we consider $k_1 = k_2 = 10^2\ [T]^{-1}$, $\overline{D} = 10^{-1}\ [L]^{1+\nu}[T]^{-1}$, and $D = 1\ [L]^2[T]^{-1}$ [201].

shown that the solutions for the scenarios worked out in Section 6.1 are given in terms of the power laws. Here, we analyse the changes in the solutions when $\mu \neq 2$, i.e., when the spatial derivative is fractional. The first problem to be treated concerns Eq. (6.1) without the external force, but with a diffusion coefficient given by Eq. (3.104). For this case, Eq. (6.1) may be rewritten as

$$\frac{\partial}{\partial t}\rho(x,t) = \frac{\partial}{\partial |x|}\left\{D|x|^{-\theta}\frac{\partial^{\mu-1}}{\partial |x|^{\mu-1}}\left[\rho(x,t)\right]^{\nu}\right\}. \tag{6.46}$$

For a suitable choice of the parameters, the equations worked out in Refs. [187, 189] are reobtained from Eq. (6.46). In order to investigate the solutions of Eq. (6.46), we also use the procedure employed in the last section, which is based on Eq. (6.2). For the fractional differential spatial operator, we consider the fractional Riemann–Liouville operator, defined in Eq. (2.8), which is very helpful in obtaining exact solutions. To do this, we work with the positive x-axis, and, later on, we will use symmetry arguments to extend the results to the entire real axis (in other words, this is equivalent to working with $\partial^{\mu}_{|x|}$) [187, 188]. We also explore the following property of the fractional derivatives:

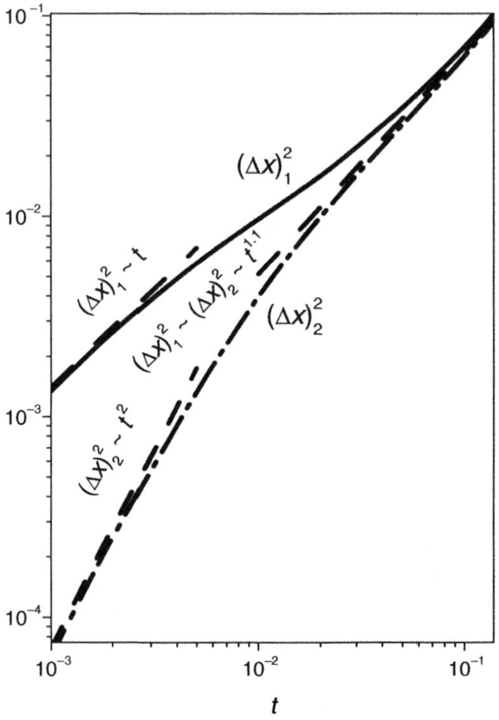

Figure 6.7 The mean square displacement obtained from Eqs. (6.29) and (6.30) for $\Psi(\rho_1) = 1 + \left(\overline{D}/D\right)\rho_1^{\nu-1}$, with $\nu = 0.8$. The curves were drawn for $\rho_1(x,0) = \delta(x)$, $\rho_2(x,0) = 0$, $k_1 = k_2 = 10^2$ $[T]^{-1}$, $\overline{D} = 1$ $[L]^{1+\nu}[T]^{-1}$, and $D = 10^{-1}$ $[L]^2[T]^{-1}$ [201].

$$\frac{d^\delta}{dx^\delta} G(ax) = a^\delta \frac{d^\delta}{d\overline{z}^\delta} G(\overline{z}), \qquad \delta \in \mathbb{R}, \tag{6.47}$$

where $\overline{z} = ax$, with a being a constant. This property is valid for all the fractional differential operators considered here. By substituting Eq. (6.2) into Eq. (6.46), after some calculation, we obtain

$$\Phi(t) = \left[\overline{k_1} Dt + \overline{k_2}\right]^{\frac{1}{\mu+\nu+\theta-1}}, \tag{6.48}$$

with $\overline{k_1} \equiv \overline{k}(1-\theta-\nu-\mu)$ and $\overline{k_2}$ is an arbitrary constant connected with the initial condition of the system. For $\rho(z)$, where $|z| = |x|/\Phi(t)$, we have to solve the following equation:

$$\frac{d}{d|z|}\left\{|z|^{-\theta}\frac{d^{\mu-1}}{d|z|^{\mu-1}}\left[\tilde{\rho}(z)\right]^\nu\right\} = \overline{k}\frac{d}{dz}\left[z\tilde{\rho}(z)\right], \tag{6.49}$$

which, after one integration, may be cast in the form

$$\frac{d^{\mu-1}}{d|z|^{\mu-1}}\left[\tilde{\rho}(z)\right]^\nu = \overline{k}|z|^{1+\theta}\tilde{\rho}(z) + C, \tag{6.50}$$

where C is an integration constant. To find a solution for Eq. (6.50) that satisfies the boundary conditions, $\tilde{\rho}(\pm\infty) = 0$, we propose the “ansatz”

$$\tilde{\rho}(z) = N z^{\alpha/\nu}(a + bz)^{\beta/\nu}$$

with the parameters to be chosen in such a way that the solution verifies Eq. (6.50) for $C = 0$. To proceed further, we may use the property

$$D_x^{\delta}\left[x^{\alpha}(a+bx)^{\beta}\right] = a^{\delta}\frac{\Gamma[1+\alpha]}{\Gamma[1+\alpha-\delta]}x^{\alpha-\delta}(a+bx)^{\beta-\delta}, \qquad (6.51)$$

with $D_x^{\delta} \equiv d^{\delta}/dx^{\delta}$, $\delta \equiv \alpha + \beta + 1$, and b is an arbitrary constant which will be considered as ± 1 depending on the values of the parameters α, β, and ν. The use of this property restricts the broadness of the solution due to the constraints imposed on the parameters; however, it enables us to find a formal (or analytical) solution to Eq. (6.50) and, consequently, to Eq. (6.46). Substitution of the ansatz in Eq. (6.50), and using Eq. (6.51), yields

$$\alpha = \frac{(2-\mu)(\mu+\theta)}{1-2\mu-\theta},$$
$$\beta = -\frac{(\mu-1)(\mu-2)}{1-2\mu-\theta}, \quad \text{and}$$
$$\nu = \frac{2-\mu}{1+\mu+\theta}. \qquad (6.52)$$

This result for parameters α, β, and ν has as particular cases the ones obtained in Ref. [187] for $\theta = 0$. By using Eq. (6.52), the solution for Eq. (6.50) can be written as

$$\rho(x,t) = \frac{N}{\left(|\bar{k}_1|t\right)^{\frac{1+\mu+\theta}{\mu^2+2\mu\theta-\mu+\theta^2-1}}}\left[\frac{z^{(\mu+\theta)(1+\mu+\theta)}}{(a+bz)^{(1-\mu)(1+\mu+\theta)}}\right]^{\frac{1}{1-2\mu-\theta}}, \qquad (6.53)$$

with

$$N = \left[\bar{k}\frac{\Gamma(-\beta)}{\Gamma(1+\alpha)}\right]^{\frac{1+\mu+\theta}{1-2\mu-\theta}} \quad \text{and} \quad z \equiv x\left(|\bar{k}_1|t\right)^{-\frac{1+\mu+\theta}{1+\mu^2+2\mu\theta-\mu+\theta^2}}, \qquad (6.54)$$

where we have chosen, for simplicity, $\bar{k}_2 = 0$.

From Eq. (6.53), we may analyse different values for the parameters μ and θ and their consequences on the system. For illustrative purposes, we consider two representative situations: (i) $-\infty < \mu < -1-\theta$, with $\theta \geq 0$, and (ii) $0 < \mu < 1/2$, with $0 \leq \theta < 1/2-\mu$. For the first case, we may choose, without loss of generality, $a = 1$ and $b = -1$. In this case, the normalisation condition is given by

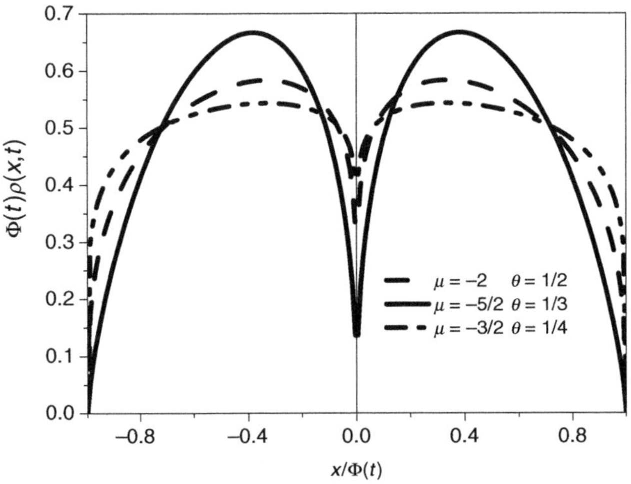

Figure 6.8 $\Phi(t)\rho(x,t)$ versus $x/\Phi(t)$, as predicted by Eq. (6.53), for typical values of μ and θ satisfying the condition $-\infty < \mu < -1-\theta$ and $\theta \geq 0$, with $a = 1$ and $b = -1$, and using the normalisation given by Eq. (6.56).

$$\mathcal{N}\int_{-1}^{1}\left[\frac{z^{(\mu+\theta)(1+\mu+\theta)}}{(1-z)^{(1-\mu)(1+\mu+\theta)}}\right]^{\frac{1}{1-2\mu-\theta}} dz = 1, \tag{6.55}$$

yielding

$$\mathcal{N} = \frac{\Gamma\left[1-\mu-\theta\right]}{2\Gamma\left[\frac{\mu^2+\mu\theta-2\theta-2\mu}{1-2\mu-\theta}\right]\Gamma\left[\frac{1-\mu+\mu^2+\theta^2+2\mu\theta}{1-2\mu-\theta}\right]}. \tag{6.56}$$

The behaviour of Eq. (6.53) is shown in Fig. 6.8, using the normalisation given by Eq. (6.56). It shows also that the distribution is compact and restricted to the spatial region defined by the condition $z \leq 1$. For $0 < \mu < 1/2$, with $0 \leq \theta < 1/2 - \mu$, where $b = 1$, the normalisation condition (6.55) yields

$$\mathcal{N} = \frac{\Gamma\left[\frac{(1-\mu)(1+\mu+\theta)}{1-2\mu-\theta}\right]}{2\Gamma\left(1+\frac{(\mu+\theta)(1+\mu+\theta)}{1-2\mu-\theta}\right)\Gamma\left(\mu+\theta\right)}. \tag{6.57}$$

Fig. 6.9 exhibits the behaviour predicted by Eq. (6.53), with the normalisation given by Eq. (6.57), for $b = 1$. The form of the distribution for these cases is different; i.e., one is compact and the other exhibits a long-tailed behaviour

Particular choices of the parameters are also possible, leading to other meaningful scenarios. For example, consider the case represented by $\mu = 0$ and $\mu = 1$ that

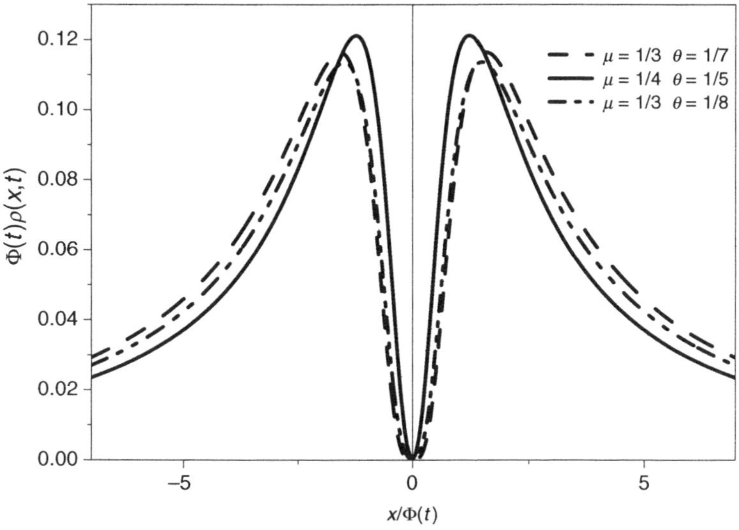

Figure 6.9 $\Phi(t)\rho(x,t)$ versus $x/\Phi(t)$, in order to illustrate Eq. (6.53) for typical values of μ and θ such that $0 < \mu < 1/2$ and $0 \leq \theta < 1/2 - \mu$, with $a = 1$ and $b = 1$.

admits simple solutions. For $\mu = 0$ and ν arbitrary with $C = 0$, Eq. (6.50) may be written as

$$\bar{k}z^{1+\theta}\tilde{\rho}(z) = \int_0^z d\bar{z}\left[\tilde{\rho}\left(\bar{z}\right)\right]^{\nu}, \tag{6.58}$$

and the solution is given by

$$\tilde{\rho}(z) \propto \frac{1}{z^{1+\theta}}\left[1 + \tilde{C}z^{1-\nu(1+\theta)}\right]^{1/(1-\nu)}, \tag{6.59}$$

where $\tilde{C}$ is a constant. For $\mu = 1$, Eq. (6.50) yields:

$$\bar{k}z\tilde{\rho}\left(z\right) = z^{-\theta}\left[\tilde{\rho}(z)\right]^{\nu} + C, \tag{6.60}$$

with $\tilde{\rho}(z)$ being implicitly determined. For the case $C = 0$, a unnormalised solution may be found, and, in particular, it is given by

$$\tilde{\rho}\left(z\right) \propto z^{(1+\theta)/(\nu-1)}. \tag{6.61}$$

The connection of the solutions in this section with distributions emerging in the context of generalised thermostatistic formalism is possible when we consider the asymptotic behaviour, i.e., the behaviour for large x. By comparing the asymptotic results obtained for the solutions found here with the ones obtained for the distributions of generalised formalism [197], namely

$$\rho(x) \sim \frac{1}{|x|^{2/(q-1)}}, \tag{6.62}$$

we conclude that

$$q = \frac{3+\mu+\theta}{1+\mu+\theta}. \tag{6.63}$$

Thus, the distribution obtained for the nonlinear fractional diffusion Eq. (6.1) is asymptotically connected with the q-distributions present in the generalised formalism.

In summary, the solutions handled in this section were obtained by considering a scaled dependence on t and x variables, described by Eq. (6.2). This procedure simplifies the calculations and allows us to reduce the partial nonlinear differential equations to ordinary differential equations, as we have seen. These equations in turn may be solved by considering suitable boundary conditions and the fact that $\int_{-\infty}^{\infty} dx\rho(x,t)$ is a conserved quantity. For the case $\mu = 2$, the solutions may be connected with the q-exponentials which suggests that the S_q entropy proposed by Tsallis [207] in 1988 may be useful to give a thermostatistics basis for these solutions. When $\mu \neq 2$, we have also obtained exact solutions which, depending on the values of the parameters, may exhibit compact or long-tailed behaviour. In this latter case, they can be connected in the asymptotic limit with the q-exponentials. Thus, a relation between the entropic index q and the fractional index μ may be established.

6.4 d-Dimensional Fractional Diffusion Equations

In this section, to obtain analytical solutions, the situations studied before are generalised to d-dimensions in the case of radial symmetry. We consider both the linear and nonlinear fractional diffusion equations written in terms of a space- and time-dependent diffusion coefficient and in the presence of external forces.

6.4.1 Linear Case

The first case to be analysed refers to the following linear fractional diffusion equations with radial symmetry [208]:

$$\frac{\partial^\gamma}{\partial t^\gamma}\rho(r,t) = \int_0^t dt' \frac{1}{r^{d-1}}\frac{\partial}{\partial r}\left\{r^{d-1}\left[D(r,t-t')\frac{\partial}{\partial r}\rho(r,t')\right]\right\} - \frac{1}{r^{d-1}}\frac{\partial}{\partial r}\left[r^{d-1}F(r)\rho(r,t)\right], \tag{6.64}$$

where, as before, $F(r)$ is the external force and d is the spatial dimensionality of the system. The boundary condition is defined on a finite interval, i.e., the system is confined, with $0 \leq r \leq a$. Later on, we will consider the semi-infinite case, i.e., when $a \to \infty$. Equation (6.64) is solved in the presence of a variable diffusion coefficient and external forces. When $\gamma = 1$, with $D(r,t) = D\delta(t)$, the usual diffusion problem is reobtained.

For simplicity, we first treat Eq. (6.64) in the absence of external forces when the diffusion coefficient is given by

$$D(r,t) = \frac{Dt^{\delta-1}}{\Gamma(\delta)}, \quad \delta \in \mathbb{R}.$$

In this case, Eq. (6.64) becomes

$$\frac{\partial^{\gamma}}{\partial t^{\gamma}}\rho(r,t) = \frac{D}{\Gamma(\delta)\, r^{d-1}} \int_0^t dt' \, (t-t')^{\delta-1} \frac{\partial}{\partial r}\left\{ r^{d-1} \frac{\partial}{\partial r}\rho(r,t') \right\}. \tag{6.65}$$

The solution will be subjected to the Dirichlet boundary condition, i.e., $\rho(a,t) = 0$. It is

$$\rho(r,t) = \int_0^a d\xi \xi^{d-1} \tilde{\rho}(\xi) G(r,\xi,t), \tag{6.66}$$

with

$$G(r,\xi,t) = \frac{2}{a^2} \sum_{n=0}^{\infty} \frac{\xi^{\frac{2-d}{2}} r^{\frac{2-d}{2}}}{\left\{ \mathrm{J}_{\frac{d}{2}}(\lambda_n a) \right\}^2} \mathrm{J}_{\frac{d-2}{2}}(\lambda_n \xi)\, \mathrm{J}_{\frac{d-2}{2}}(\lambda_n r), \\ \times \mathrm{E}_{\gamma+\delta}(-\lambda_n^2 D t^{\gamma+\delta}), \tag{6.67}$$

for $d \geq 2$. In Eq. (6.67), $\mathrm{J}_{\nu}(x)$ is the Bessel function of the first kind, defined in Section 1.2.2, and the values of λ_n are obtained by solving the following eigenvalue equation:

$$\mathrm{J}_{\frac{d-2}{2}}(\lambda_n a) = 0, \tag{6.68}$$

and using the initial condition, $\rho(r,0) = \tilde{\rho}(r)$. In Eq. (6.67), $G(r,\xi,t)$ is the Green's function that contains the dynamical aspects of the system for its time evolution from the initial condition. Its spatial behaviour is exhibited in Fig. 6.10 for particular values of the parameters. The Mittag-Leffler function in Eq. (6.67) is connected with the changes produced by the fractional derivative, whereas the time dependence of $D(r,t)$ on the waiting time distribution function is more akin to this process, which in turn is connected with an anomalous diffusion process. For $\gamma + \delta = 1$, we reobtain the solution of the d-dimensional usual diffusion equation with radial symmetry.

The next step is to incorporate the following spatial dependence in the diffusion coefficient:

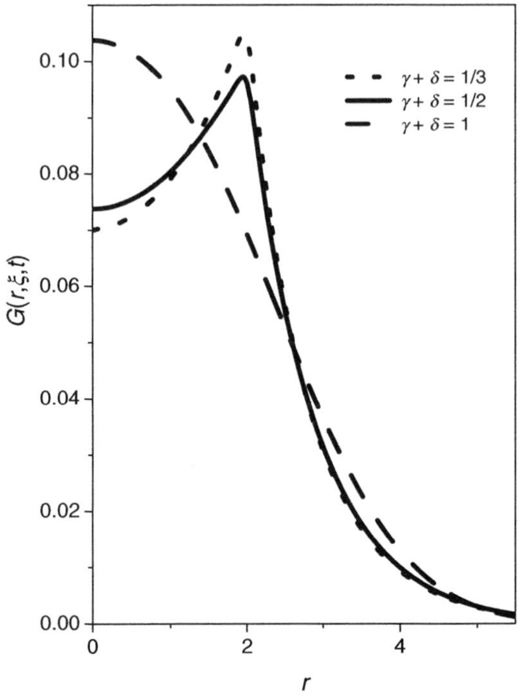

Figure 6.10 $G(r,\xi,t)$ versus r for typical values of $\gamma+\delta$, when $a = 10.0$, $t = 1.0$, $\xi = 2.0$, $D = 1.0$, and $d = 3$ [208]. Modified from Physica A, **360**/2, E. K. Lenzi, R. S. Mendes, G. Gonçalves, M. K. Lenzi, L. R. da Silva, Fractional diffusion equation and Green function approach: Exact solutions, 215–226. Copyright (2006), with permission from Elsevier.

$$D(r) = \frac{Dr^{-\theta}t^{\delta-1}}{\Gamma(\delta)}. \tag{6.69}$$

The equation to be solved is thus

$$\frac{\partial^{\gamma}}{\partial t^{\gamma}}\rho(r,t) = \frac{D}{\Gamma(\delta)\, r^{d-1}} \int_0^t dt'\,(t-t')^{\delta-1}\frac{\partial}{\partial r}\left\{r^{d-1-\theta}\frac{\partial}{\partial r}\,\rho(r,t')\right\}. \tag{6.70}$$

The solution may be written as

$$\rho(r,t) = \int_0^a d\xi\,\xi^{d-1}\tilde{\rho}(\xi)G(r,\xi,t), \quad \text{with}$$

$$G(r,\xi,t) = \frac{2+\theta}{a^{2+\theta}}\sum_{n=0}^{\infty}\frac{\xi^{\frac{1}{2}(2+\theta-d)}r^{\frac{1}{2}(2+\theta-d)}}{\left\{\mathrm{J}_{\frac{d}{2+\theta}}\left(\frac{2\lambda_n}{2+\theta}a^{\frac{1}{2}(2+\theta)}\right)\right\}^2}E_{\gamma+\delta}(-D\lambda_n^2 t^{\gamma+\delta})$$
$$\times\,\mathrm{J}_{\frac{d-2-\theta}{2+\theta}}\left(\frac{2\lambda_n\xi^{\frac{2+\theta}{2}}}{2+\theta}\right)\mathrm{J}_{\frac{d-2-\theta}{2+\theta}}\left(\frac{2\lambda_n r^{\frac{2+\theta}{2}}}{2+\theta}\right), \tag{6.71}$$

where $d \geq 2+\theta$, with λ_n being determined by the following eigenvalue equation:

$$\mathrm{J}_{\frac{d}{2+\theta}-1}\left[\frac{2\lambda_n}{2+\theta}a^{\frac{1}{2}(2+\theta)}\right]=0. \tag{6.72}$$

To tackle this case, we may also consider a mixed boundary condition, in the form:

$$\partial_r\rho + H\rho|_{r=a}=0, \tag{6.73}$$

where H is a constant. The condition (6.73) is commonly employed to investigate the heat conduction. The solution satisfying this boundary condition may be written as

$$\rho(r,t)=\int_0^a d\xi\,\xi^{d-1}\tilde{\rho}(\xi)G(r,\xi,t), \quad \text{with}$$

$$G(r,\xi,t)=\frac{2+\theta}{a^{2+\theta}}\sum_{n=0}^{\infty}\frac{\xi^{\frac{1}{2}(2+\theta-d)}r^{\frac{1}{2}(2+\theta-d)}}{\overline{D}_n\left\{\mathrm{J}_{\frac{d}{2+\theta}}\left(\frac{2\lambda_n}{2+\theta}a^{\frac{1}{2}(2+\theta)}\right)\right\}^2}E_{\gamma+\delta}(-D\lambda_n^2t^{\gamma+\delta})$$
$$\times\,\mathrm{J}_{\frac{d-2-\theta}{2+\theta}}\left(\frac{2\lambda_n\xi^{\frac{2+\theta}{2}}}{2+\theta}\right)\mathrm{J}_{\frac{d-2-\theta}{2+\theta}}\left(\frac{2\lambda_n r^{\frac{2+\theta}{2}}}{2+\theta}\right), \tag{6.74}$$

where

$$\overline{D}_n=\left[\left(\frac{\lambda_n}{H}\right)^2a^{\theta}+1-\frac{d-(2+\theta)}{Ha}\right].$$

The result obtained above may be extended further if we consider a system in a semi-infinite space, by taking the limit $a\to\infty$. To find the solution, we use the following integral transform:

$$\rho(r,t)=\int_0^{\infty}dk\;C(k,t)\Psi(r,k), \quad \text{with}$$

$$\Psi(r,k)=r^{\frac{2+\theta-d}{2}}\mathrm{J}_{\frac{d-2-\theta}{2+\theta}}\left(\frac{2kr^{\frac{2+\theta}{2}}}{2+\theta}\right), \tag{6.75}$$

where $C(k,t)$ is a time-dependent function to be determined. The kernels in Eq. (6.75) employ the eigenfunctions of the spatial operator of the diffusion equation to be solved. By substituting Eq. (6.75) into Eq. (6.70), we obtain

$$\frac{d^{\gamma}}{dt^{\gamma}}C(k,t)=-\frac{Dk^2}{\Gamma(\delta)}\int_0^t dt'\,(t-t')^{\delta-1}C(k,t). \tag{6.76}$$

The solution for Eq. (6.76) is

$$C(k,t)=C(k,0)E_{\gamma+\delta}(-k^2Dt^{\gamma+\delta}),$$

where $C(k,0)$ is to be determined by the initial condition. For an arbitrary initial condition as, for instance, $\rho(r,0)=\tilde{\rho}(r)$, we obtain

$$C(k,0)=\frac{2k}{2+\theta}\int_0^\infty d\xi\,\xi^{d-1}\tilde{\rho}(\xi)\Psi(\xi,k). \tag{6.77}$$

The solution (6.75) becomes

$$\rho(r,t)=\int_0^\infty d\xi\,\xi^{d-1}\tilde{\rho}(\xi)G(r,\xi,t), \qquad \text{where}$$

$$G(r,\xi,t)=\frac{2}{2+\theta}\int_0^\infty dk k\Psi(\xi,k)\Psi(r,k)E_{\gamma+\delta}(-k^2Dt^{\gamma+\delta}). \tag{6.78}$$

For $\delta=0$, this equation recovers results presented in Ref. [209], and for $\gamma+\delta=1$ it can be simplified by means of the identity:

$$\int_0^\infty dk\,k\,\mathrm{J}_\nu(\alpha k)\mathrm{J}_\nu(\beta k)e^{-a^2k^2}=\frac{1}{2a^2}e^{-\frac{\beta^2+\alpha^2}{4a^2}}\mathrm{I}_\nu\left(\frac{\alpha\beta}{2a^2}\right). \tag{6.79}$$

Applying Eq. (6.79) to Eq. (6.78), we obtain

$$G(r,\xi,t)=\frac{e^{-\frac{r^{2+\theta}+\xi^{2+\theta}}{(2+\theta)^2Dt}}}{(2+\theta)(\xi r)^{\frac{d-2-\theta}{2}}Dt}\mathrm{I}_{\frac{d-2-\theta}{2+\theta}}\left[\frac{2(\xi r)^{\frac{2+\theta}{2}}}{(2+\theta)^2Dt}\right], \tag{6.80}$$

where $\mathrm{I}_\nu(x)$ is the modified Bessel function, defined in Section 1.2.2.

Let us now consider the system subjected to an external force. Specifically, consider

$$F(r)=\mathcal{K}r^{\epsilon}, \quad \text{when} \quad D(r,t)=Dr^{-\theta}\delta(t), \tag{6.81}$$

with $\epsilon=-1-\theta$. For this case, there is not a stationary solution. The equation which governs the system is

$$\frac{\partial^\gamma}{\partial t^\gamma}\rho(r,t)=\frac{D}{r^{d-1}}\frac{\partial}{\partial r}\left\{r^{d-1-\theta}\frac{\partial}{\partial r}\rho(r,t')\right\}-\frac{\mathcal{K}}{r^{d-1}}\frac{\partial}{\partial r}\left\{r^{d-1+\epsilon}\rho(r,t)\right\} \tag{6.82}$$

To search for a solution to this equation, we consider first a limited region of space and, after that, an extended semi-infinite region. Thus, the solution satisfying the condition $\rho(a,t)=0$ is

$$\rho(r,t) = \int_0^a d\xi \xi^{d-1-\frac{\mathcal{K}}{D}} \tilde{\rho}(\xi) G(r,\xi,t), \quad \text{with}$$

$$G(r,\xi,t) = \frac{2+\theta}{a^{2+\theta}} \sum_{n=0}^{\infty} \frac{\xi^{\frac{1}{2}(2+\theta-d)+\frac{\mathcal{K}}{2D}} r^{\frac{1}{2}(2+\theta-d)+\frac{\mathcal{K}}{2D}}}{\left\{ \mathrm{J}_{\overline{\nu}+1}\left[\frac{2\lambda_n}{2+\theta} a^{\frac{1}{2}(2+\theta)}\right]\right\}^2} \mathrm{E}_\gamma(-D\lambda_n^2 t^\gamma)$$
$$\times \mathrm{J}_{\overline{\nu}}\left(\frac{2\lambda_n \xi^{\frac{2+\theta}{2}}}{2+\theta}\right) \mathrm{J}_{\overline{\nu}}\left(\frac{2\lambda_n r^{\frac{2+\theta}{2}}}{2+\theta}\right), \tag{6.83}$$

where

$$\overline{\nu} = \frac{2\sqrt{\overline{k}+k'^2}}{(2+\theta)}, \quad \overline{k} = \frac{[d-(2+\theta)]\mathcal{K}}{D}, \quad \text{and} \quad k' = \frac{1}{2}\left[d-(2+\theta)-\frac{\mathcal{K}}{D}\right]. \tag{6.84}$$

The eigenvalue λ_n is obtained from the equation

$$\mathrm{J}_{\overline{\nu}}\left(\frac{2\lambda_n}{2+\theta} a^{\frac{2+\theta}{2}}\right) = 0. \tag{6.85}$$

The solution of the problem may be also obtained for an extended semi-infinite region when $a \to \infty$. The same procedure employed for the case without external forces may be used now, by introducing the function

$$\overline{\Psi}(r,k) = r^{\frac{1}{2}(2+\theta-d)+\frac{\mathcal{K}}{2D}} \mathrm{J}_{\overline{\nu}}\left(\frac{2kr^{\frac{2+\theta}{2}}}{2+\theta}\right), \tag{6.86}$$

instead of $\Psi(r,k)$ in Eq. (6.75), and by considering a generic initial condition $\rho(x,0) = \tilde{\rho}(x)$. The solution is

$$\rho(r,t) = \int_0^\infty d\xi \xi^{d-1-\frac{\mathcal{K}}{D}} \tilde{\rho}(\xi) G(r,\xi,t), \quad \text{with}$$

$$G(r,\xi,t) = \frac{2}{2+\theta} \int_0^\infty dk k \overline{\Psi}(\xi,k) \overline{\Psi}(r,k) E_\gamma(-k^2 D t^\gamma). \tag{6.87}$$

We may also incorporate in Eq. (6.64) an external force in the form

$$F(r) = -kr + \mathcal{K} r^\epsilon,$$

with $\epsilon = -1-\theta$. For this external force there is a stationary solution depending on the values of ϵ. To obtain the solution for this case, we consider a series in terms of an eigenfunction, as follows:

$$\rho(r,t) = r^{\frac{\mathcal{K}}{D}} e^{-\frac{kr^{2+\theta}}{(2+\theta)D}} \sum_{n=0}^{\infty} \Psi_n(r)\Phi_n(t). \tag{6.88}$$

The function $\Psi_n(r)$ represents the eigenfunction of the spatial operator and $\Phi_n(t)$ is a time-dependent function to be determined. By considering the initial condition $\rho(x,0) = \tilde{\rho}(x)$, we obtain

$$\rho(r,t) = \int_0^{\infty} d\xi \xi^{d-1} \tilde{\rho}(\xi) G(r,\xi,t), \quad \text{with}$$

$$G(r,\xi,t) = r^{\frac{\mathcal{K}}{D}} e^{-\frac{kr^{2+\theta}}{(2+\theta)D}} \left[\frac{k}{(2+\theta)D}\right]^{\frac{\mathcal{K}+dD}{(2+\theta)D}} \sum_{n=0}^{\infty} \frac{(2+\theta)\Gamma(n+1)}{\Gamma\left[\frac{\mathcal{K}+dD}{(2+\theta)D}+n\right]} \mathrm{E}_{\gamma}(-\lambda_n t^{\gamma})$$
$$\times \mathrm{L}_n^{(\alpha)}\left[\frac{kr^{2+\theta}}{(2+\theta)D}\right] \mathrm{L}_n^{(\alpha)}\left[\frac{k\xi^{2+\theta}}{(2+\theta)D}\right], \tag{6.89}$$

in which

$$\alpha = \frac{(\mathcal{K}+dD)}{[(2+\theta)D]} - 1. \tag{6.90}$$

In Eq. (6.89), $\mathrm{L}_n^{(\alpha)}(x)$ are the associated Laguerre polynomials and $\lambda_n = (2+\theta)\,n\,k$ represents the eigenvalues.

6.4.2 Nonlinear Case

We consider now the nonlinear case in d-dimensions, using spatial fractional derivatives. To be specific, we focus our attention on the following equation:

$$\frac{\partial}{\partial t}\rho(r,t) = \frac{1}{r^{d-1}} \frac{\partial^{\mu'}}{\partial r^{\mu'}} \left\{ r^{d-1} D(r,t;\rho) \frac{\partial^{\mu}}{\partial r^{\mu}} [r^{-\eta} \rho(r,t)^{\nu}] \right\}$$
$$- \frac{1}{r^{d-1}} \frac{\partial}{\partial r} \left\{ r^{d-1} F(r,t)\rho(r,t) \right\} + \overline{\alpha}(t)\rho(r,t), \tag{6.91}$$

in which the diffusion coefficient will be given by

$$D(r,t,\rho) = D(t) r^{-\theta} \rho^{\gamma}. \tag{6.92}$$

In Eqs. (6.91) and (6.92), $\overline{\alpha}(t)$ represents a reaction term and μ, μ', η, γ, and θ are real coefficients. The external force to be considered has the functional form:

$$F(r,t) = -k(t) r. \tag{6.93}$$

As before, we consider for Eq. (6.91) the solutions that have the form of Eq. (6.2):

$$\overline{\rho}\,(r,t) = \frac{1}{\Phi\,(t)} \tilde{\rho}\left[\frac{r}{\Phi\,(t)}\right]. \tag{6.94}$$

This permits us to start our search for the solution of Eq. (6.91) by using the transform

$$\rho(r,t) = \frac{1}{r^{d-1}} \exp\left[\int_0^t d\tilde{t}\overline{\alpha}(\tilde{t})\right] \overline{\rho}(r,t), \tag{6.95}$$

in such a way that $\overline{\rho}(r,t)$ becomes the function to be determined. Applying this transform to Eq. (6.91) and considering, initially, the system in the absence of external forces, we obtain

$$\frac{\partial}{\partial t}\overline{\rho}(r,t) = \tilde{D}(t)\frac{\partial^{\mu'}}{\partial r^{\mu'}}\left\{r^{(1-\gamma)(d-1)-\theta}[\overline{\rho}(r,t)]^{\gamma}\frac{\partial^{\mu}}{\partial r^{\mu}}\{r^{-\eta-\nu(d-1)}[\overline{\rho}(r,t)]^{\nu}\}\right\} \tag{6.96}$$

with

$$\tilde{D}(t) = \exp\left[(\nu+\gamma-1)\int_0^t d\tilde{t}\alpha(\tilde{t})\right] D(t). \tag{6.97}$$

Substitution of Eq. (6.94) into Eq. (6.91), using $z = r/\Phi(t)$, yields the following equations to be solved:

$$\frac{d^{\mu'}}{dz^{\mu'}}\left\{z^{(1-\gamma)(d-1)-\theta}[\tilde{\rho}(z)]^{\gamma}\frac{d^{\mu}}{dz^{\mu}}\{z^{m-\nu(d-1)}\left[\tilde{\rho}\left(z\right)\right]\right\} = \overline{k}\frac{d}{dz}\left[z\tilde{\rho}\left(z\right)^{\nu}\right] \tag{6.98}$$

and

$$\dot{\Phi}(t) = -\tilde{D}(t)\overline{k}\,[\Phi(t)]^{2-\xi}\,, \tag{6.99}$$

where

$$\xi = (d-1)(\gamma+\nu-1)+\theta+\gamma+\nu-m+\mu+\mu', \tag{6.100}$$

with $\overline{k}$ being a separation constant to be determined by the normalisation condition. The solution for $\Phi(t)$ is

$$\Phi(t) = \left[\Phi(0)^{\xi-1} + (1-\xi)\overline{k}\int_0^t \tilde{D}(t')dt'\right]^{\frac{1}{\xi-1}}. \tag{6.101}$$

The presence of the linear external force Eq. (6.93) only changes the behaviour of the time-dependent function $\Phi(t)$, i.e., the time-dependent part of the problem, which in this case is determined by solving the following equation:

$$\dot{\Phi}(t) = -\overline{k}\,\tilde{D}(t)[\Phi(t)]^{2-\xi} - k(t)\Phi(t). \tag{6.102}$$

The solution is

$$\Phi(t) = \left\{ \Phi(0)^{\xi-1} + (1-\xi)\bar{k} \int_0^t d\tilde{t}\,\tilde{D}(\tilde{t})\, e^{(\xi-1)\int_0^{\tilde{t}} dt' k(t')} \right\}^{\frac{1}{\xi-1}} e^{-\int_0^t dt' k(t')}. \tag{6.103}$$

To proceed further with the search for the solutions of Eq. (6.91), we integrate Eq. (6.98) once, in order to simplify the analysis. After that, the equation to be solved will be cast in the form:

$$\frac{d^{\mu'-1}}{dz^{\mu'-1}} \left\{ z^{(d-1)(1-\gamma)-\theta} [\tilde{\rho}(z)]^{\gamma} \frac{d^{\mu}}{dz^{\mu}} \left[z^{-\eta-(d-1)\nu} [\tilde{\rho}(z)]^{\nu} \right] \right\} = \bar{k} z \tilde{\rho}(z) + C, \tag{6.104}$$

where C denotes again an integration constant. In Eq. (6.104) several problems can be analysed depending on the values of the parameters. A particular solution may present a long-tailed or a compact behaviour, as we shall show below. For simplicity, we start with the value $\mu' = 1$ and, after that, with the values $\mu' \neq 1$. When $\mu' = 1$, Eq. (6.104) reduces to

$$z^{(d-1)(1-\gamma)-\theta} [\tilde{\rho}(z)]^{\gamma} \frac{d^{\mu}}{dz^{\mu}} \left[z^{-\eta-(d-1)\nu} [\tilde{\rho}(z)]^{\nu} \right] = \bar{k} z \tilde{\rho}(z), \tag{6.105}$$

because, without loss of generality, we can choose $C = 0$. To solve Eq. (6.105) we propose the ansatz

$$\tilde{\rho}(z) = N z^{\frac{\alpha}{\nu}} (a + bz)^{\frac{\beta}{\nu}}. \tag{6.106}$$

This equation resembles the one found before in one dimension, thus suggesting the use of the property stated in Eq. (6.51). In this way as well, we find

$$\begin{aligned} \frac{\alpha}{\nu} &= \frac{(\theta+\mu+\eta+1)(2+\mu+\theta)}{(\gamma-1)(\theta+2\mu+\eta+1)} + d - 1, \\ \frac{\beta}{\nu} &= \frac{\mu(2+\theta+\mu)}{(\gamma-1)(\theta+\eta+1+2\mu)}, \quad \text{and} \\ \nu &= \frac{(1-\gamma)(1-\eta-\mu)}{(2+\mu+\theta)}, \end{aligned} \tag{6.107}$$

where N and $\bar{k}$ are determined by the normalisation condition. Summing up the previous results, the solution for Eq. (6.105) can be written as

$$\tilde{\rho}(z) = N z^{d-1} \left[z^{\theta+\mu+1+\eta} (a+bz)^{\mu} \right]^{\frac{2+\mu+\theta}{(\gamma-1)(\theta+2\mu+\eta+1)}}. \tag{6.108}$$

In Eq. (6.108), b is a constant to be taken as ± 1 depending on the case considered. The value $b = -1$ implies a solution with a compact behaviour; the value $b = 1$

implies a solution with a long-tailed behaviour. In particular, for this latter case it is possible to connect the solutions found here with the distributions emerging in the generalised thermostatistics in the asymptotic limit, as in the preceding cases analysed in this chapter.

By means of the same procedure, we shall obtain the solutions for $\mu' \neq 1$. The equation to be solved becomes now

$$\frac{d^{\mu'-1}}{dz^{\mu'-1}}\left\{z^{(d-1)(1-\gamma)-\theta}[\tilde{\rho}(z)]^{\gamma}\frac{d^{\mu}}{dz^{\mu}}\left[z^{m-(d-1)\nu}[\tilde{\rho}(z)]^{\nu}\right]\right\} = \bar{k}z\tilde{\rho}(z), \tag{6.109}$$

where, for simplicity, we fix again $C = 0$. Solving Eq. (6.109), having in mind Eq. (6.51), we obtain:

$$\widetilde{\rho}(z) = \overline{N}z^{d-1}\left[z^{\theta-m+\mu+\mu'}(1+bz)^{\mu+\mu'-1}\right]^{\frac{\theta+2\mu'+\mu}{(1-\gamma)(1-\theta-\eta-2(\mu+\mu'))}}, \tag{6.110}$$

in which

$$\gamma = \frac{d+\theta-\mu'-\mu}{\mu'+d}, \tag{6.111}$$

while $\overline{N}$ and $\overline{k}$ are defined by the normalisation condition, for $b = \pm 1$, as before. The behaviour of Eq. (6.108) is shown in Fig. 6.11 for typical values of the parameters.

To conclude, we analyse the solution of Eq. (6.109) for three particular cases corresponding to the values (i) $\mu = \mu' = 1$, (ii) $\mu' = -\mu = 1$, and (iii) $\mu' = 2$ and $\mu = 0$. We take $\overline{k} = -k'$ in order to obtain a solution which may represent the spreading of the physical system. We also express these solutions in terms of the q-exponentials of the generalised thermostatistics formalism. Substitution of the conditions required for the first case ($\mu = \mu' = 1$), in Eq. (6.109), yields

$$z^{(d-1)(1-\gamma)-\theta}[\tilde{\rho}(z)]^{\gamma}\frac{d}{dz}\left[z^{-\eta-(d-1)\nu}[\tilde{\rho}(z)]^{\nu}\right] = -k'z\tilde{\rho}(z). \tag{6.112}$$

The solution of Eq. (6.112) is

$$\tilde{\rho}(z) = z^{d-1+\frac{\eta}{\nu}}\exp_q\left[-\mathcal{K}z^{2+\theta+\frac{\eta}{\nu}}\right], \tag{6.113}$$

with

$$\mathcal{K} = \frac{k'}{\nu(2+\theta)+\frac{\eta}{\nu}}. \tag{6.114}$$

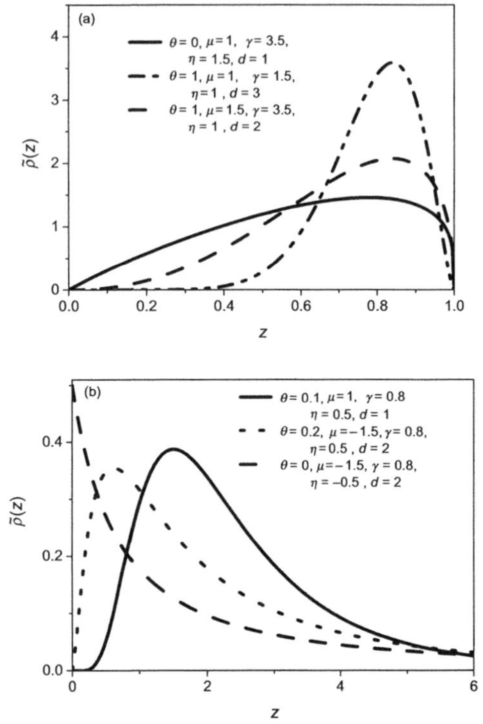

Figure 6.11 The solution constructed from Eq. (6.108) versus z for typical values of μ, γ, θ, and η. Depending on the values of these parameters, we may encounter a compact (a) ($b = -1$) or a long-tailed behaviour (b) ($b = 1$).

For the second case ($\mu' = -\mu = 1$), we may rewrite Eq. (6.109) as

$$z^{(d-1)(1-\gamma)-\theta}[\tilde{\rho}(z)]^{\gamma} \int_0^z dz z^{-\eta-(d-1)\nu}[\tilde{\rho}(z)]^{\nu} = -k' z \tilde{\rho}(z), \tag{6.115}$$

and the solution is

$$\tilde{\rho}(z) = z^{d-1+\frac{\theta+1}{\gamma-1}} \exp_q \left[-z^{1-\eta+\frac{(\theta+1)}{\gamma-1}\nu} / \mathcal{K} \right], \tag{6.116}$$

with

$$\mathcal{K} = k'[(1-\gamma)(1+m) - \nu(\theta+1)] \quad \text{and} \quad q = \nu + \gamma. \tag{6.117}$$

Finally, for the third case ($\mu' = 2$ and $\mu = 0$), Eq. (6.109) is reduced to

$$\frac{d}{dz}\left\{ z^{(d-1)(1-\gamma-\nu)-\theta-\eta}[\tilde{\rho}(z)]^{\gamma+\nu} \right\} = -k' z \tilde{\rho}(z), \tag{6.118}$$

and the solution may be given in terms of the q-exponential function as in the above cases. It is

$$\widetilde{\rho}(z) = z^{d-1+\frac{\eta+\theta-(d-1)}{\gamma+\nu}} \exp_q\left[-\mathcal{K}z^{2+(d-1)\frac{\gamma+\nu-1}{\gamma+\nu}}\right], \tag{6.119}$$

with

$$\mathcal{K} = \frac{k'}{(d-1)(\gamma+\nu-1)+2(\gamma+\nu)} \quad \text{and} \quad q = 2-\nu-\gamma. \tag{6.120}$$

The solutions found for these particular values of the parameters may exhibit a long-tailed or a compact behaviour, depending on the values of ν, γ, θ, and η.

7

Anomalous Diffusion

Anisotropic Case

The preceding chapters dealt with the fractional diffusion equation with spatial and temporal fractional derivatives, diffusion coefficients with space and time dependencies, external forces, and surface effects in finite length situations. Remarkable consequences appear also when we consider the diffusion process in the presence of anisotropy.

To analyse the anisotropic case, we first face a problem in which suspended or dispersed particles diffuse through an anisotropic semi-infinite medium. The process is described in the framework of the usual diffusion equation, but anomalous diffusion behaviour arises in the system because the phenomenon of adsorption–desorption of particles occurs at the interface, and the conservation of the number of particles in the system has to be imposed.

The second problem is to consider a fractional diffusion equation subjected to an anisotropy, with a nonsingular spatial and temporal diffusion coefficient. We will show that the distribution governed by the equation is not separable in terms of space and time variables as in the usual diffusion, which is an unexpected behaviour since the fractional operator is linear. As a specific application, the chapter closes with the search for the solutions to the comb model with integer and fractional derivatives, and also with a drift term. This model is a simplified picture of highly disordered systems and can be connected with a rich class of diffusive processes due to geometric constraints.

7.1 The Adsorption–Desorption Process in Anisotropic Media

We consider first the diffusion problem in a semi-infinite anisotropic medium in contact with a solid substrate at which an adsorption–desorption process takes place [114, 210]. Initially, a defined number of particles is suspended or dispersed in the medium and an anisotropic diffusive process starts. The particles reaching the solid substrate can be adsorbed and desorbed in such a way that the kinetics of this process is governed by a typical balance equation characterising a chemical reaction of first kind (Langmuir approximation) as the one considered in Section 5.4. The conservation of the number of particles is then invoked and the profiles of the surface as well as of the bulk density of particles are analytically obtained by means of Laplace–Fourier techniques. The results for the momentum distribution show that the system exhibits anomalous diffusion [211] behaviour, according to the values of the characteristic times entering the problem. More precisely, not only can a single subdiffusive or superdiffusive motion be found, but the system presents a multiple behaviour that includes both modes of subdiffusion and superdiffusion and, for large times, the normal diffusive behaviour. A system presenting similar behaviour can be found in the dynamics of vesicles driven by adhesion gradients of a Langmuir monolayer [212]. The same theoretical framework can be relevant to describe the diffusion process of suspended or dispersed particles in an anisotropic media like liquid crystals [213] and effects of interfaces on diffusion [214]. Dispersions of particles in an anisotropic host medium like nematic liquid crystals are responsible for a series of different physical scenarios [213]. Other systems recently considered are the suspension of magnetic grains [215], silica spheres [216], latex particles [217], and oil droplets [218]. All the bulk effects in these systems have been object of attention in the last few years.

To mathematically formulate the problem we consider a typical geometry for the sample such that the Cartesian reference frame has the z-axis perpendicular to the bounding surface, located at $z = 0$ [210]. If the system is a nematic liquid crystal, we can consider that in this geometry the sample is homeotropically oriented [152]. In the general case of other host media, it is enough to consider an anisotropic diffusion coefficient, and the bulk density of particles $\rho(x, y, z; t)$ will be governed by the anisotropic diffusion equation

$$\frac{\partial}{\partial t}\rho(x,y,z;t) = D_{\perp}\left[\frac{\partial^2}{\partial x^2}\rho(x,y,z;t) + \frac{\partial^2}{\partial y^2}\rho(x,y,z;t)\right] + D_{\parallel}\frac{\partial^2}{\partial z^2}\rho(x,y,z;t), \tag{7.1}$$

where $D_{xx} = D_{yy} = D_{\perp}$ and $D_{zz} = D_{\parallel}$ are, respectively, the diffusion coefficients connected with the x, y, and z directions. The bulk density of particles is subjected to the boundary conditions:

$$\rho(\pm\infty, y, z; t) = 0, \quad \rho(x, \pm\infty, z; t) = 0, \quad \text{and} \quad \rho(x, y, \infty; t) = 0. \tag{7.2}$$

The boundary condition of $\rho(x, y, z; t)$ on the surface $z = 0$ is defined in terms of the surface density of particles, $\sigma(x, y; t)$, by the kinetic equation

$$\frac{d}{dt}\sigma(x,y;t) + \frac{1}{\tau}\sigma(x,y;t) = \kappa\rho(x,y,0;t), \tag{7.3}$$

where κ and τ are parameters describing the adsorption phenomenon [152]. For simplicity, we consider an arbitrary initial condition which is normalised, i.e., $\rho(x,y,z;0) = \widetilde{\rho}(x,y,z)$ such that

$$\int_{-\infty}^{\infty} dx \int_{-\infty}^{\infty} dy \int_{0}^{\infty} dz \rho(x,y,z;0) = 1.$$

This quantity, $\widetilde{\rho}(x,y,z)$, may represent a definite amount of particles introduced in the host sample, at a given position. The other condition to be imposed on the system is a consequence of the first one, actually, the conservation of the number of particles:

$$\int_{-\infty}^{\infty} dx \int_{-\infty}^{\infty} dy \sigma(x,y;t) + \int_{-\infty}^{\infty} dx \int_{-\infty}^{\infty} dy \int_{0}^{\infty} dz \rho(x,y,z;t) = 1, \tag{7.4}$$

and, for simplicity, we assume that $\sigma(x,y;0) = 0$. The first term of Eq. (7.4) gives the quantity of particles adsorbed by the surface, whereas the second term gives the quantity of particles dispersed in the bulk, which is nothing but the survival probability $\mathcal{S}(t)$, defined in Eq. (4.54). To solve Eq. (7.1) subject to the conditions stated above, we use the Green's function approach and integral (Laplace and Fourier) transforms.

We start by applying the Laplace transform to the variable t and the Fourier transform to the variables x and y in Eq. (7.1). We obtain

$$D_{\|} \frac{d^2}{dz^2} \rho(k_x,k_y,z;s) - \left[s + D_{\perp}\left(k_x^2 + k_y^2\right)\right] \rho(k_x,k_y,z;s) = -\widetilde{\rho}(k_x,k_y,z). \tag{7.5}$$

The solution of Eq. (7.5) is formally given by

$$\begin{aligned} \rho(k_x,k_y,z;s) = & -\int_0^{\infty} d\overline{z}\, \widetilde{\rho}(k_x,k_y,\overline{z}) G(k_x,k_y,z,\overline{z};s) \\ & - D_{\|} \rho(k_x,k_y,0;s) \frac{d}{d\overline{z}} G(k_x,k_y,z,\overline{z};s)\bigg|_{\overline{z}=0}, \end{aligned} \tag{7.6}$$

with the Green's function obtained from

$$D_{\|} \frac{d^2}{dz^2} G(k_x,k_y,z,\overline{z};s) - \left[s + D_{\perp}\left(k_x^2 + k_y^2\right)\right] G(k_x,k_y,z,\overline{z};s) = \delta(z - \overline{z}) \tag{7.7}$$

and subject to the conditions

$$G(k_x,k_y,0,\overline{z};s) = G(k_x,k_y,\infty,\overline{z};s) = 0.$$

By using the sine-Fourier transform, as defined in Eq. (1.10), it is possible to show that the solution of Eq. (7.7) is

$$G(k_x,k_y,z,\bar{z};s) = -\frac{2}{\pi}\int_0^\infty dk_z \frac{\sin(k_z z)\sin(k_z\bar{z})}{s + D_\parallel k_z^2 + D_\perp\left(k_x^2+k_y^2\right)}. \tag{7.8}$$

By performing the inverse Fourier and Laplace transforms, Eq. (7.8) may be simplified to

$$G(x,y,z,\bar{z};t) = -\frac{1}{\sqrt{4\pi D_\parallel t}}\left[e^{-\frac{(z-\bar{z})^2}{4D_\parallel t}} - e^{-\frac{(z+\bar{z})^2}{4D_\parallel t}}\right]\frac{e^{-\frac{x^2+y^2}{4D_\perp t}}}{4\pi D_\perp t}. \tag{7.9}$$

Thus, the solution becomes

$$\begin{aligned}\rho(x,y,z;t) = &-\int_{-\infty}^{\infty} d\bar{x}\int_{-\infty}^{\infty} d\bar{y}\int_0^\infty d\bar{z}\,\tilde{\rho}(\bar{x},\bar{y},\bar{z})G(x-\bar{x},y-\bar{y},z,\bar{z};t)\\ &- D_\parallel\int_{-\infty}^{\infty} d\bar{x}\int_{-\infty}^{\infty} d\bar{y}\int_0^t d\bar{t}\rho(\bar{x},\bar{y},0;\bar{t})\\ &\times\left.\frac{d}{d\bar{z}}G(x-\bar{x},y-\bar{y},z,\bar{z};t-\bar{t})\right|_{\bar{z}=0}.\end{aligned} \tag{7.10}$$

To obtain the time dependence of the quantity of adsorbed particles at the surface, i.e.,

$$\bar{\sigma}(t) = \int_{-\infty}^{\infty} dx\int_{-\infty}^{\infty} dy\sigma(x,y;t),$$

we consider Eqs. (7.3), (7.4), and (7.10). After some calculation, using the Laplace transform, it is possible to show that

$$\bar{\sigma}(s) = \int_{-\infty}^{\infty} dx\int_{-\infty}^{\infty} dy\int_0^\infty dz\,\frac{\kappa\tilde{\rho}(x,y,z)e^{-\sqrt{\frac{s}{D_\parallel}}z}}{\kappa s + \sqrt{sD_\parallel}\left(s+\frac{1}{\tau}\right)} \tag{7.11}$$

and, consequently, that

$$\bar{\sigma}(t) = \int_{-\infty}^{\infty} dx\int_{-\infty}^{\infty} dy\int_0^\infty dz\tilde{\rho}(x,y,z)\int_0^t d\bar{t}\,\frac{\Phi(\bar{t})e^{-\frac{z^2}{4D_\parallel(t-\bar{t})}}}{\sqrt{\pi(t-\bar{t})}}, \tag{7.12}$$

with

$$\Phi(t) = \frac{\tau_R}{\sqrt{\tau(\tau_R^2 - 1)}} \left[\gamma_+ e^{\gamma_+^2 t/\tau} \operatorname{erfc}\left(\gamma_+\sqrt{t/\tau}\right) - \gamma_- e^{\gamma_-^2 t/\tau} \operatorname{erfc}\left(\gamma_-\sqrt{t/\tau}\right)\right]. \tag{7.13}$$

In Eq. (7.13), $\gamma_\pm = \tau_R \pm \sqrt{\tau_R^2 - 1}$ and $\operatorname{erfc}(x)$ is the complementary error function, defined in Eq. (1.116). Actually, $\overline{\sigma}(t)$ represents the number of particles on the surface, since we have performed an integration over x and y. In the preceding expressions, we have introduced the quantity

$$\tau_R = \frac{\kappa}{2}\sqrt{\frac{\tau}{D_\parallel}}, \tag{7.14}$$

which involves the characteristic parameters connected with the dynamics of the diffusion process and with the kinetics of the adsorption phenomenon in the sample.

In Fig. 7.1, the bulk density of particles integrated over x and y is shown as a function of $z/\kappa\tau$ for three different times, illustrating the spreading of the distribution

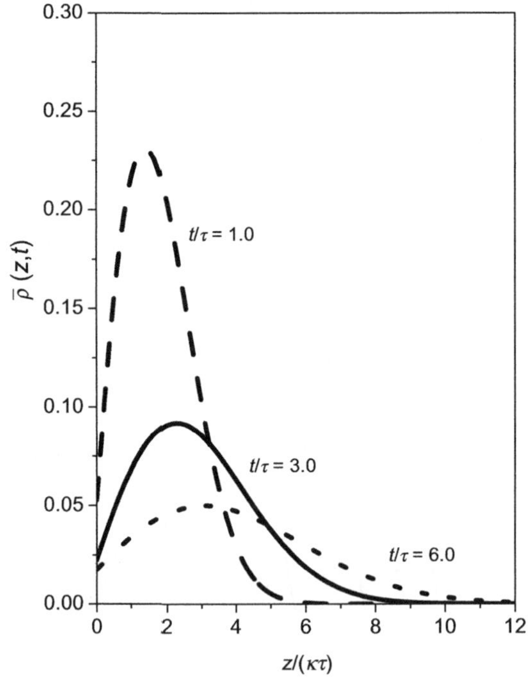

Figure 7.1 Profile of the bulk density of particles $\overline{\rho}(z,t) = \int_{-\infty}^{\infty} dx \int_{-\infty}^{\infty} dy \rho(x,y,z;t)$ versus z, for three different values of the ratio t/τ, when the initial condition is $\widetilde{\rho}(x,y,z) = \delta(x)\delta(y)\delta(z-1)$. For illustrative purposes, the curves were drawn for $D_\parallel = 1$, $\kappa = 5$, and $\tau = 10$ ($\tau_R \approx 7.9$) [210]. Modified from E. K. Lenzi, L. R. Evangelista, G. Barbero, and F. Mantegazza, Anomalous diffusion and the adsorption–desorption process in anisotropic media, EPL **85** (2009) 28004.

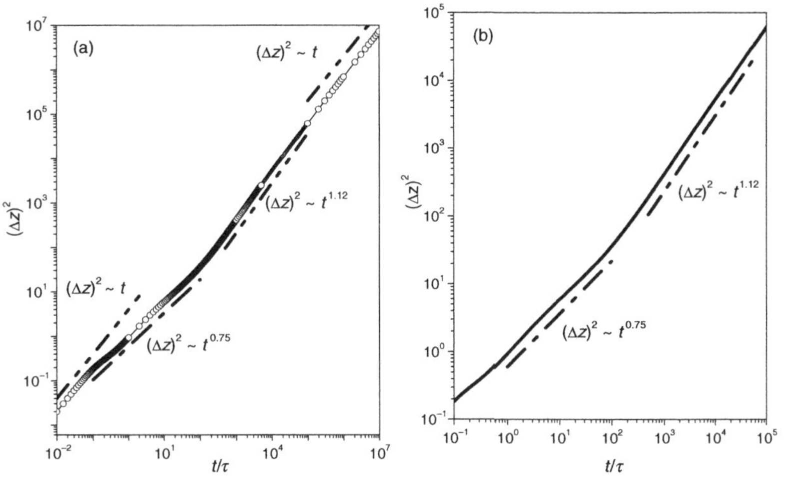

Figure 7.2 (a) $(\Delta z)^2$ versus t/τ, for the initial condition $\widetilde{\rho}(x,y,z) = \delta(x)\delta(y)\delta(z-1)$, showing the existence of different diffusive regimes in the system. (b) The times for which the anomalous regime is present. The straight lines are approximations used to show the different diffusive regimes in the time scale considered. The curves were drawn for $D_{\|} = 1$, $\kappa = 5$, and $\tau = 10$ ($\tau_R \approx 7.9$) [210]. Modified from E. K. Lenzi, L. R. Evangelista, G. Barbero, and F. Mantegazza, Anomalous diffusion and the adsorption–desorption process in anisotropic media, EPL **85** (2009) 28004.

in connection with the lowering of the bulk density near the surface limiting the samples.

In Fig. 7.2, the dispersion relation for the z-direction of the distribution function, i.e., $(\Delta z)^2 = \langle (z - \langle z \rangle)^2 \rangle$, is shown as a function of time (solid line with circles). It is remarkable that this simple system can exhibit anomalous diffusion behaviour, which becomes evident when we draw comparative straight lines characterising the different diffusion regimes.

After starting the diffusion process, which is initially usual for very short times, the system exhibits a subdiffusive regime for which $(\Delta z)^2 \propto t^{0.75}$ (dashed line). This initial regime indicates that the adsorption process on the surface plays the main role for initial times. After that, the process becomes superdiffusive with $(\Delta z)^2 \propto t^{1.12}$ (dotted line). The second regime reflects the effect of the desorption process of the particles by the surface located at $z = 0$, which dominates the solution for intermediate times. For long enough times, the system presents a normal diffusive behaviour $(\Delta z)^2 \propto t$ (dashed-dotted line). The last behaviour may be understood as a consequence of the absence of adsorption or desorption. This feature may be verified in Fig. 7.3, which for long times shows that the particles initially adsorbed by the surface are essentially desorbed and remain in the bulk.

This anomalous behaviour, in particular the presence of different diffusive regimes, is also a consequence of the conservation of the number of particles,

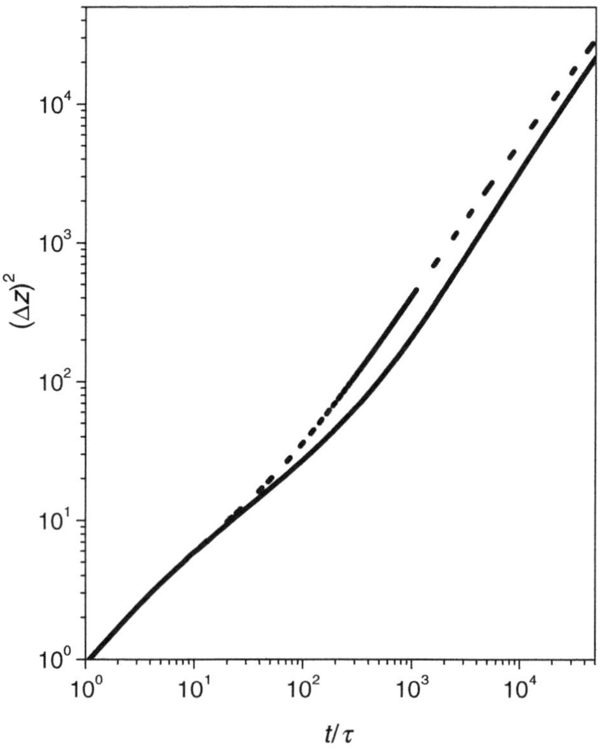

Figure 7.3 The same as in Fig. 7.2 in order to show the role of the parameter τ as a crossover time in two illustrative situations: $\tau = 10$ ($\tau_R \approx 7.9$) (dotted line) and $\tau = 50$ ($\tau_R \approx 18$) (solid line) [210]. Modified from E. K. Lenzi, L. R. Evangelista, G. Barbero, and F. Mantegazza, Anomalous diffusion and the adsorption–desorption process in anisotropic media, EPL **85** (2009) 28004.

Eq. (7.4), imposed on the system. In the model, the number of particles inserted in the system is fixed and the adsorption–desorption process, together with the necessity to conserve the total number of particles, leads to different diffusive regimes according to the time scale considered. As we have seen throughout this book, models based on fractional diffusion equations are used to investigate anomalous diffusion and may be the mathematical basis to explain the existence of different diffusive regimes [102, 219]. Indeed, the anomalous diffusion produced by these models is essentially due to the presence of fractional operators and not explicitly due to the interaction between the surface and the bulk (adsorption–desorption) as we have found in this section for a system with a fixed number of particles.

In Fig. 7.3, one observes a crossover time which is essentially governed by τ. This result is a manifestation of the importance of the desorption process, whose characteristic time is τ, in the anomalous behaviour of the diffusion. The onset of this anomalous behaviour is sensitive to the values of the model parameters.

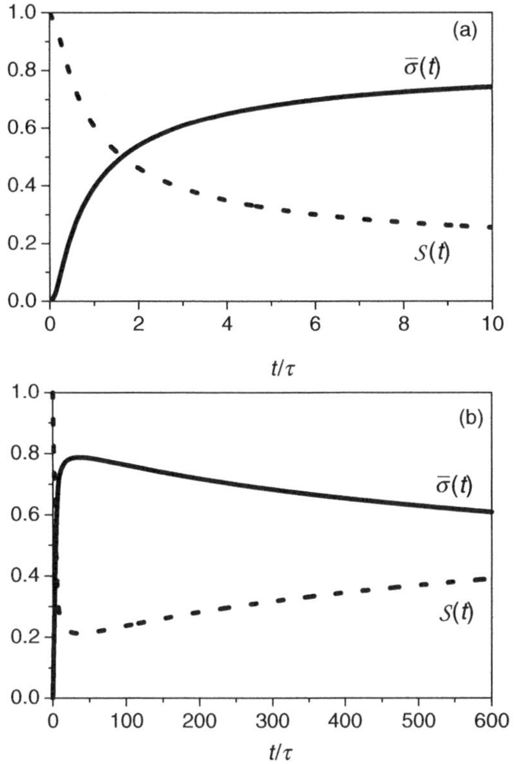

Figure 7.4 Number of adsorbed particles $\overline{\sigma}(t)$, as given by Eq. (7.12) (solid line), and survival probability $\mathcal{S}(t)$ (dotted line) versus t for the initial condition $\widetilde{\rho}(x,y,z) = \delta(x)\delta(y)\delta(z-1)$. The curves were drawn for $D_{\parallel} = 1$, $\kappa = 5$, and $\tau = 10$ ($\tau_R \approx 7.9$) [210]. Modified from E. K. Lenzi, L. R. Evangelista, G. Barbero, and F. Mantegazza, Anomalous diffusion and the adsorption–desorption process in anisotropic media, EPL **85** (2009) 28004.

A simple criterion to choose these values is that the characteristic times may be such that the diffusion process is followed by the adsorption and then by the desorption process.

Regarding the number of particles at the surface, here represented by $\overline{\sigma}(t)$, in Fig. 7.4 one can find two typical behaviours according to the time scale considered. In Fig. 7.4a, this quantity is exhibited for $t < \tau$, where τ is a characteristic time connected with the desorption process. This means that, after the diffusion takes place in the bulk, the surface adsorbs and the number of particles on it rapidly grows (solid line). In the same figure, the survival probability is shown as a function of time. This quantity informs us that the particles dispersed in the bulk are initially adsorbed by the surface $z = 0$ and after are desorbed into the bulk. The behaviour is in agreement with the one shown in panel (a), exhibiting a decreasing behaviour. For $t \gg \tau$, the situation is different, as can be seen in Fig. 7.4b, where both

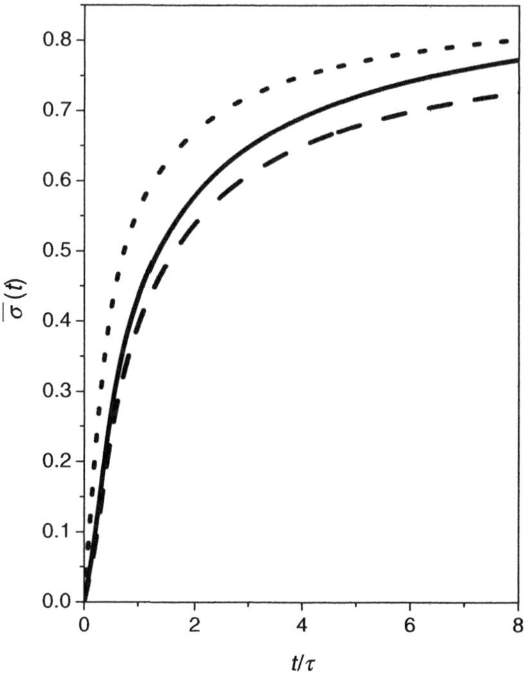

Figure 7.5 Number of adsorbed particles $\overline{\sigma}(t)$ as given by Eq. (7.12) versus t/τ for different sets of parameters: $D_{\parallel} = 3.0$, $\kappa = 7.0$, and $\tau = 15.0$ ($\tau_R = 7.8$) (dotted line); $D_{\parallel} = 1.0$, $\kappa = 5.0$, and $\tau = 10.0$ ($\tau_R = 7.9$) (dashed line); and $D_{\parallel} = 1.0$, $\kappa = 10.0$, and $\tau = 10.0$ ($\tau_R = 15.8$) (solid line) [210]. Modified from E. K. Lenzi, L. R. Evangelista, G. Barbero, and F. Mantegazza, Anomalous diffusion and the adsorption–desorption process in anisotropic media, EPL **85** (2009) 28004.

quantities are shown for a large time interval. After rapidly increasing, $\overline{\sigma}(t)$ tends to decrease while $\mathcal{S}(t)$ shows the inverse behaviour. The asymptotic limit for long time is

$$\overline{\sigma}(t) \sim \tau_{\mathrm{R}}\sqrt{\tau/\pi t}, \qquad t \gg \tau.$$

Finally, in Fig. 7.5 the short time behaviour of $\overline{\sigma}(t)$ illustrates its sensitivity to the values of the model parameters. Notice that from Eq. (7.11) or Eq. (7.12) only $D_{\parallel}$ contributes to $\overline{\sigma}(t)$. This means that anomalous diffusion may also occur even in the case of isotropic diffusion constant.

This permits us to conclude that after a short time interval, in which the diffusion process together with the adsorption process works, the number of particles on the surface increases; after that, however, when a large number of particles have been adsorbed, the desorption process is operating and the bulk again becomes more populated. This behaviour is a consequence of the boundary conditions imposed on the system and the absence of external forces. Different boundary conditions (see, for example, the cases treated in Refs. [152, 220]) or external forces applied to the system may produce a stationary solution for the density of particles, $\rho(x, y, z; t)$, and consequently a stationary rate of adsorption–desorption process on the surface $z = 0$.

The anomalous diffusion can be found in those semi-infinite anisotropic systems for which the diffusive behaviour is accompanied by the adsorption–desorption process at a uniform interface, when the conservation of the number of particles is imposed on the whole system. Since this mechanism is general, the theoretical framework presented here can find applications in a wide variety of systems, in particular, in colloidal suspensions or any kind of dispersed particles that undergo diffusion through a host system. This happens because the flat interface, at which the adsorption–desorption phenomenon takes place, breaks the translational invariance and together with the adsorption–desorption process may represent the disappearing or appearing of a particle in a given system where particles can be confined in some phases. In some colloidal systems, as the concentration in a dense colloidal suspension is increased, particles become confined in transient cages formed by their neighbors [221, 222]. This prevents them from diffusing freely throughout the sample. Such an inhibition of the molecular motion is responsible for a subdiffusion process whose mechanisms are in many respects similar to the ones described in the theoretical framework we have discussed in this section.

We notice that a system composed of two adsorbing surfaces, i.e., in a slab shape, in one dimension does not exhibit anomalous behaviour in the case of neutral particles [151, 220]. This is not true for charged particles, as we will discuss in Chapters 9 and 10. This reinforces the idea that to find anomalous diffusion behaviour in the system we are analysing here, it is necessary to consider a semi-infinite system limited by a surface in which the adsorption–desorption process is governed by a kinetic equation when the conservation of particle is imposed on the system. In fact, as stated above, for long enough times, $\overline{\sigma}(t) \propto 1/\sqrt{t}$. This means that, when the particles are inserted in the system, the diffusion process starts, and, for short time intervals, the number of particles increases at the surface. However, in view of the desorption process and the semi-infinite character of the medium, for long enough times the overwhelming majority of the particles can be found again in the bulk. For this system, there is no stationary solution for the bulk density of particles, as is found in slab-shaped systems.

7.2 Fractional Diffusion Equations in Anisotropic Media

We consider again the anisotropic diffusion process but now described by means of fractional derivatives. Specifically, we handle the following fractional diffusion equation in d-dimensions [223]:

$$\frac{\partial^\gamma}{\partial t^\gamma}\rho(\overline{x},t) = \sum_{i,j=1}^{d}\int_0^t dt'\frac{\partial}{\partial x_i}\left\{D_{ij}(\overline{x},t-t')\frac{\partial}{\partial x_j}[\rho(\overline{x},t')]\right\}$$
$$-\sum_{i=1}^{d}\frac{\partial}{\partial x_i}[F_i(\overline{x})\rho(\overline{x},t)], \tag{7.15}$$

where $\overline{x} = (x_1, x_2, ..., x_d)$, $F_i(\overline{x})$ is an external force, $D_{ij}(\overline{x};t)$ are the diffusion coefficients, and the fractional operator is that of Caputo (see Section 2.3). We have also taken into account for Eq. (7.15) the boundary condition

$$\lim_{|\overline{x}|\to\infty}\rho(\overline{x},t) = 0,$$

in order to assure that

$$\int_{-\infty}^{\infty}\prod_{i=1}^{d}dx_i\rho(\overline{x},t)$$

is time independent. We analyse Eq. (7.15) in two representative scenarios: first, with a diffusion coefficient with a nonsingular spatial and temporal dependence, in the presence of external forces; subsequently, we consider it with a singular dependence on the diffusion coefficient, i.e., the comb model [224], with the time fractional derivative replaced by a fractional derivative in space. Finally, a drift term is incorporated in the comb model and the whole situation is reanalysed.

When the diffusion coefficient is $D_{ij}(\overline{x},t) = D_i(t)\delta_{ij}$, in the absence of external forces, Eq. (7.15) becomes

$$\frac{\partial^\gamma}{\partial t^\gamma}\rho(\overline{x},t) = \sum_{i=1}^{d}\int_0^t dt' D_i(t-t')\frac{\partial^2}{\partial x_i^2}\rho(\overline{x},t'). \tag{7.16}$$

By means of this equation, for a suitable kernel in the diffusive term, it is possible to introduce a finite phase velocity, which is not present in the usual diffusion equation, and a fractional diffusion equation of distributed order may be obtained. Before analysing the solutions of Eq. (7.15), we investigate the behaviour of the second moment $\langle x_i^2\rangle$ in order to gain some information about the diffusive process. The spreading of the system is connected with the second moment (or variance) and consequently this calculation is useful to characterise the diffusive process. After some calculation, it is possible to show that

$$\langle x_i^2\rangle = \frac{2}{\Gamma(1+\gamma)}\int_0^t dt'(t-t')^\gamma D_i(t'), \tag{7.17}$$

for the initial condition $\rho(\overline{x},0) = \Pi_{i=1}^d\delta(x_i)$. Equation (7.17) shows that the spreadings of Eq. (7.16) are uncoupled; i.e., the spreading of the distribution in one direction is independent of the spreading of the distribution in the other ones. Depending on the form of $D_i(t)$, different behaviours for $\langle x_i^2 \rangle$ may be obtained in connection with subdiffusive or superdiffusive processes. In particular, for $D_i(t) = D_i\delta(t)$, we reobtain the one-dimensional case analysed in Ref. [225]; for $\gamma = 1$, with $D_i(t) = D_i\delta(t)$, the usual case is found.

To search for the solution, we apply the Laplace transform to Eq. (7.16), simplifying it to

$$s^\gamma \rho(\overline{x}, s) - s^{\gamma-1}\rho(\overline{x}, 0) = \sum_{i=1}^{d} D_i(s)\frac{\partial^2}{\partial x_i^2}\rho(\overline{x}, s). \tag{7.18}$$

By means of standard procedures, it is possible to find a solution for Eq. (7.18) in the form

$$\rho(\overline{x}, s) = \frac{s^{\frac{\gamma}{4}(2+d)-1}}{\left[2\pi \overline{D}_{1\cdots d}(s)\right]^{\frac{d}{2}}}\left[\sum_{i=1}^{d}\frac{x_i^2}{D_i(s)}\right]^{\frac{1}{2}-\frac{d}{4}} \mathrm{K}_{\frac{d}{2}-1}\left\{s^{\frac{\gamma}{2}}\left[\sum_{i=1}^{d}\frac{x_i^2}{D_i(s)}\right]^{\frac{1}{2}}\right\}, \tag{7.19}$$

for the initial condition $\rho(\overline{x}, 0) = \prod_{i=1}^{d}\delta(x_i)$, where

$$\overline{D}_{1\ldots d}(s) = \prod_{i=1}^{d}[D_i(s)]^{\frac{1}{d}},$$

and $\mathrm{K}_\nu(x)$ is the modified Bessel function, defined in Section 1.2.2. To obtain the inverse Laplace transform, we have to take a specific diffusion coefficient, since for an arbitrary one it is a hard task. Some meaningful possibilities for the diffusion coefficient exist.

For instance, we may consider

$$D_i(s) = D_i s^{-\alpha} \quad \Longleftrightarrow \quad D_{ij}(\overline{x}, t) = D_i\frac{\delta_{ij}t^{\alpha-1}}{\Gamma(\alpha)}.$$

Thus, we obtain

$$\rho(\overline{x}, t) = \frac{1}{\left(4\pi\overline{D}_{1\cdots d}t^{\alpha+\gamma}\right)^{\frac{d}{2}}}\mathrm{H}^{2\,0}_{1\,2}\left[\sum_{i=1}^{d}\frac{x_i^2}{4D_i t^{\gamma+\alpha}}\middle|\begin{matrix}\left(1-\frac{d}{2}(\gamma+\alpha),\gamma+\alpha\right)\\ \\ (1-\frac{d}{2},1)\ \ (0,1)\end{matrix}\right]. \tag{7.20}$$

Equation (7.20) predicts an anomalous spreading, evidenced by the behaviour of the second moment, given by

$$\langle x_i^2\rangle \propto t^{\gamma+\alpha},$$

where $\gamma + \alpha < 1$, $\gamma + \alpha = 1$ or $\gamma + \alpha > 1$, represents, respectively a sub-, a normal, or a superdiffusive process.

It is possible to incorporate a spatial dependence into the diffusion coefficient of the problem treated before. A typical dependence is represented by

$$D_{ij}(\overline{x},t) = D_i t^{\alpha-1}\frac{\delta_{ij}|x_i|^{-\theta_i}}{\Gamma(\alpha)}.$$

In this case, Eq. (7.15) can be rewritten as

$$\frac{\partial^{\gamma}}{\partial t^{\gamma}}\rho(\overline{x},t) = \frac{1}{\Gamma(\alpha)}\sum_{i=1}^{d}\int_0^t dt'(t-t')^{\alpha-1}\frac{\partial}{\partial x_i}\left[|x_i|^{-\theta_i}\frac{\partial}{\partial x_i}\rho(\overline{x},t')\right]. \tag{7.21}$$

By means of calculations similar to the ones performed before, we find the solution as

$$\rho(\overline{x},t) = \prod_{i=1}^{d}\frac{2+\theta_i}{2\Gamma\left(\frac{1}{2+\theta_i}\right)} \times \left[\frac{1}{(2+\theta_i)^2 D_i t^{\gamma+\alpha}}\right]^{\frac{1}{2+\theta_i}} \mathrm{H}^{2\,0}_{1\,2}\left[\sum_{i=1}^{d}\frac{|x_i|^{2+\theta_i}}{4D_i t^{\gamma+\alpha}}\left|\begin{matrix}(1-\xi(\gamma+\alpha),\gamma+\alpha)\\ (1-\xi,1)\ (0,1)\end{matrix}\right.\right], \tag{7.22}$$

where $\xi = \sum_{i=1}^{d} 1/(2+\theta_i)$.

Figures 7.6 and 7.7 illustrate $\gamma+\alpha = 1/3$,

$$r^2 = \sum_{i=1}^{N}\frac{|x_i|^{2+\theta_i}}{4D_i},$$

and

$$\overline{N} = \prod_{i=1}^{N}(2+\theta_i)\frac{1}{2\left[(2+\theta_i)^2 D_i\right]^{\frac{1}{2+\theta_i}}\Gamma\left(\frac{1}{2+\theta_i}\right)},$$

which was introduced for compactness. Equation (7.22) represents a result more general than the one obtained in Ref. [198], because it incorporates anisotropic effects and a time dependence in the diffusion coefficient. It is also more general than the ones presented in Ref. [226] for the two-dimensional case. Note that Eq. (7.22) may exhibit an anomalous behaviour at the origin depending on the dimension d and on the values of the parameters θ_i. This can be analysed by using the series form of the H-function of Fox in Eq. (7.22) which, for small values, reads

$$\rho(\overline{x},t) \sim \frac{r^{-2(\xi-1)}}{t^{(\gamma+\alpha)\xi}}, \qquad \text{for} \quad \xi > 1,$$

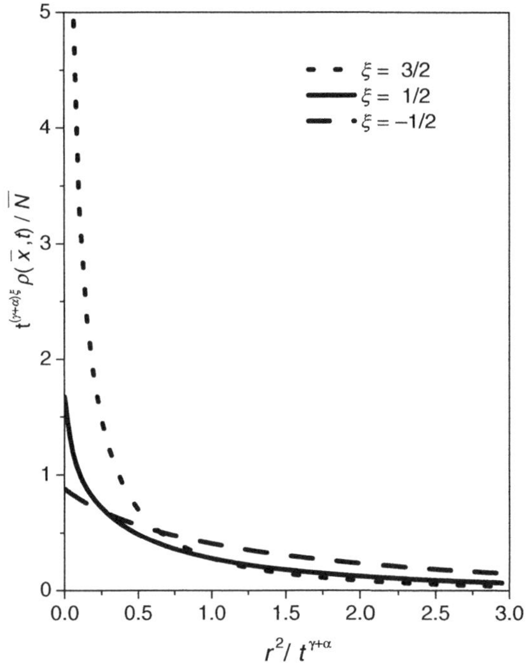

Figure 7.6 $t^{(\gamma+\alpha)\xi}\rho(\overline{x},t)/\overline{N}$ versus $r^2/t^{\gamma+\alpha}$ for typical values of ξ [223]. Modified from Physics Letters A, **347**/4–6, M. F. de Andrade, E. K. Lenzi, L. R. Evangelista, R. S. Mendes, and L. C. Malacarne, Anomalous diffusion and fractional diffusion equation: anisotropic media and external forces, 160–169. Copyright (2005), with permission from Elsevier.

where

$$r^2 = \sum_{i=1}^{d} \frac{|x_i|^{2+\theta_i}}{(2+\theta_i)^2 D_i t^{\gamma}},$$

and has a divergent behaviour at the origin ($r = 0$).

Another challenging scenario appears when the condition $\lim_{|\overline{x}|\to 0}\rho(\overline{x},t) = 0$ is incorporated in the previous problem. The solution of Eq. (7.21), subjected to the initial condition $\rho(\overline{x},0) = \prod_{i=1}^{d}\delta(x_i - \xi_i)$, is

$$\rho(\overline{x},t) = \prod_{i=1}^{N} \frac{2(x_i\xi_i)^{\frac{1+\theta}{2}}}{2+\theta_i} \int_0^{\infty} \cdots \int_0^{\infty} \prod_{i=1}^{N} dk_i k_i \mathrm{J}_{\frac{1+\theta_i}{2+\theta_i}}\left(\frac{2k_i\xi_i^{\frac{2+\theta_i}{2}}}{2+\theta_i}\right) \times \mathrm{J}_{\frac{1+\theta_i}{2+\theta_i}}\left(\frac{2k_i x_i^{\frac{2+\theta_i}{2}}}{2+\theta}\right) \mathrm{E}_{\gamma+\alpha}\left(-\sum_{i=1}^{N} k_i^2 D_i t^{\gamma+\alpha}\right), \tag{7.23}$$

whose behaviour is shown in Fig. 7.8 for typical values of t. For the particular case $\gamma + \alpha = 1$, by using some identities, Eq. (7.23), which is written in terms of Bessel and Mittag-Leffler functions, can be rewritten as

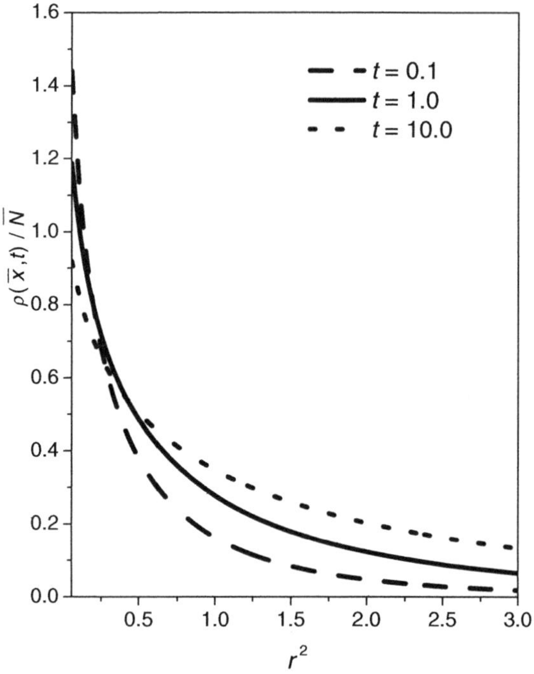

Figure 7.7 $\rho(\overline{x},t)/\overline{N}$ versus r^2 for typical values of t and $\gamma + \alpha = 1/3$ [223]. Modified from Physics Letters A, **347**/4–6, M. F. de Andrade, E. K. Lenzi, L. R. Evangelista, R. S. Mendes, and L. C. Malacarne, Anomalous diffusion and fractional diffusion equation: anisotropic media and external forces, 160–169. Copyright (2005), with permission from Elsevier.

$$\rho(\overline{x},t) = \prod_{i=1}^{d} \frac{(\xi_i x_i)^{\frac{1+\theta_i}{2}}}{(2+\theta_i)D_i t}\, e^{-\frac{x_i^{2+\theta_i}+\xi_i^{2+\theta_i}}{(2+\theta_i)^2 D_i t}}\, \mathrm{I}_{\frac{1+\theta_i}{2+\theta_i}}\left[\frac{2(\xi_i x_i)^{\frac{2+\theta_i}{2}}}{(2+\theta_i)^2 D_i t}\right], \tag{7.24}$$

where $\mathrm{I}_\nu(x)$ is modified Bessel function defined in Section 1.2.2.

For this problem, we can incorporate in Eq. (7.15) the external force and the diffusion coefficient below:

$$F_i(\overline{x}) = (2+\theta_i)\frac{\mathcal{K}_i}{x_i}|x_i|^{-\theta_i} \quad \text{and} \quad D_{ij}(\overline{x},t) = D_i|x_i|^{-\theta_i}\delta(t)\delta_{ij}.$$

The potential corresponding to this external force generalises the logarithmic potential used, for example, to establish the connection between the fractional diffusion coefficient and the generalised mobility [166]. The resulting equation to be solved will be

$$\frac{\partial^\gamma}{\partial t^\gamma}\rho(\overline{x},t) = \sum_{i=1}^{d}\frac{\partial}{\partial x_i}\left\{D_i|x_i|^{-\theta_i}\frac{\partial}{\partial x_i}\rho(\overline{x},t)\right\} - \sum_{i=1}^{d}\frac{\partial}{\partial x_i}\left[(2+\theta_i)\frac{\mathcal{K}_i}{x_i}|x_i|^{-\theta_i}\rho(\overline{x},t)\right]. \tag{7.25}$$

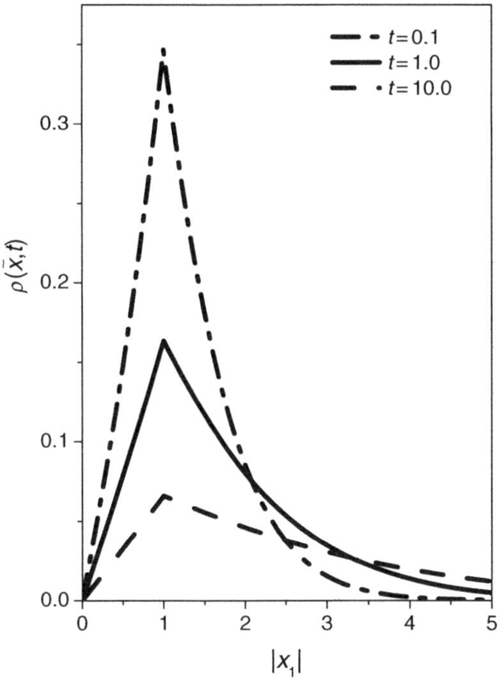

Figure 7.8 $\rho(\bar{x},t)$ versus $|x_1|$ for typical values of t by considering, for simplicity, the one-dimensional case of Eq. (7.23), $\gamma = 1/2$, $\theta_1 = 0$, $\xi_1 = 1$, and $D_1 = 1$ [223]. Modified from Physics Letters A, **347**/4–6, M. F. de Andrade, E. K. Lenzi, L. R. Evangelista, R. S. Mendes, and L. C. Malacarne, Anomalous diffusion and fractional diffusion equation: anisotropic media and external forces, 160–169. Copyright (2005), with permission from Elsevier.

The second moment,

$$\langle x_i^2 \rangle \sim t^{\gamma/(2+\theta_i)},$$

is such that $\rho(\bar{x}, t \to \infty) \to 0$. This implies that Eq. (7.25) in this case has no stationary solution. After some calculation, it is possible to show that the solution is

$$\rho(\bar{x},t) = \prod_{i=1}^{d} \frac{2+\theta_i}{2\Gamma\left(\frac{1}{2+\theta_i} + \frac{\mathcal{K}_i}{D_i}\right)} \left[\frac{1}{(2+\theta_i)^2 D_i t^\gamma}\right]^{\frac{1}{2+\theta_i}} \left[\frac{4\chi_i(t)}{(2+\theta_i)^2}\right]^{\frac{\mathcal{K}_i}{D_i}}$$
$$\times \mathrm{H}_{1\,2}^{2\,0}\left[\sum_{i=1}^{N} \chi_i(t) \middle|_{(1-\Delta,1)\ (0,1)}^{(1-\gamma\Delta,\gamma)}\right], \tag{7.26}$$

where $\Delta = \sum_{i=1}^{d} \mathcal{K}_i/D_i + \xi$ and, for compactness, we have introduced the quantity $\chi_i(t) = |x_i|^{2+\theta_i}/(4D_i t^\gamma)$. The asymptotic behaviour of Eq. (7.26) is

$$\rho(\overline{x},t) \sim \prod_{i=1}^{d} \frac{2+\theta_i}{2\Gamma\left(\frac{1}{2+\theta_i}+\frac{\mathcal{K}_i}{D_i}\right)} \left[\frac{1}{(2+\theta_i)^2 D_i t^{\gamma}}\right]^{\frac{1}{2+\theta_i}} \left[\frac{4\chi_i(t)}{(2+\theta_i)^2}\right]^{\frac{\mathcal{K}_i}{D_i}}$$
$$\times \left[\sum_{i=1}^{d} \chi_i(t)\right]^{\frac{(\gamma-1)\Delta}{2-\gamma}} \exp\left\{(\gamma-2)\gamma^{\frac{\gamma}{2-\gamma}} \left[\sum_{i=1}^{d} \chi_i(t)\right]^{\frac{1}{2-\gamma}}\right\}. \qquad (7.27)$$

The problem we are analysing may present a stationary solution if we add to the external force an extra term $-kx$, which produces a potential exhibiting a minimum and, consequently, confines the distribution. The external force thus becomes

$$F_i(\overline{x}) = -k_i x_i + \frac{\mathcal{K}_i}{x_i}|x_i|^{-\theta_i},$$

which may be obtained from a power-law potential with a second-order term. The equation to be solved is now

$$\frac{\partial^{\gamma}}{\partial t^{\gamma}}\rho(\overline{x},t) = \sum_{i=1}^{d} \frac{\partial}{\partial x_i}\left\{D_i|x_i|^{-\theta_i}\frac{\partial}{\partial x_i}\rho(\overline{x},t) + \left[k_i x_i - \frac{\mathcal{K}_i}{x_i}|x_i|^{-\theta_i}\right]\rho(\overline{x},t)\right\}. \qquad (7.28)$$

We assume that the solution may be written as

$$\rho(\overline{x},t) = \sum_{n_1=0}^{\infty}\cdots\sum_{n_d=0}^{\infty} \Psi_{n_1\cdots n_d}(\overline{x})\phi_{n_1\cdots n_d}(t), \qquad (7.29)$$

where $\Psi_{n_1\ldots n_d}(\overline{x})$ is the eigenfunction of the spatial operator of Eq. (7.28), to be determined, and $\phi_{n_1\ldots n_d}(t)$ is a time-dependent function determined from the initial condition. Substitution of Eq. (7.29) into Eq. (7.28) yields

$$\frac{d^{\gamma}}{dt^{\gamma}}\phi_{n_1\cdots n_d}(t) = -\lambda_{n_1\cdots n_d}\phi_{n_1\cdots n_d}(t) \qquad (7.30)$$

and

$$\sum_{i=1}^{d} \frac{\partial}{\partial x_i}\left\{D_i|x_i|^{-\theta_i}\frac{\partial}{\partial x_i}\Psi(\overline{x}) + \left[k_i x_i - \frac{\mathcal{K}_i}{x_i}|x_i|^{-\theta_i}\right]\Psi(\overline{x})\right\} = -\lambda_{n_1\cdots n_d}\Psi(\overline{x}). \qquad (7.31)$$

The solution of Eq. (7.30) is given in terms of the Mittag-Leffler function:

$$\phi_{n_1\cdots n_d}(t) = \phi_{n_1\cdots n_d}(0)\, \mathrm{E}_{\gamma}(-\lambda_{n_1\cdots n_d}t^{\gamma}), \qquad (7.32)$$

with

$$\lambda_{n_1\ldots n_d} = (2+\theta_i)k_1 n_1 + \cdots + (2+\theta_d)k_d n_d.$$

On the other hand, the solution for the eigenfunction $\Psi_{n_1 \ldots n_d}(\overline{x})$ of the spatial operator may be expressed in terms of the associated Laguerre polynomials, i.e.,

$$\Psi_{n_1 \cdots n_d}(\overline{x}) = \prod_{i=1}^{d} \frac{(2+\theta_i)\Gamma(n_i+1)}{2\Gamma\left(\frac{\mathcal{K}_i + D_i}{(2+\theta_i)D_i} + n_i\right)} \left[\frac{k_i}{(2+\theta_i)D_i}\right]^{\frac{\mathcal{K}_i+D_i}{(2+\theta_i)D_i}} |x_i|^{\frac{\mathcal{K}_i}{D_i}} \times e^{-\frac{k_i|x_i|^{2+\theta_i}}{(2+\theta_i)D_i}} \mathrm{L}_{n_i}^{(\overline{\alpha}_i)}\left[\frac{k_i|x_i|^{2+\theta_i}}{(2+\theta_i)D_i}\right], \tag{7.33}$$

where

$$\overline{\alpha}_i = \left[\frac{\mathcal{K}_i + D_i}{(2+\theta_i)D_i}\right] - 1.$$

If we consider the initial condition $\rho(\overline{x}, 0) = \prod_{i=1}^{d} \delta(x_i - \xi_i)$, then we obtain the following solution for Eq. (7.28):

$$\rho(\overline{x}, t) = \sum_{n_1=0}^{\infty} \cdots \sum_{n_d=0}^{\infty} \prod_{i=1}^{d} \frac{(2+\theta_i)\Gamma(n_i+1)}{2\Gamma\left[\frac{\mathcal{K}_i+D_i}{(2+\theta_i)D_i} + n_i\right]} \left[\frac{k_i}{(2+\theta_i)D_i}\right]^{\frac{\mathcal{K}_i+D_i}{(2+\theta_i)D_i}} e^{-\frac{k_i|x_i|^{2+\theta_i}}{(2+\theta_i)D_i}} \times |x_i|^{\frac{\mathcal{K}_i}{D_i}} \mathrm{L}_{n_i}^{(\overline{\alpha}_i)}\left[\frac{k_i|x_i|^{2+\theta_i}}{(2+\theta_i)D_i}\right] \mathrm{L}_{n_i}^{(\overline{\alpha}_i)}\left[\frac{k_i|\xi_i|^{2+\theta_i}}{(2+\theta_i)D_i}\right] \mathrm{E}_\gamma(-\lambda_{n_1 \cdots n_d} t^\gamma). \tag{7.34}$$

This problem admits also the following stationary solution:

$$\rho(\overline{x}) = \prod_{i=1}^{d} \frac{2+\theta_i}{2\Gamma\left[\frac{\mathcal{K}_i+D_i}{(2+\theta_i)D_i}\right]} \left[\frac{k_i}{(2+\theta_i)D_i}\right]^{\frac{\mathcal{K}_i+D_i}{(2+\theta_i)D_i}} |x_i|^{\frac{\mathcal{K}_i}{D_i}} e^{-\frac{k_i|x_i|^{2+\theta_i}}{(2+\theta_i)D_i}}, \tag{7.35}$$

which coincides with the usual one. The behaviour of Eq. (7.34) is shown in Figs. 7.9 and 7.10. From Fig. 7.9, we notice that Eq. (7.34) relaxes to Eq. (7.35) when long times are considered. This indicates that the fractional time derivative of the diffusion equation, as in the one-dimensional case, produces an anomalous relaxation of the initial condition to the usual stationary solution.

The problem can be solved when an adsorbent term $\alpha|x_i|^{\eta_i}\rho(\overline{x}, t)$, with $\eta_i = 2+\theta_i$, is also taken into account. This term can be connected with a reaction process occurring in the bulk system. If this is assumed, then the diffusion equation becomes

$$\frac{\partial^\gamma}{\partial t^\gamma}\rho(\overline{x}, t) = \sum_{i=1}^{N} \frac{\partial}{\partial x_i}\left\{D_i|x_i|^{-\theta_i}\frac{\partial}{\partial x_i}\rho(\overline{x}, t)\right\} - \sum_{i=1}^{d} \alpha|x_i|^{2+\theta_i}\rho(\overline{x}, t) - \sum_{i=1}^{d} \frac{\partial}{\partial x_i}\left[\left(-k_i x_i + \frac{\mathcal{K}_i}{x_i}|x_i|^{-\theta_i}\right)\rho(\overline{x}, t)\right]. \tag{7.36}$$

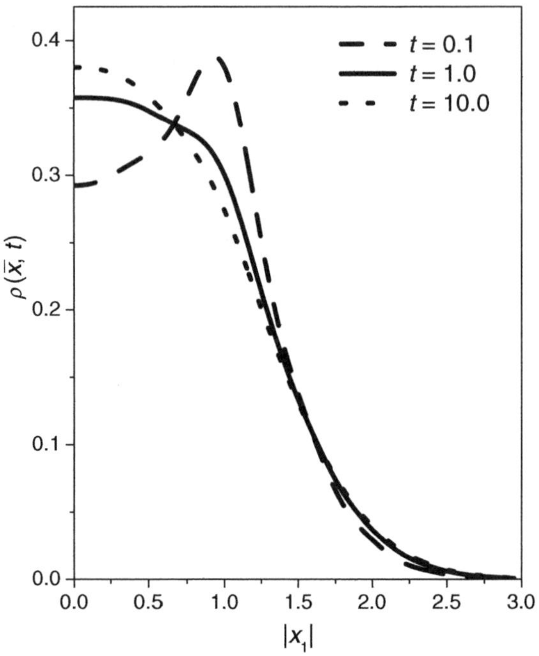

Figure 7.9 $\rho(\bar{x},t)$ versus $|x_1|$ for the one-dimensional case of Eq. (7.34), $\gamma = 1/2$, $\mathcal{K}_1 = 0$, $\theta_1 = 1$, $\xi_1 = 1$, $k_1 = 1$, and $D_1 = 1$. Modified from Physics Letters A, **347**/4–6, M. F. de Andrade, E. K. Lenzi, L. R. Evangelista, R. S. Mendes, and L. C. Malacarne, Anomalous diffusion and fractional diffusion equation: anisotropic media and external forces, 160–169. Copyright (2005), with permission from Elsevier.

For the initial condition $\rho(\bar{x},0) = \prod_{i-1}^{d} \delta(x_i - \xi_i)$, the solution is

$$\rho(\bar{x},t) = \sum_{n_1=0}^{\infty} \cdots \sum_{n_d=0}^{\infty} \prod_{i=1}^{d} |x_i|^{\frac{\mathcal{K}_i}{D_i}} e^{-\frac{k_i(|x_i|^{2+\theta_i} - |\xi_i|^{2+\theta_i})}{2(2+\theta_i)D_i}} e^{-\frac{\sqrt{k_i^2+4\alpha D_i}}{2(2+\theta_i)D_i}(|x_i|^{2+\theta_i}+|\xi_i|^{2+\theta_i})}$$

$$\times \left[\frac{\sqrt{k_i^2 + 4\alpha D_i}}{(2+\theta_i)D_i}\right]^{\frac{\mathcal{K}_i+D_i}{(2+\theta_i)D_i}} \frac{(2+\theta_i)\Gamma(n_i+1)}{2\Gamma\left[\frac{\mathcal{K}_i+D_i}{(2+\theta_i)D_i} + n_i\right]} \mathrm{L}_{n_i}^{(\overline{\alpha}_i)}\left[\frac{\sqrt{k^2+4\alpha D}}{(2+\theta_i)D_i}|x_i|^{2+\theta}\right]$$

$$\times \mathrm{L}_{n_i}^{(\overline{\alpha}_i)}\left[\frac{\sqrt{k_i^2+4\alpha D}}{(2+\theta)D_i}|\xi_i|^{2+\theta}\right] \mathrm{E}_\gamma(-\overline{\lambda}_{n_1\cdots n_d}t^\gamma), \tag{7.37}$$

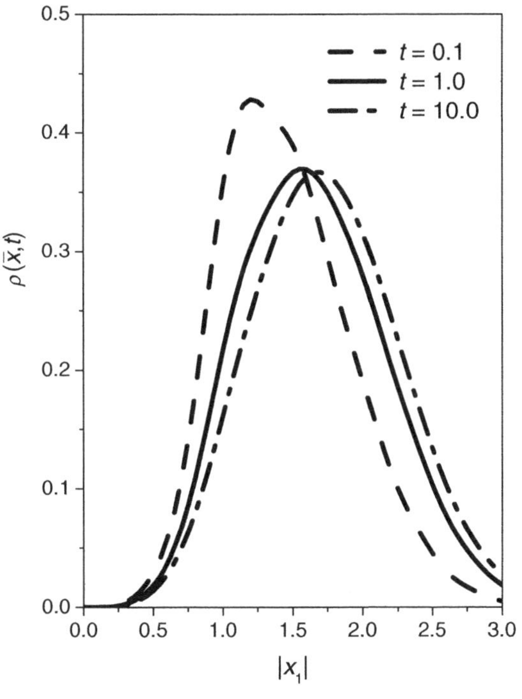

Figure 7.10 The same as in Fig. 7.9 but for $\mathcal{K}_1 = 4$ [223]. Modified from Physics Letters A, **347**/4–6, M. F. de Andrade, E. K. Lenzi, L. R. Evangelista, R. S. Mendes, and L. C. Malacarne, Anomalous diffusion and fractional diffusion equation: anisotropic media and external forces, 160–169. Copyright (2005), with permission from Elsevier.

where

$$\overline{\lambda}_{n_1 \ldots n_d} = \sum_{i=1}^{d} \sqrt{k_i^2 + 4\overline{\alpha}_i D_i}(2+\theta_i) \left\{ n_i + \frac{\mathcal{K}_i + D_i}{2D_i(2+\theta_i)} - k_i \frac{\mathcal{K}_i + D_i}{[2(2+\theta_i)D_i\sqrt{k_i^2 + 4\overline{\alpha}_i D_i}]} \right\}.$$

The solution (7.37) represents a very general expression also embodying a reaction phenomenon in the anomalous diffusion process.

It is worth mentioning that anisotropic situations such as the ones worked out in this section could also be considered in the context of the nonlinear diffusion equation (porous media equation), discussed in the previous chapter. In this context, an interesting feature is the transverse effect [227] manifested by the solutions when external forces are applied to the system.

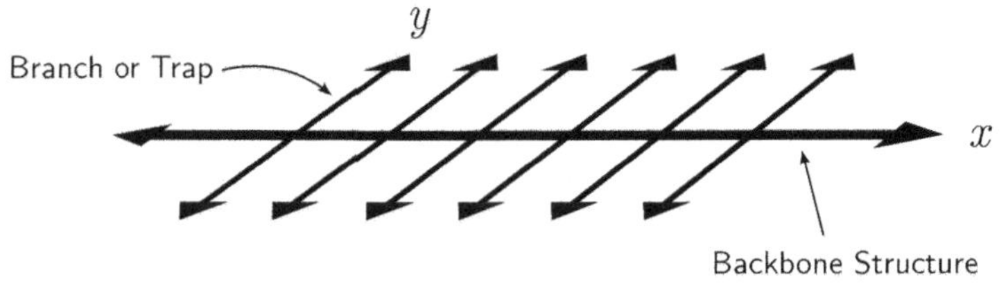

Figure 7.11 The backbone structure used in the comb model. The diffusion in the x-direction only occurs when $y = 0$ and the diffusion in the y-direction is perpendicular to the x-axis [239]. Reprinted with permission from E. K. Lenzi, L. R. da Silva, A. A. Tateishi, M. K. Lenzi, and H. V. Ribeiro, Physical Review E, **87**, 012121 (2013). Copyright (2013) by the American Physical Society.

7.3 The Comb Model

In this section, we consider an important two-dimensional problem with anisotropy on the diffusion coefficients, one of which is singular, represented by the Dirac delta function. This model is known as comb-model (see Fig. 7.11) and has been analysed in several contexts, such as in a porous medium related to exploration of low-dimensional percolation clusters [228–232] and in quantum dynamics [233, 234]. It has also been used to model the problem of flow transfer in disordered systems [235] and electrophoresis, tumour development [236, 237], and anomalous transport [238], among others.

We first investigate the solutions of the equation used to describe the comb model by considering a standard form for it, i.e., by employing differential operators of integer order. A remarkable aspect of this approach is the relation between the diffusion along the x-direction and the fractional time derivative of order $1/2$. Anyway, we may go further by considering the following diffusion equation:

$$\frac{\partial}{\partial t}\rho(x,y;t) = \int_0^t dt'\, D_y(t-t')\frac{\partial^2}{\partial y^2}\rho(x,y;t') + \delta(y)\int_0^t dt'\, D_x(t-t')\frac{\partial^\mu}{\partial|x|^\mu}\rho(x,y;t'), \tag{7.38}$$

where $D_y(t)$ and $D_x(t)$ are time-dependent diffusion coefficients and the fractional derivative applied to the spatial variable is the Riesz–Weyl operator defined in Section 2.4. Note that Eq. (7.38) has essentially the same form as Eq. (7.16) except for the presence of the spatial fractional derivatives. Our goal in this section is to investigate Eq. (7.38) by taking into account the boundary and initial conditions in the form

$$\rho(\pm\infty,y,t) = 0, \quad \rho(x,\pm\infty,t) = 0, \quad \text{and} \quad \rho(x,y,0) = \widehat{\rho}(x,y), \tag{7.39}$$

where $\widehat{\rho}(x, y)$ indicates how the system is initially distributed. In addition, normalisation is imposed such that

$$\int_{-\infty}^{\infty} dx \int_{-\infty}^{\infty} dy \widehat{\rho}(x, y) = 1. \tag{7.40}$$

These conditions are similar to the ones defined in Eq. (7.2). Equation (7.38) will be initially handled by assuming $D_y(t) = D_y\delta(t)$, $D_x(t) = D_x\delta(t)$, and $\mu = 2$. The result obtained for this case shows the effect produced on the solution by an arbitrary initial condition. The spreading of the distribution for initial times indicates that the influence of the initial condition is decisive.

In order to connect with the fractional time derivative, we consider that the diffusion coefficients are represented by power laws in the Laplace space, i.e., that they are given by

$$D_y(s) = s^{1-\gamma_y} D_y \quad \text{and} \quad D_x(s) = s^{1-\gamma_x} D_x, \tag{7.41}$$

which corresponds, for example, to the asymptotic limit of long time of the coefficients

$$D_y(t) \propto (\tau + t)^{\gamma_y - 2} D_y \quad \text{and} \quad D_x(t) = (\tau + t)^{1-\gamma_x} D_x \tag{7.42}$$

in the time domain. Likewise, it is also useful to consider $D_y(t)$ and $D_x(t)$ expressed in terms of Mittag-Leffler functions because they present the same asymptotic behaviour.

Other time dependencies of the diffusion coefficients are possible, but the calculations may be cumbersome.

7.3.1 Integer Order Derivatives

We consider first Eq. (7.38) for $D_y(t) = D_y\delta(t)$, $D_x(t) = D_x\delta(t)$, and $\mu = 2$. Thus, it may be rewritten as [239]

$$\frac{\partial}{\partial t}\rho(x, y, t) = D_y \frac{\partial^2}{\partial y^2}\rho(x, y, t) + D_x\delta(y)\frac{\partial^2}{\partial x^2}\rho(x, y, t). \tag{7.43}$$

The singular term in Eq. (7.43) is responsible for the fact that the diffusion of the system along the x-direction only occurs in the line $y = 0$, i.e., the diffusion coefficient $D_x \neq 0$ only in $y = 0$. A direct consequence of this restriction on the system is the appearance of an anomalous diffusion (actually a subdiffusion) along this direction. This may be analytically verified by evaluating the mean square displacement of the solution on the variable x.

To do this, we search for a solution to Eq. (7.43) satisfying the boundary conditions of Eq. (7.39), when the normalisation Eq. (7.40) is imposed. The solutions to this equation are obtained here by taking a general initial condition into account, which is an extension of some particular forms of the initial condition that have been considered elswhere [240–243]. By applying the Fourier transform to the variable x and the Laplace transform to the variable t in Eq. (7.43), we obtain

$$D_y \frac{d^2}{dy^2}\rho(k_x, y, s) - \left[s + D_x k_x^2 \delta(y)\right]\rho(k_x, y, s) = -\rho(k_x, y, 0). \tag{7.44}$$

This equation may be solved by using the Green's function approach, which yields

$$\rho(k_x, y, s) = -\int_{-\infty}^{\infty} d\bar{y}\widehat{\rho}(k_x, y)G(k_x, y, \bar{y}, s), \tag{7.45}$$

with the Green's function governed by the equation

$$D_y \frac{d^2}{dy^2}G(k_x, y, \bar{y}, s) - \left[s + D_x k_x^2 \delta(y)\right]G(k_x, y, \bar{y}, s) = \delta\left(y - \bar{y}\right) \tag{7.46}$$

and subjected to the boundary condition $G(k_x, \pm\infty, \bar{y}, s) = 0$. After some calculation, it is possible to show that the solution of Eq. (7.46) is

$$\begin{aligned} G(k_x, y, \bar{y}, s) = &-\frac{1}{2\sqrt{sD_y}}\left[e^{-\sqrt{\frac{s}{D_y}}\,|y-\bar{y}|} - e^{-\sqrt{\frac{s}{D_y}}\,(|y|+|\bar{y}|)}\right] \\ &-\frac{e^{-\sqrt{\frac{s}{D_y}}\,(|y|+|\bar{y}|)}}{2\sqrt{sD_y} - D_x k_x^2}. \end{aligned} \tag{7.47}$$

By performing the inverse of Laplace and Fourier transforms in Eq. (7.47), we have

$$\begin{aligned} G(x, y, \bar{y}, t) = &-\frac{\delta(x)}{\sqrt{4\pi D_y t}}\left(e^{-\frac{(y-\bar{y})^2}{4D_y t}} - e^{-\frac{(|y|+|\bar{y}|)^2}{4D_y t}}\right) \\ &-\frac{1}{\sqrt{8D_x\sqrt{D_y}}}\left(\frac{|y|+|\bar{y}|}{\sqrt{4\pi D_y}}\right)\int_0^t d\bar{t}\frac{e^{-\frac{(|y|+|\bar{y}|)^2}{4D_y(t-\bar{t})}}}{\left[\left(t-\bar{t}\right)\bar{t}^{\frac{1}{2}}\right]^{\frac{3}{2}}} \\ &\times \mathrm{H}_{1,1}^{1,0}\left[\sqrt{\frac{2\sqrt{D_y}}{D_x\sqrt{\bar{t}}}}\,|x|\left|\begin{matrix}\left(\frac{1}{4},\frac{1}{4}\right)\\(0,1)\end{matrix}\right.\right]. \end{aligned} \tag{7.48}$$

Again, we notice that the presence of the H-function of Fox in Eq. (7.48) indicates that the system described by the comb structure exhibits anomalous diffusion. This anomalous behaviour in the x-direction may be shown by analysing the dispersion relation $\sigma_x^2 = \langle x^2\rangle - \langle x\rangle^2$, which presents a subdiffusive behaviour, as we will show in Eq. (7.50). By applying the inverse Fourier transform to Eq. (7.45) and using Eq. (7.48) it is possible to find the distribution $\rho(x, y; t)$ as

$$\rho(x, y; t) = \int_{-\infty}^{\infty} d\bar{y} \frac{\widehat{\rho}(x, \bar{y})}{\sqrt{4\pi D_y t}} \left[e^{-\frac{(y-\bar{y})^2}{4D_y t}} - e^{-\frac{(|y|+|\bar{y}|)^2}{4D_y t}} \right]$$
$$- \int_{-\infty}^{\infty} d\bar{x} \int_{-\infty}^{\infty} d\bar{y} \int_{0}^{t} d\bar{t} \, \frac{\widehat{\rho}(\bar{x}, \bar{y})}{\sqrt{8D_x\sqrt{D_y}}} \left(\frac{|y| + |\bar{y}|}{\sqrt{4\pi D_y}} \right) \frac{e^{-\frac{(|y|+|\bar{y}|)^2}{4D_y \bar{t}}}}{\left[(t-\bar{t})\, \bar{t}^{\frac{1}{2}} \right]^{\frac{3}{2}}}$$
$$\times \mathrm{H}_{1,1}^{1,0} \left[\sqrt{\frac{2}{D_x}} \sqrt{\frac{D_y}{\bar{t}}} \, |x - \bar{x}| \left| \begin{matrix} \left(\frac{1}{4}, \frac{1}{4}\right) \\ (0,1) \end{matrix} \right. \right]. \tag{7.49}$$

The first term in Eq. (7.49) is essentially due to the arbitrary form of the initial condition; depending on the form of the initial condition, it vanishes and it is possible to reobtain the results of Ref. [243]. This additional term in $\rho(x, y; t)$ may also play an important role if one is interested in investigating situations characterised by the system not initially localised at the point $(x, y) = (0, 0)$, for example, in drug delivery, in flow transfer in disordered systems, in tumour development, or in contaminant diffusion in heterogeneous media. By using the above equation it is possible to find the dispersion relation for x- and y-directions, which are useful to understand the diffusive process produced by the comb model. For the initial condition $\rho(x, y; 0) = \delta(x - \widetilde{x})\delta(y - \widetilde{y})$, we can show that the dispersion relation connected with the x-direction is

$$\sigma_x^2 = 2D_x \sqrt{\frac{t}{\pi D_y}} \, e^{-\frac{\widetilde{y}^2}{4D_y t}} + |\widetilde{y}| \frac{D_x}{D_y} \mathrm{erfc} \left(\frac{|\widetilde{y}|}{2\sqrt{D_y t}} \right), \tag{7.50}$$

whereas the dispersion relation for the y-variable is

$$\sigma_y^2 = 2\, D_y\, t,$$

indicating that the diffusive behaviour in this direction is the usual one.

We now analyse the solutions of Eq. (7.38) when the diffusion coefficients are given by (7.41) and (7.42), but still with $\mu = 2$. As underlined above, this kind of time dependence of the diffusion coefficients is associated with the fractional time derivatives employed to analyse the subdiffusive processes. Thus, Eq. (7.38) can be rewritten as

$$\frac{\partial}{\partial t} \rho(x, y; t) = \int_{0}^{t} dt' \, D_y(t - t') \frac{\partial^2}{\partial y^2} \rho(x, y; t')$$
$$+ \delta(y) \int_{0}^{t} dt' \, D_x(t - t') \frac{\partial^2}{\partial x^2} \rho(x, y; t'), \tag{7.51}$$

with $D_y(t)$ and $D_x(t)$ as defined above. By taking the previous boundary and initial conditions into account, it is possible to show that the solution in the Laplace space is

$$\rho(x,y,s) = -\int_{-\infty}^{\infty} d\overline{x} \int_{-\infty}^{\infty} d\overline{y}\widehat{\rho}(\overline{x},\overline{y})\widetilde{G}(x,\overline{x},y,\overline{y},s), \tag{7.52}$$

with the Green's function satisfying the equation

$$D_y(s)\frac{\partial^2}{\partial y^2}\widetilde{G}(x,\overline{x},y,\overline{y},s) + \delta(y)D_x(s)\frac{\partial^2}{\partial x^2}\widetilde{G}(x,\overline{x},y,\overline{y},s) - s\widetilde{G}(x,\overline{x},y,\overline{y},s) = \delta(y-\overline{y})\,\delta(x-\overline{x}) \tag{7.53}$$

and subject to the conditions $\widetilde{G}(x,\overline{x},\pm\infty,\overline{y},s) = 0$ and $\widetilde{G}(\pm\infty,\overline{x},\overline{y},y,s) = 0$. After some calculation, the solution of the above equation can be written as

$$\widetilde{G}(x,\overline{x},y,\overline{y};s) = -\frac{\delta(x-\overline{x})}{2\sqrt{sD_y(s)}}\left[e^{-\sqrt{\frac{s}{D_y(s)}}|y-\overline{y}|} - e^{-\sqrt{\frac{s}{D_y(s)}}(|y|-|\overline{y}|)}\right] - \frac{e^{-\sqrt{\frac{2\sqrt{sD_y(s)}}{D_x(s)}}|x-\overline{x}|}}{\sqrt{8D_x(s)\sqrt{sD_y(s)}}}e^{-\sqrt{\frac{s}{D_y(s)}}(|y|+|\overline{y}|)}. \tag{7.54}$$

By using Eqs. (7.52) and (7.54), we can find the dispersion relations σ_x and σ_y in the Laplace space by considering the initial condition $\widehat{\rho}(x,y) = \delta(x-\widetilde{x})\delta(y-\widetilde{y})$. For $\sigma_x^2(s)$, we have

$$\sigma_x^2(s) = \frac{D_x(s)}{\sqrt{s^3 D_y(s)}}e^{-\sqrt{\frac{s}{D_y(s)}}|\widetilde{y}|}, \tag{7.55}$$

and, for $\sigma_y^2(s)$, we obtain

$$\sigma_y^2(s) = \frac{2D_y(s)}{s^2}.$$

These results for $\sigma_x^2(s)$ and $\sigma_y^2(s)$ show that, depending on the forms of $D_x(s)$ and $D_y(s)$, the spreading of the distribution may exhibit different diffusive behaviours. By performing the inverse Laplace transform in $\sigma_x^2(s)$ and $\sigma_y^2(s)$ and taking the time dependence required above for the diffusion coefficients into account, we obtain

$$\sigma_x^2(t) = \frac{D_x t^{\gamma_x}}{\sqrt{D_y t^{\gamma_y}}}\mathrm{H}_{1,1}^{1,0}\left[\frac{|\widetilde{y}|}{\sqrt{D_y t^{\gamma_y}}}\middle|\begin{matrix}(1+\gamma_x-\frac{\gamma_y}{2},\frac{\gamma_y}{2})\\(0,1)\end{matrix}\right] \tag{7.56a}$$

and

$$\sigma_y^2(t) = \frac{2D_y t^{\gamma_y}}{\Gamma(1+\gamma_y)}. \tag{7.56b}$$

The dispersion relations demonstrate that the solution has an anomalous dispersion in both x and y directions and depends on the parameters γ_x and γ_y. The dispersion relation obtained for the y-direction is the same as the one obtained for the fractional diffusion equations. Moreover, the asymptotic behaviour of $\sigma_x^2(t)$ is $t^{\gamma_x-\gamma_y/2}$, when $t\to\infty$. The inverse Laplace transform of the Green's function is

$$\begin{aligned}\widetilde{G}(x,\bar{x},y,\bar{y};t) = &-\frac{\delta(x-\bar{x})}{\sqrt{4\pi D_y t^{\gamma_y}}}\left\{\mathrm{H}^{1,0}_{1,1}\left[\frac{(y-\bar{y})^2}{\sqrt{D_y t^{\gamma_y}}}\left|\begin{matrix}\left(1-\frac{\gamma_y}{2},\frac{\gamma_y}{2}\right)\\(0,1)\end{matrix}\right.\right]\right.\\ &\left.-\mathrm{H}^{1,0}_{1,1}\left[\frac{(|y|+|\bar{y}|)^2}{\sqrt{D_y t^{\gamma_y}}}\left|\begin{matrix}\left(1-\frac{\gamma_y}{2},\frac{\gamma_y}{2}\right)\\(0,1)\end{matrix}\right.\right]\right\}\\ &-\int_0^t \frac{d\bar{t}}{(t-\bar{t})\sqrt{8D_x\bar{t}^{\gamma_x}\sqrt{D_y\bar{t}^{\gamma_y}}}}\mathrm{H}^{1,0}_{1,1}\left[\sqrt{\frac{2}{D_x\bar{t}^{\gamma_x}}\sqrt{D_y\bar{t}^{\gamma_y}}}|x-\bar{x}|\left|\begin{matrix}(\beta,\xi)\\(0,1)\end{matrix}\right.\right]\\ &\times\mathrm{H}^{1,0}_{1,1}\left[\frac{|y|+|\bar{y}|}{\sqrt{D_y\bar{t}^{\gamma_y}}}\left|\begin{matrix}\left(0,\frac{\gamma_y}{2}\right)\\(0,1)\end{matrix}\right.\right], \end{aligned}\tag{7.57}$$

with $\beta = 1-\gamma_x/2-\gamma_y/4$ and $\xi = \gamma_x/2-\gamma_y/4$. The next step is to incorporate spatial fractional derivatives, i.e., $\mu\neq 2$, in our analysis. For these cases, as we shall show below, it is possible to obtain the exact solution by using the Green's function approach and the dispersion relation, when it is defined.

7.3.2 Spatial Fractional Derivatives

Let us now consider Eq. (7.51), with $\mu\neq 2$. Applying the previous procedure one can show that the solution has the general form

$$\rho(x,y;t) = \int_{-\infty}^{\infty} d\bar{x}\int_{-\infty}^{\infty} d\bar{y}\widehat{\rho}(\bar{x},\bar{y})\overline{G}(x-\bar{x},y,\bar{y};t),$$

with the Green's function being defined as

$$\begin{aligned}\overline{G}(x,y,\bar{y},t) = &-\frac{\delta(x)}{\sqrt{4\pi D_y t^{\gamma_y}}}\left\{\mathrm{H}^{1,0}_{1,1}\left[\frac{(y-\bar{y})^2}{\sqrt{D_y t^{\gamma_y}}}\left|\begin{matrix}\left(1-\frac{\gamma_y}{2},\frac{\gamma_y}{2}\right)\\(0,1)\end{matrix}\right.\right]\right.\\ &\left.-\mathrm{H}^{1,0}_{1,1}\left[\frac{(|y|+|\bar{y}|)^2}{\sqrt{D_y t^{\gamma_y}}}\left|\begin{matrix}\left(1-\frac{\gamma_y}{2},\frac{\gamma_y}{2}\right)\\(0,1)\end{matrix}\right.\right]\right\}\\ &-\frac{1}{\pi|x|}\int_0^t \frac{d\bar{t}}{(t-\bar{t})\bar{t}^{\frac{\gamma_y}{2}}}\mathrm{H}^{2,1}_{2,3}\left[\frac{2\sqrt{D_y\bar{t}^{\gamma_y}}}{D_x\bar{t}^{\gamma_x}}|x|^{\mu}\left|\begin{matrix}(1,1),(\bar{\beta},\bar{\xi})\\\left(\frac{1}{2},\frac{\mu}{2}\right),(1,1),\left(1,\frac{\mu}{2}\right)\end{matrix}\right.\right]\\ &\times\mathrm{H}^{1,0}_{1,1}\left[\frac{|y|+|\bar{y}|}{\sqrt{D_y(t-\bar{t})^{\gamma_y}}}\left|\begin{matrix}\left(0,\frac{\gamma_y}{2}\right)\\(0,1)\end{matrix}\right.\right], \end{aligned}\tag{7.58}$$

with $\overline{\beta} = 1 - \gamma_y/2$ and $\overline{\xi} = \gamma_x - \gamma_y/2$. This solution is indeed very general; it incorporates both time dependencies in the diffusion coefficients and also fractional spatial derivatives in the diffusion process.

In a few words, a non-Markovian Fokker–Planck equation was analytically solved, taking different and complex scenarios into account and focusing on the comb-like structure. The solutions were obtained by considering the presence of time-dependent diffusion coefficients, fractional spatial derivatives, and a general initial condition, without external forces. In the next section, we add an external force to this system; i.e., we add a drift term to Eq. (7.43) and search for a solution to it satisfying the boundary conditions (7.39) and Eq. (7.40), i.e., the normalisation requirement.

7.3.3 Integer Order: External Forces

In this section, we investigate the effects produced by an external force on the diffusive processes subjected to the structure represented in Fig. (7.11). Specifically, we consider the following Fokker–Planck equation [239]:

$$\frac{\partial}{\partial t}\rho(x,y;t) = D_y\frac{\partial^2}{\partial y^2}\rho(x,y;t) + D_x\delta(y)\left[\frac{\partial^2}{\partial x^2} - \overline{v}_x\frac{\partial}{\partial x}\right]\rho(x,y;t) \\ - \nabla\cdot[\vec{v}\rho(x,y;t)], \tag{7.59}$$

with $\vec{v} = (v_x, v_y)$, where v_x, v_y, and $\overline{v}_x$ are constants. Equation (7.59) is an extension of Eq. (7.43) that incorporates the drift term $\vec{F} = [v_x + \delta(y)\overline{v}_x, v_y]$, which accounts for an external force acting on the system. The boundary and initial conditions used to investigate the solutions of Eq. (7.59) are given by Eq. (7.39); i.e., the backbone and the branches of the comb are not limited, and the solution is normalised. The calculations will show that the presence of the external force changes the diffusive process and may introduce different diffusive regimes in the system, depending on the choice of the parameters v_x, v_y, and $\overline{v}_x$.

Let us first consider that the drift forces act outside the backbone structure, i.e., $v_x \neq 0$, $v_y \neq 0$, with $\overline{v}_x = 0$. Equation (7.59) subjected to these conditions can be rewritten as

$$\frac{\partial}{\partial t}\rho(x,y;t) = D_y\frac{\partial^2}{\partial y^2}\rho(x,y;t) + \delta(y)D_x\frac{\partial^2}{\partial x^2}\rho(x,y;t) - \nabla\cdot\left[\vec{v}\rho(x,y;t)\right]. \tag{7.60}$$

To obtain the solution for Eq. (7.60), we use again the Laplace and Fourier transforms. Indeed, by applying the Laplace transform to the t-variable of Eq. (7.60), we obtain

$$D_y\frac{\partial^2}{\partial y^2}\rho(x,y;s) + \delta(y)D_x\frac{\partial^2}{\partial x^2}\rho(x,y;s) - \nabla\cdot(\overline{v}\rho(x,y;s)) = s\rho(x,y;s) - \widehat{\rho}(x,y). \tag{7.61}$$

This equation can be simplified further by applying the Fourier transform to the x-variable, yielding the differential equation:

$$D_y \frac{\partial^2}{\partial y^2} \rho(k_x, y; s) - \left[s + v_y \frac{\partial}{\partial y} + \delta(y) D_x k_x^2 + i k_x v_x \right] \rho(k_x, y; s) = -\widehat{\rho}(k_x, y), \quad (7.62)$$

which can be solved by using the Green's function approach. The solution of Eq. (7.62) is

$$\rho(k_x, y; s) = -\int_{-\infty}^{\infty} d\bar{y} \widehat{\rho}(k_x, y) G(k_x, y, \bar{y}; s), \quad (7.63)$$

with the Green's function to be obtained from the equation

$$\left[D_y \frac{\partial^2}{\partial y^2} - v_y \frac{\partial}{\partial y} - \delta(y) D_x k_x^2 - i k_x v_x - s \right] G(k_x, y, \bar{y}; s) = \delta(y - \bar{y}), \quad (7.64)$$

subject to the Dirichlet boundary condition, i.e., $G(k_x, \pm\infty, \bar{y}; s) = 0$. Equation (7.64) may also be solved by using the Fourier transform with respect to variable y. The solution of Eq. (7.64) in the Fourier space is

$$G(k_x, k_y, \bar{y}; s) = -\frac{e^{-i k_y \bar{y}}}{D_y k_y^2 + i k_y v_y + i k_x v_x + s} - \frac{D_x k_x^2}{D_y k_y^2 + i k_y v_y + i k_x v_x + s} G(k_x, 0, \bar{y}; s). \quad (7.65)$$

After some calculation, it is possible to show that

$$G(k_x, 0, \bar{y}; s) = -\frac{e^{-\frac{v_y}{2D_y}\bar{y}} e^{-\frac{\sqrt{\beta}}{2D_y}|\bar{y}|}}{\sqrt{\beta} + D_x k_x^2}, \quad (7.66)$$

with $\beta = v_y^2 + 4D_y s + 4iD_y k_x v_x$. By substituting Eq. (7.66) into Eq. (7.65) and performing the inverse of Laplace transform, we obtain

$$G(k_x, y, \bar{y}; t) = -e^{\frac{v_y}{2D_y}(y-\bar{y}) - \frac{v_y^2}{4D_y}t - i k_x v_x t} \left\{ \frac{1}{\sqrt{4\pi D_y t}} \left(e^{-\frac{(|y|+|\bar{y}|)^2}{4D_y t}} - e^{-\frac{(y-\bar{y})^2}{4D_y t}} \right) \right.$$
$$\left. + \frac{1}{2D_y}(|y| + |\bar{y}|) \int_0^t d\bar{t} \frac{e^{-\frac{(|y|+|\bar{y}|)^2}{4D_y(t-\bar{t})}}}{\sqrt{4\pi \bar{t}(t-\bar{t})^3}} \mathrm{E}_{\frac{1}{2},\frac{1}{2}}\left(-\frac{D_x k_x^2}{2\sqrt{D_y}} \sqrt{\bar{t}} \right) \right\}. \quad (7.67)$$

We notice, *en passant*, the presence of the generalised Mittag-Leffler function in the last part of previous equation. Performing the inverse Fourier transform on the x-variable and considering some identities of the H-function of Fox [30, 42] (see also Chapter 1), it is possible to show that the Green's function is

$$G(x,y,\bar{y};t) = -e^{\frac{v_y}{2D_y}(y-\bar{y})}e^{-\frac{v_y^2}{4D_y}t}\overline{G}'(x,y,\bar{y};t), \tag{7.68}$$

where

$$\begin{aligned}\overline{G}'(x,y,\bar{y};t) &= \frac{1}{\sqrt{4\pi D_y t}}\delta(x - v_x t)\left(e^{-\frac{(y-\bar{y})^2}{4D_y t}} - e^{-\frac{(|y|+|\bar{y}|)^2}{4D_y t}}\right)\\ &+ \frac{1}{\sqrt{8D_x\sqrt{D_y}}}\frac{|y|+|\bar{y}|}{\sqrt{4\pi D_y}}\\ &\times \int_0^t d\bar{t}\frac{e^{-\frac{(|y|+|\bar{y}|)^2}{4D_y(t-\bar{t})}}}{[(t-\bar{t})\bar{t}^{\frac{1}{2}}]^{\frac{3}{2}}}\mathrm{H}_{1,1}^{1,0}\left[\sqrt{\frac{2}{D_x}}\sqrt{\frac{D_y}{\bar{t}}}\,|x - v_x\bar{t}|\left|\begin{matrix}\left(\frac{1}{4},\frac{1}{4}\right)\\(0,1)\end{matrix}\right.\right].\end{aligned} \tag{7.69}$$

We now address our attention to the relaxation of the system in order to characterise the effect produced by the drift forces. To perform this analysis, we consider the mean square displacement for the x- and y-directions, for simplicity, by taking into account the initial condition $\hat{\rho}(x,y) = \delta(x)\delta(y-\tilde{y})$. After some calculation, we obtain

$$\begin{aligned}\sigma_x^2(t) &= \langle (x - \langle x\rangle)^2\rangle\\ &= 2D_x\, e^{-\frac{v_y}{2D_y}\tilde{y}}\int_0^t \frac{d\tau}{\sqrt{4\pi D_y\tau}}e^{-\frac{v_y^2}{4D_y}\tau}e^{-\frac{\tilde{y}^2}{4D_y\tau}}\end{aligned} \tag{7.70}$$

and $\sigma_y^2(t) = 2D_y t$. Equation (7.70) shows that the drift force v_x does not have influence on the backbone structure and that, for $v_y = 0$, we reobtain the mean square displacement presented in Ref. [224]. For $v_y \neq 0$, we observe a kind of confined diffusive regime for long times, also called a saturation regime, as shown in Fig. 7.12. This kind of behaviour has been reported in Brownian dynamics simulations of a single polymer [244], in time series of continuous-time random walk and fractional Brownian motion [245], and also in living cells [246–249], where the crowded environment of the cytoplasm and constrained diffusion are possible physical mechanisms of the anomalous diffusion. Moreover, once the comb model was proposed to mimic percolation-like structures, it is remarkable that the behaviour of spreading obtained here is in good agreement with the diffusion reported in percolation clusters below the criticality, i.e., $p < p_c$ (where p_c is the critical probability threshold of percolation transition) and the clusters are considered finite [250]. Hence, due to the constant force acting in the y-axis the system remains confined in the branches; i.e., the particles fall into a trap of the labyrinth and do not return to the backbone. In this manner, we verify that the drift term in the y-direction changes the spreading of the system in the backbone and leads it to a stationary solution in the x-direction.

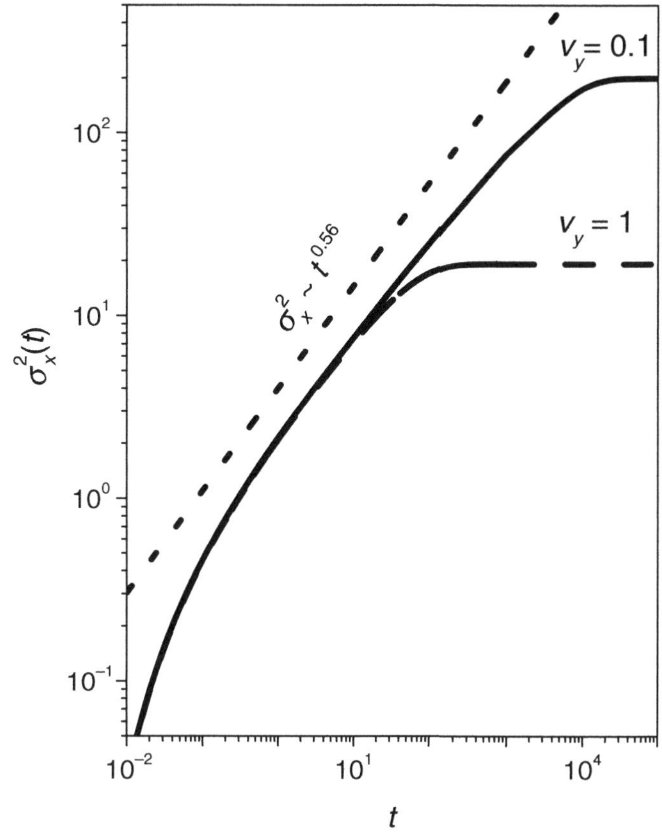

Figure 7.12 The mean square displacement with $v_x = 0$, $D_y = 20$, $D_x = 10$, and $\tilde{y} = 0.8$. The solid line corresponds to the case $v_y = 0.1$ and the dashed line to the case $v_y = 1$. The saturation in the x-direction is reached faster for large values of v_y and slower for small values of v_y. The dotted line was incorporated in order to show the subdiffusive behaviour presented by the system before reaching the stationary state [239]. Modified from E. K. Lenzi, L. R. da Silva, A. A. Tateishi, M. K. Lenzi, and H. V. Ribeiro, Physical Review E, **87**, 012121 (2013). Copyright (2013) by the American Physical Society.

Let us now incorporate a drift force acting on the backbone structure, i.e., $\overline{v}_x \neq 0$, in the x-direction; i.e., we now search for solutions for $\overline{v}_x \neq 0$, $v_y \neq 0$, and $v_x \neq 0$. In this case, Eq. (7.59) is rewritten as

$$\frac{\partial}{\partial t}\rho(x,y;t) = D_y \frac{\partial^2}{\partial y^2}\rho(x,y;t) + \delta(y)\left[D_x \frac{\partial^2}{\partial x^2} - \overline{v}_x \frac{\partial}{\partial x}\right]\rho(x,y;t) \\ - v_y \frac{\partial}{\partial y}\rho(x,y;t) - v_x \frac{\partial}{\partial x}\rho(x,y;t). \tag{7.71}$$

To solve this equation, it is possible to employ the same procedure used to obtain the solution of Eq. (7.60). Thus, we start by applying the Laplace transform to the

t-variable and the Fourier transform to the x-variable in Eq. (7.71). We obtain the following equation:

$$D_y \frac{\partial^2}{\partial y^2} \rho(k_x, y; s) - h[k_x, y; s] \rho(k_x, y; s) = -\widehat{\rho}(k_x, y), \tag{7.72}$$

with the solution given by Eq. (7.63) and the Green's function governed by the equation:

$$D_y \frac{\partial^2}{\partial y^2} G(k_x, y; s) - h[k_x, y; s] G(k_x, y, \bar{y}; s) = \delta\left(y - \bar{y}\right), \tag{7.73}$$

where

$$h[k_x, y; s] = s + v_y \frac{\partial}{\partial y} + \delta(y) \left(D_x k_x^2 + i k_x \bar{v}_x\right) + i k_x v_x,$$

subject to the boundary conditions $G(k_x, \pm\infty, \bar{y}; s) = 0$. By now using the Fourier transform with respect to variable y and performing some calculations, it is possible to show that the solution of Eq. (7.73) in the Fourier space is

$$\begin{aligned} G(k_x, k_y, \bar{y}; s) = &- \frac{e^{-i k_y \bar{y}}}{D_y k_y^2 + i k_y v_y + i k_x v_x + s} \\ &- \frac{D_x k_x^2 + i k_x \bar{v}_x}{D_y k_y^2 + i k_y v_y + i k_x v_x + s} G(k_x, 0, \bar{y}; s), \end{aligned} \tag{7.74}$$

with

$$G(k_x, 0, \bar{y}; s) = -\frac{e^{-\frac{v_y}{2D_y}\bar{y}} e^{-\frac{\sqrt{\beta}}{2D_y}|\bar{y}|}}{\sqrt{\beta} + D_x k_x^2 + i k_x \bar{v}_x}. \tag{7.75}$$

Applying the inverses of Laplace and Fourier transforms to the previous equations, we obtain

$$G(x, y, \bar{y}; t) = -e^{-\frac{v_y^2}{4D_y} t} e^{\frac{v_y}{2D_y}(y - \bar{y})} \widetilde{G}(x, y, \bar{y}; t), \tag{7.76}$$

where

$$\begin{aligned} \widetilde{G}(x, y, \bar{y}; t) = &\frac{1}{\sqrt{4\pi D_y t}} \delta(x - v_x t) \left[e^{-\frac{(y-\bar{y})^2}{4D_y t}} - e^{-\frac{(|y| + |\bar{y}|)^2}{4D_y t}} \right] \\ &+ \frac{1}{t} \int_0^\infty du \left(|y| + |\bar{y}| + 2D_y u\right) G_y\left(|y|, |\bar{y}|, 2D_y u; u\right) \\ &\times G_x\left(x, -\bar{v}_x u, -v_x t; t\right) \end{aligned} \tag{7.77}$$

and

$$G_\alpha(x,y,z;u) = \frac{1}{\sqrt{4\pi D_\alpha u}} e^{-\frac{1}{4D_\alpha u}(x+y+z)^2}.$$

As we have done in the previous case, the next step is to analyse the behaviour of the mean square displacement in the x- and y-directions to investigate the effect of the external force on the relaxation of the system. We obtain

$$\langle x \rangle = v_x\, t + \frac{\overline{v}_x\, e^{-\frac{v_y}{2D_y}\tilde{y}}}{\sqrt{4\pi D_y}} \int_0^t \frac{d\bar{t}}{\sqrt{\bar{t}}} e^{-\frac{v_y^2}{4D_y}\bar{t}} e^{-\frac{\tilde{y}^2}{4D_y \bar{t}}}, \tag{7.78}$$

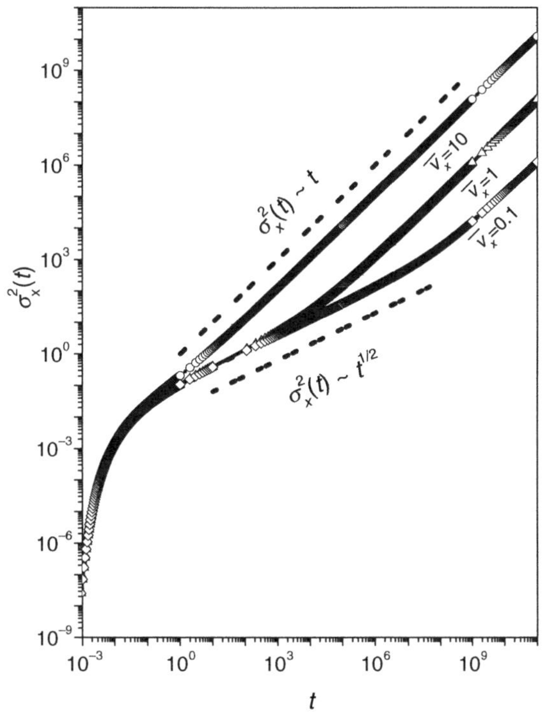

Figure 7.13 The mean square displacement for $v_y = 0$, $D_y = 20$, $D_x = 10$, and $\tilde{y} = 0.8$. The straight dotted lines were added to the figure in order to show the usual and the subdiffusive behaviour which can be manifested by the mean square displacement obtained from Eqs. (7.78) and (7.79). The circle, triangle, and diamond symbols correspond to the cases $\overline{v}_x = 10$, $\overline{v}_x = 1$, and $\overline{v}_x = 0.1$, respectively [239]. Modified from E. K. Lenzi, L. R. da Silva, A. A. Tateishi, M. K. Lenzi, and H. V. Ribeiro, Physical Review E, **87**, 012121 (2013). Copyright (2013) by the American Physical Society.

and

$$\langle x^2 \rangle = v_x^2 t^2 + \int_0^t \frac{d\bar{t}}{\sqrt{\pi D_y \bar{t}}} \left(2 v_x \bar{v}_x (t - \bar{t}) + D_x\right) e^{-\frac{(\tilde{y}+v_y\bar{t})^2}{4D_y\bar{t}}}$$

$$+ \frac{\bar{v}_x^2}{2D_y} \int_0^t d\bar{t} e^{-\frac{v_y}{2D_y}\tilde{y}} e^{-\frac{v_y^2}{4D_y}\bar{t}} \text{rfc}\left(-\frac{|\tilde{y}|}{2\sqrt{D_y \bar{t}}}\right). \tag{7.79}$$

Figure 7.13 illustrates the behaviour of the mean square displacement obtained for the previous equations by considering $v_y = 0$ and $v_x = 0$. A remarkable feature is the presence of two different diffusive regimes after an initial transient. One of these regimes is subdiffusive and the other one is usual. The presence of the subdiffusive regime depends on the values of $\bar{v}_x$; e.g., $\bar{v}_x \gg 1$ yields an usual behaviour, while $\bar{v}_x \ll 1$ yields a subdiffusive behaviour. In addition, the same dynamical crossover between subdiffusion and normal diffusion is reported in

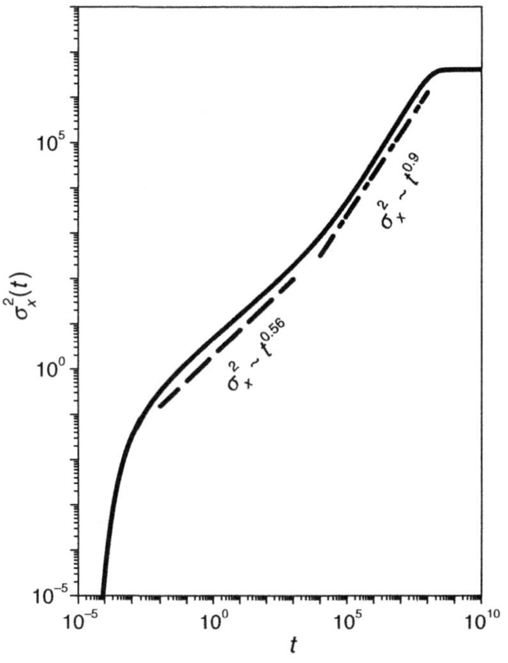

Figure 7.14 The time behaviour of the mean square displacement for $v_y = 5\ 10^{-4}$, $v_x = 0, \bar{v}_x = 1, D_y = 5, D_x = 10$ and $\tilde{y} = 0.1$. The dashed and dashed-dotted lines were incorporated to show the subdiffusive behaviours presented for the system before getting the stationary state [239]. Modified from E. K. Lenzi, L. R. da Silva, A. A. Tateishi, M. K. Lenzi, and H. V. Ribeiro, Physical Review E, **87**, 012121 (2013). Copyright (2013) by the American Physical Society.

Ref. [251], where the authors studied the random walk in a comb lattice by numerical simulation. This kind of crossover is also found in polymer physics [252]: the dynamics of a tagged monomer in a polymer must be anomalous until the terminal relaxation time, and this anomalous dynamics is connected with the mean relaxation response of the polymers to local strains [253]. In particular, for phantom Rouse polymers the mean square displacement of a tagged monomer behaves as $t^{1/2}$ until the terminal relaxation time τ, and only after that time does the dynamics of the polymer become diffusive [254]. Moreover, in the theory of percolation clusters the diffusion is usual when the percolation happens, i.e., $p > p_c$, and the clusters are considered infinite [250].

In Fig. 7.14, we incorporate the drift term in the y-direction, i.e., $v_y \neq 0$, and a stationary behaviour is obtained for the distribution in the x-direction, since $\sigma_x^2(t)$ is constant for long times as in the previous case dealing with the drift term outside the backbone structure.

8
Fractional Schrödinger Equations

In this chapter, we investigate the solutions to a Schrödinger equation with nonlocal terms and the connections with the anomalous diffusion phenomenon. We consider first a usual one-dimensional equation, with a time-dependent potential in the nonlocal term, showing anomalous spreading of the solution and a non-Markovian behaviour. Extensions of the Schrödinger equation encompassing fractional derivatives in the spatial and temporal variables are an elegant way to tackle nonlocal and non-Markovian effects. For an arbitrary initial condition, the time-dependent solutions of a fractional Schrödinger equation in the presence of delta potentials are also determined. We use the continuous-time random walk (CTRW) approach to obtain the fractional Schrödinger equation accounting for nonlocal effects. This formulation is applied to the problem of a free particle in the half-space. Time-dependent solutions are also found for Schrödinger-like equations with memory kernels, distributed order memory kernels, and nonlocal terms. It is shown that the results may be transformed into those for the probability distribution function of a diffusion-like equation with memory kernel. Finally, effects of the spatial fractional derivatives on the energy spectra are discussed in connection with their potential relevance for the description of thermal properties of glassy systems at very low temperature.

8.1 The Schrödinger Equation and Anomalous Behaviour

Let us start our discussion by reviewing some formal aspects of the time-dependent Schrödinger equation. It is defined as:

$$\imath\hbar\frac{\partial}{\partial t}\Psi = \hat{H}\Psi, \tag{8.1}$$

where $\Psi(\mathbf{r}, t)$ is the wave function, $\hbar = h/(2\pi)$ is the Planck constant, and $\hat{H}$ is the Hamiltonian which describes the time evolution of a quantum-mechanical system. It has the form

$$\hat{H}\Psi(\mathbf{r},t) = -\frac{\hbar^2}{2M}\nabla^2\Psi(\mathbf{r},t) + V(\mathbf{r})\Psi(\mathbf{r},t), \tag{8.2}$$

where M is the effective mass and $V(\mathbf{r})$ is the potential energy function, for simplicity, considered here as time independent. To obtain the time-independent Schrödinger equation and consequently to determine the eigenvalues E and eigenfunctions Ψ, according to the eigenvalue equation $\hat{H}\Psi = E\Psi$, we may employ the separation of variables ansatz $\Psi(\mathbf{r},t) = \Phi(\mathbf{r})T(t)$ in the time-dependent Schrödinger equation. We obtain

$$-\frac{\hbar^2}{2M}\nabla^2\Phi(\mathbf{r}) + V(\mathbf{r})\Phi(\mathbf{r}) = E\Phi(\mathbf{r}). \tag{8.3}$$

The time-dependent part of the solution is $T(t) = e^{-\imath\frac{E}{\hbar}t}$. Thus, a plane wave form

$$\Psi(\mathbf{r},t) = \Phi(\mathbf{r})e^{-\imath\frac{E}{\hbar}t}$$

is also obtained, and the probability distribution function $|\Psi(\mathbf{r},t)|^2$ is time independent, since

$$|\Psi(\mathbf{r},t)|^2 = \Psi(\mathbf{r},t)\Psi^*(\mathbf{r},t) = |\Phi(\mathbf{r})|^2.$$

The solution of Eq. (8.1) for an arbitrary initial condition can be represented as a linear combination of the particular solutions, i.e.,

$$\Psi(\mathbf{r},t) = \sum_n c_n\Phi_n(\mathbf{r})e^{-\imath\frac{E_n}{\hbar}t}.$$

The transition from the Schrödinger equation to the diffusion equation $\partial_t\Psi(x,t) = D\nabla^2\Psi(x,t)$ may be done by using an imaginary diffusion coefficient $D \to \imath\hbar/(2M)$ [255], or by analytical continuation $\imath t \to t$, where $D \to \hbar/(2M)$ is the diffusion coefficient [256]. An effective Hamiltonian taking into account the interaction between components of the system may be considered [257] and a nonlocal term of the form

$$\int_0^t d\tau \int_{-\infty}^{\infty} d\xi U(x-\xi,t-\tau)\Psi(x,\tau)$$

may be added to the one-dimensional Hamiltonian leading to the Schrödinger equation [258], i.e.,

$$\hat{H}\Psi(x,t) = -\frac{\hbar^2}{2M}\nabla^2\Psi(x,t) + V(x,t)\Psi(x,t) + \int_0^t d\tau \int_{-\infty}^{\infty} d\xi\, U(x-\xi,t-\tau)\Psi(x,\tau). \tag{8.4}$$

Similar nonlocal terms have been added to a diffusion equation showing that the model can be used to generate various anomalous diffusive behaviours [259].

We focus our attention on the following one-dimensional Schrödinger equation:

$$i\hbar\frac{\partial}{\partial t}\Psi(x,t) = -\frac{\hbar^2}{2m}\frac{\partial^2}{\partial x^2}\Psi(x,t) + \int_0^t d\bar{t}\int_{-\infty}^{\infty} d\bar{x}\, U(x-\bar{x}, t-\bar{t})\Psi(\bar{x},\bar{t}). \tag{8.5}$$

The potential in the nonlocal part may also be related to the fractional derivatives depending on the functional form of $U(x,t)$, and is such that for $U(x,t) = 0$ the usual Schrödinger equation for the free particle may be reobtained. The behaviour of the solution, governed by the kinetic term and the nonlocal potential, may present different spreading regimes depending on the choice performed for $U(x,t)$. One of these is dominated by a Gaussian-like behaviour due to the kinetic term and the other one may be directly dependent on $U(x,t)$. In particular, for

$$U(x,t) \propto \frac{\delta(t)}{|x|^{1+\alpha}}, \qquad 0 < \alpha < 1, \tag{8.6}$$

the solution is asymptotically governed by a power-law behaviour for long times, resembling the Lévy distribution. As we have discussed before, situations characterised by two regimes have been reported in several contexts as, for instance, long-range interaction systems [149] and enhanced diffusion in active intracellular transport [147, 148], among others.

Another remarkable feature of this formalism concerns the continuity equation satisfied by Eq. (8.5), when $U(x,t) = U^*(x,t)$, which has the form [260]

$$i\hbar\frac{\partial}{\partial t}P(x,t) = -\frac{\hbar^2}{2m}\frac{\partial}{\partial x}J(x,t) + \int_0^t d\bar{t}\int_{-\infty}^{\infty} d\bar{x} U(x-\bar{x}, t-\bar{t})\, K(x,\bar{x};t,\bar{t}), \tag{8.7}$$

where

$$P(x,t) = \Psi(x,t)\Psi^*(x,t),$$

$$J(x,t) = \Psi^*(x,t)\frac{\partial}{\partial x}\Psi(x,t) - \Psi(x,t)\frac{\partial}{\partial x}\Psi^*(x,t),$$

and

$$K(x,\bar{x};t,\bar{t}) = \Psi(\bar{x},\bar{t})\Psi^*(x,t) - \Psi(x,t)\Psi^*(\bar{x},\bar{t}).$$

For $U(x,t) = \delta(t)\overline{U}(x)$, with the function $\overline{U}(x)$ real and symmetric, it can be shown that $\int_{-\infty}^{\infty} dx P(x,t)$ is a constant, if the current density satisfies the boundary condition $J(\pm\infty,t) = 0$. This result obtained from Eq. (8.7) implies that this form of $U(x,t)$ keeps the probability constant, i.e., time independent, as verified for the usual case. Other functional forms of $U(x,t)$ may not fulfil this requirement, which is crucial to preserve the probabilistic interpretation of the wave function $\Psi(x,t)$.

Let us investigate the solutions of Eq. (8.5) by considering specific forms of the function $U(x,t)$ in the nonlocal term. We analyse three particular cases. The first one is given by Eq. (8.6) which, in the Fourier–Laplace space, assumes the form

$$U(k,s) = \widetilde{U}|k|^{\alpha}, \tag{8.8}$$

where $\widetilde{U}$ is a constant. This potential has a long-tailed behaviour resembling the asymptotic behaviour of the Lévy distribution. It is responsible for the existence of two different regimes in the solution; these can be verified by analysing the Green's function for short and long times, as we will show later on. The second case,

$$U(x,t) \propto \delta(t)\overline{U}(x), \tag{8.9}$$

is relevant to the investigation of the general solution for a nonlocal potential with an arbitrary dependence on the spatial function. The third case is

$$U(x,t) \propto \frac{t^{\gamma-1}}{|x|^{1+\alpha}}, \qquad 0 < \alpha < 1, \tag{8.10}$$

which in the Fourier–Laplace space becomes

$$U(k,s) = \widehat{U}s^{-\gamma}|k|^{\alpha}, \tag{8.11}$$

where $\widehat{U}$ is a constant. This potential incorporates a time dependence in the nonlocal potential, discussed in the first case, which may be related to a fractional derivative for a suitable value of γ.

To proceed, we apply the Fourier transform to Eq. (8.5) in order to obtain the integral equation

$$i\hbar\frac{\partial}{\partial t}\Psi(k,t) = \frac{\hbar^2k^2}{2m}\Psi(k,t) + \int_0^t d\bar{t}\, U(k,t-\bar{t})\Psi(k,\bar{t}). \tag{8.12}$$

This integral equation may also be simplified by using the Laplace transform, leading to the following algebraic equation:

$$i\hbar\left[s\Psi(k,s) - \Psi(k,0)\right] = \frac{\hbar^2k^2}{2m}\Psi(k,s) + U(k,s)\Psi(k,s), \tag{8.13}$$

whose solution is given by

$$\Psi(k,s) = \Psi(k,0)G(k,s), \tag{8.14a}$$

where

$$G(k,s) = \frac{1}{s + i\hbar k^2/(2m) + iU(k,s)/\hbar}. \tag{8.14b}$$

In Eqs. (8.14a) and (8.14b), $\Psi(k,0)$ is the Fourier transform of the initial condition and $G(k,s)$ is the Green's function of Eq. (8.5) in the Fourier–Laplace space. We notice that Eq. (8.14b) reduces to the usual case, i.e., the Green's function of the free particle case, if we put $U(k,s)=0$. Using (8.6) and (8.8), we obtain

$$\Psi(k,s)=\frac{\Psi(k,0)}{s+i\hbar k^2/(2m)+i\widetilde{U}/\hbar|k|^\alpha}. \tag{8.15}$$

The inverse Laplace transform of Eq. (8.15) is

$$\Psi(k,t)=\Psi(k,0)\exp\left[-\frac{it}{\hbar}\left(\frac{\hbar^2k^2}{2m}+\widetilde{U}|k|^\alpha\right)\right]. \tag{8.16}$$

Finally, the inverse Fourier transform of Eq. (8.16) yields, in terms of the H-function of Fox, the formal solution of the problem in the form:

$$\Psi(x,t)=\int_{-\infty}^{\infty}d\bar{x}\,G(x-\bar{x},t)\Psi(\bar{x},0), \tag{8.17a}$$

with

$$G(x,t)=\frac{1}{2|x|}\sum_{n=0}^{\infty}\frac{\tilde{A}_n(t)}{n!}\mathrm{H}^{1,1}_{2,2}\left[\sqrt{\frac{2m}{ith}}|x|\left|\begin{matrix}\left(1-\frac{n\alpha}{2},\frac{1}{2}\right),\left(1,\frac{1}{2}\right)\\(1,1),\left(1,\frac{1}{2}\right)\end{matrix}\right.\right], \tag{8.17b}$$

where

$$\tilde{A}_n(t)=\left[\frac{\widetilde{U}t}{i\hbar}\left(\frac{2m}{ith}\right)^{\frac{\alpha}{2}}\right]^n.$$

In Fig. 8.1, some examples of the behaviour of $|\Psi(x,t)|^2$ versus x are shown, for typical values of t. We notice an anomalous spreading of the initial condition, i.e., a delocalisation of the initial condition (Gaussian packet) as the time evolves.

Figure 8.2 illustrates the behaviour of $|\Psi(x,t)|^2$ versus x for different values of α. The presence of H-function of Fox in the solution is evidence that the external potential in Eq. (8.5) produces an anomalous spreading of the solution. A remarkable feature of the solution is the existence of two regimes; one of them is characterised by the usual behaviour (i.e., Gaussian), whereas the other is dominated asymptotically by a power-law behaviour. It is possible to verify from Eq. (8.17b) that, for short times,

$$G(x,t)\sim\sqrt{\frac{m}{2\pi ith}}\,e^{\frac{imx^2}{2th}}+\cdots, \tag{8.18}$$

whereas for long times

$$G(x,t)\sim\frac{1}{\alpha|x|}\mathrm{H}^{1,1}_{2,2}\left[\left(\frac{\hbar}{i\widetilde{U}t}\right)^{\frac{1}{\alpha}}|x|\left|\begin{matrix}\left(1,\frac{1}{\alpha}\right),\left(1,\frac{1}{2}\right)\\(1,1),\left(1,\frac{1}{2}\right)\end{matrix}\right.\right]+\cdots, \tag{8.19}$$

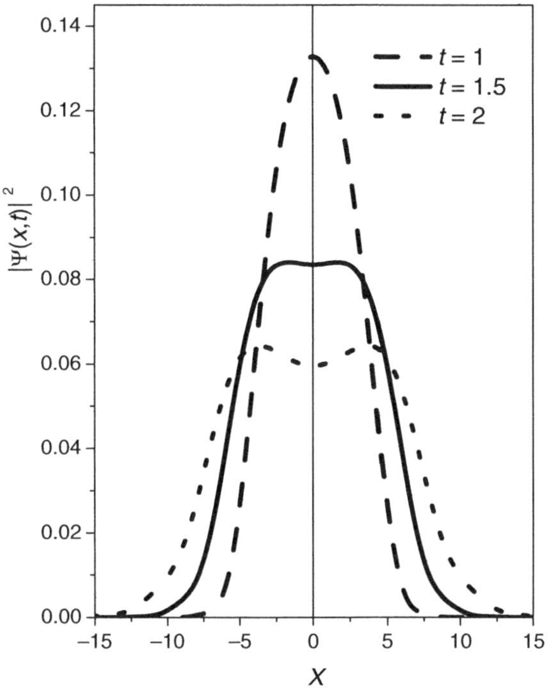

Figure 8.1 $|\Psi(x,t)|^2$ versus x for $\alpha = 3/2$, $\hbar = 1$, $m = 1$, $U = 1$, and the initial condition $\Psi(x,0) = e^{-x^2/2}/\pi^{1/4}$ (Gaussian packet) [260]. Modified from E. K. Lenzi, B. F. de Oliveira, L. R. da Silva, and L. R. Evangelista, Solutions for a Schrödinger equation with a nonlocal term, Journal of Mathematical Physics **49**, 032108 (2008), with the permission of AIP Publishing.

which resembles the form of the Lévy distribution found in anomalous diffusion processes, as mentioned before. The asymptotic behaviour of Eq. (8.19) in the limit of $|x| \to \infty$ is given by

$$G(x,t) \sim i\frac{\widetilde{U}t}{\alpha\hbar|x|^{1+\alpha}}.$$

To proceed further, we consider the second case, characterised by Eq. (8.9), with $\bar{U}(x)$ arbitrary and symmetric to preserve the probabilistic interpretation of the solution, i.e., to assure that $\int_{-\infty}^{\infty} dxP(x,t)$ is constant. By using Eq. (8.14a), with $U(k,s) = \bar{U}(k)$, and applying the inverse Laplace transform, we obtain

$$\Psi(k,t) = \Psi(k,0)\exp\left[-\frac{it}{\hbar}\left(\frac{\hbar^2k^2}{2m} + \bar{U}(k)\right)\right], \tag{8.20}$$

as expected. Employing the inverse Fourier transform and the convolution theorem to Eq. (8.20), the solution is

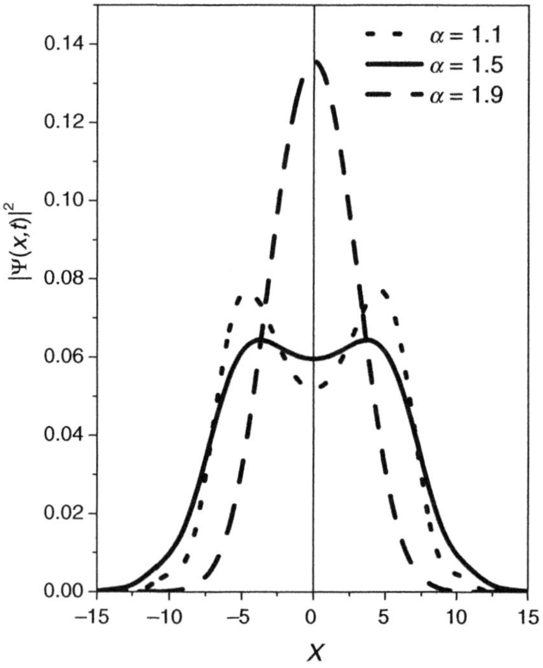

Figure 8.2 $|\Psi(x,t)|^2$ versus x for different values of α by considering, for illustrative purposes, $t = 2$, $\hbar = 1$, $m = 1$, $U = 1$, and the initial condition $\Psi(x,0) = e^{-x^2/2}/\pi^{1/4}$ [260]. Modified from E. K. Lenzi, B. F. de Oliveira, L. R. da Silva, and L. R. Evangelista, Solutions for a Schrödinger equation with a nonlocal term, Journal of Mathematical Physics **49**, 032108 (2008), with the permission of AIP Publishing.

$$\Psi(x,t) = \int_{-\infty}^{\infty} d\bar{x}\Psi(\bar{x},0)G(x-\bar{x},t), \tag{8.21a}$$

where

$$G(x,t) = \bar{G}(x,t) + \sum_{n=1}^{\infty}\frac{1}{n!}\left(-i\frac{t}{\hbar}\right)^n \int_{-\infty}^{\infty} dx_n\bar{U}(x-x_n)\cdots$$
$$\times \int_{-\infty}^{\infty} dx_1\bar{U}(x_2-x_1)\bar{G}(x_1,t), \tag{8.21b}$$

in which $\Psi(x,0)$ is the initial condition and

$$\bar{G}(x,t) = \sqrt{\frac{m}{2\pi i t\hbar}}\, e^{\frac{imx^2}{2t\hbar}} \tag{8.22}$$

is the usual Green's function for the free particle case.

Let us now incorporate a time dependence in $U(x,t)$ defined by Eq. (8.10). In this case, Eq. (8.14a) becomes

$$\Psi(k,s) = \frac{\Psi(k,0)}{s + i\hbar k^2/(2m) + i(\widehat{U}/\hbar)s^{-\gamma}|k|^{\alpha}}. \tag{8.23}$$

We underline that Eq. (8.23) reduces to the previous case, when $\gamma = 0$, and to the usual one, when $\widehat{U} = 0$. This time dependence of the nonlocal term introduces a memory effect in a way similar to what happens with fractional diffusion equations. Thus, the solution obtained for this case (and for the previous one) has a non-Markovian character, in contrast to what happens in the usual case.

To find the solution for the potential represented by Eq. (8.10), we start by using the inverse Laplace transform. After some calculation it is possible to show that (see also Refs. [261, 262])

$$\Psi(k,t) = \Psi(k,0)\sum_{n=0}^{\infty}\frac{1}{n!}\left(-i\frac{\widehat{U}}{\hbar}t^{1+\gamma}|k|^{\alpha}\right)^{n} \mathrm{E}^{(n)}_{1,1+n\gamma}\left(-i\frac{\hbar k^2}{2m}t\right), \tag{8.24}$$

where $E^{(n)}_{\alpha,\beta}(x)$ is the nth-derivative of the generalised Mittag-Leffler function. The inverse Fourier transform of Eq. (8.24) is

$$\Psi(x,t) = \int_{-\infty}^{\infty} d\overline{x}\Psi(\overline{x},0)G(x-\overline{x},t), \tag{8.25a}$$

where

$$G(x,t) = \frac{1}{2|x|}\sum_{n=0}^{\infty}\frac{t^n}{n!}\widehat{A}_n(t)\mathrm{H}^{2,1}_{3,3}\left[\sqrt{\frac{2m}{ith}}|x|\left|\begin{matrix}\left(1-\frac{n\alpha}{2},\frac{1}{2}\right),\left(1+\left(1+\gamma-\frac{\alpha}{2}\right)n,\frac{1}{2}\right),\left(1,\frac{1}{2}\right)\\(1,1),\left(1-\left(1-\frac{\alpha}{2}\right)n,\frac{1}{2}\right),\left(1,\frac{1}{2}\right)\end{matrix}\right.\right], \tag{8.25b}$$

in which

$$\widehat{A}_n(t) = \left[\frac{\widehat{U}}{i\hbar}\left(\frac{2m}{ith}\right)^{\frac{\alpha}{2}}t^{\gamma}\right]^n.$$

The general solution that takes into account an arbitrary form of the nonlocal potential can be found by using the inverse Laplace transform and its corresponding convolution theorem. Some routine calculations yield

$$\Psi(x,t) = \int_{-\infty}^{\infty} d\overline{x}\Psi(\overline{x},0)G(x-\overline{x},t), \tag{8.26a}$$

where

$$G(x,t) = \overline{G}(x,t) + \sum_{n=1}^{\infty} \frac{1}{n!}\left(-\frac{i}{\hbar}\right)^n \int_{-\infty}^{\infty} dx_n \int_0^t dt_n \overline{U}(x - x_n, t - t_n) \cdots$$

$$\times \int_{-\infty}^{\infty} dx_1 \int_0^{t_2} dt_1 t_1^n \overline{U}(x_2 - x_1, t_2 - t_1)\overline{G}(x_1, t_1), \tag{8.26b}$$

with $\Psi(x,0)$ being the initial condition and $\overline{G}(x,t)$ given by Eq. (8.22). Equation (8.26b) generalises Eq. (8.21b) and may be used to obtain approximated solutions when the nonlocal term can be considered as a perturbation to the usual solution.

8.2 Time-Dependent Solutions

In this section, we investigate, for an arbitrary initial condition, the time-dependent solutions of a fractional Schrödinger equation, in the presence of delta potentials, by again using the Green's function approach. This problem may be relevant because the solutions show an anomalous spreading asymptotically characterised by a power-law behaviour which, in turn, is governed by the order of the fractional spatial operator in the Schrödinger equation. The key point is again to extend the Schrödinger equation in order to incorporate nonlocal and non-Markovian characteristics [263–266]. This has been also done by using an extension of the Feynman path integral [267] and in several other contexts [258, 260, 268–275]. In the light of these approaches, here we discuss a problem to obtain time-dependent solutions for a fractional Schrödinger equation in the simplified case of a potential in the form of a delta function.

The investigation starts by considering the one-dimensional fractional Schrödinger equation [265, 266] with $V(x) = U\delta(x)$, where U is a constant, and subjected to the boundary condition $\Psi(\pm\infty, t) = 0$ and to an initial condition $\Psi(x,0) = \Phi(x)$, where $\Phi(x)$ is an arbitrary function. For this case, the fractional Schrödinger equation can be written as [276]

$$i\hbar\frac{\partial}{\partial t}\Psi(x,t) = D_\mu \left(-\hbar^2\nabla^2\right)^{\frac{\mu}{2}} \Psi(x,t) + U\delta(x)\Psi(x,t), \tag{8.27}$$

where D_μ is a constant, and the spatial operator in the Fourier space, such that

$$\mathcal{F}\{\Psi(x,t)\} = \int_{-\infty}^{\infty} dx\, e^{-i\frac{p}{\hbar}x}\Psi(x,t) = \overline{\Psi}(p,t)$$

and

$$\mathcal{F}^{-1}\{\overline{\Psi}(p,t)\} = \frac{1}{2\pi\hbar}\int_{-\infty}^{\infty} dp\, e^{i\frac{p}{\hbar}x}\overline{\Psi}(p,t) = \Psi(x,t),$$

may be written as

$$\left(-\hbar^2\nabla^2\right)^{\frac{\mu}{2}}\Psi(x,t) = \frac{1}{2\pi\hbar}\int_{-\infty}^{\infty} dp e^{-i\frac{p}{\hbar}x}|p|^{\mu}\overline{\Psi}(p,t), \tag{8.28}$$

which corresponds to Riesz–Weyl operator introduced in Section 2.4. The usual Laplacian operator is reobtained for $\mu = 2$, i.e., by choosing $D_{\mu=2} = \hbar^2/(2m)$. The discussion involving the presence of the fractional time derivative will be performed later on, after the development of the solution for the potential formed by two delta functions.

The solutions and properties of Eq. (8.27) have already been investigated in different contexts [272, 277, 278]. Here, we focus our attention on the time-dependent solutions and, consequently, on the influence of the spatial fractional derivative on the spreading of the solution, obtained with the Green's function approach. In this framework, the solution of Eq. (8.27) can be written as

$$\Psi(x,t) = \int_0^t dt' \int_{-\infty}^{\infty} dx' G(x,x';t,t')\Phi(x), \tag{8.29}$$

with the Green's function being governed by the following equation:

$$i\hbar\frac{\partial}{\partial t}G(x,x';t,t') - \mathcal{H}G(x,x';t,t') = i\hbar\delta(x-x')\delta(t-t'), \tag{8.30}$$

where

$$\mathcal{H}G(x,x';t,t') = D_\mu\left(-\hbar^2\nabla^2\right)^{\frac{\mu}{2}}G(x,x';t,t') + U\delta(x)G(x,x';t,t'), \tag{8.31}$$

with $G(\pm\infty,x';t,t') = 0$ and $G(x,x';t,t') = 0$, for $t < t'$ (causality condition). Applying the Laplace and Fourier transforms to Eq. (8.30), we obtain

$$\widetilde{\overline{G}}(p,x';s,t') = \widetilde{\overline{G}}_f(p,s)e^{-i\frac{p}{\hbar}x'}e^{-st'} + U\widetilde{\overline{G}}_f(p,s)\widetilde{G}(0,x';s,t'), \tag{8.32}$$

where

$$\widetilde{\overline{G}}_f(p,s) = \frac{1}{s + iD_\mu|p|^\mu/\hbar}. \tag{8.33}$$

Equation (8.33) corresponds to the Green's function of the cases for which the fractional Schrödinger equation is considered in absence of a potential term, i.e., $U(x,t) = 0$, in a unbounded region, and for $\mu = 2$ it reduces to the usual form of the propagator in the Laplace–Fourier space [263–266]. After some calculation, it is possible to show that

$$\widetilde{G}(0,x';s,t') = \frac{e^{-st'}}{1 - U\widetilde{G}_f(0,s)}\widetilde{G}_f(x',s) \tag{8.34a}$$

and

$$\widetilde{G}_f(0,s) = \frac{1}{\mu}\left(\frac{\hbar}{iD_\mu}\right)^{\frac{1}{\mu}} \frac{s^{\frac{1}{\mu}-1}}{\sin(\pi/\mu)}. \tag{8.34b}$$

By substituting Eqs. (8.33), (8.34a), and (8.34b) into Eq. (8.32), and performing the inverse Laplace and Fourier transforms, we obtain

$$G(x,x';t,t') = G_f(x-x',t)\theta(t-t')U\theta(t-t')\int_0^t d\eta G_f(x,t-\eta)$$
$$\times\left[G_f(x',\eta)\int_0^\eta d\xi\,\Delta_U(\eta-\xi)G_f(x',\xi)\right], \tag{8.35}$$

with

$$G_f(x,t) = \frac{1}{\mu|x|}\mathrm{H}_{2,2}^{1,1}\left[\frac{|x|}{(D_\mu it/\hbar)^{\frac{1}{\mu}}}\middle|\begin{matrix}\left(1,\frac{1}{\mu}\right)\left(1,\frac{1}{2}\right)\\ (1,1)\quad\left(1,\frac{1}{2}\right)\end{matrix}\right] \tag{8.36}$$

and

$$\Delta_U(t) = U\mathcal{E}t^{-1/\mu}\mathrm{E}_{1-1/\mu,1-1/\mu}\left(U\mathcal{E}t^{1-1/\mu}\right), \tag{8.37}$$

where

$$\mathcal{E} = \frac{1}{\mu}\left(\frac{\hbar}{iD_\mu}\right)^{\frac{1}{\mu}}\frac{1}{\sin(\pi/\mu)}. \tag{8.38}$$

As we have seen before, the H-function of Fox, present in Eq. (8.36), is asymptotically governed by a power-law behaviour, in contrast to the behaviour of the nonfractional case that is characterised by an exponential relaxation. It is possible to show that

$$G_f(x,t) \sim \frac{iD_\mu t}{\left(\mu\hbar|x|^{1+\mu}\right)}, \qquad \mu \neq 1.$$

The generalised Mittag-Leffler function, $\mathrm{E}_{\alpha,\beta}(x)$ in Eq. (8.37), also has a power-law behaviour in the asymptotic limit of long times. These characteristics incorporated in Eq. (8.35) produce a behaviour that is different from the usual one.

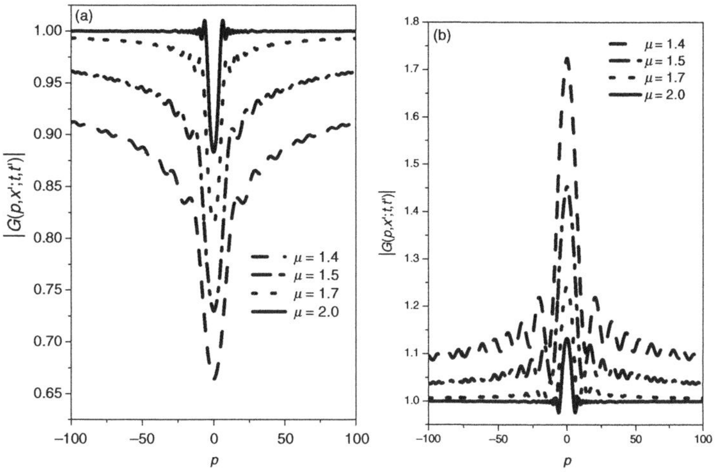

Figure 8.3 $G(p, x'; t, t')$ given by Eq. (8.35) versus p in the repulsive ($U = 1$) (a) and attractive ($U = -1$) (b) cases for different values of μ and, for simplicity, $x' = 0$, $t = 0.1$, $\hbar = 1$, and $D_\mu = 1$ [276]. Modified from E. K. Lenzi, H. V. Ribeiro, M. A. F. dos Santos, R. Rossato, and R. S. Mendes, Time dependent solutions for a fractional Schrödinger equation with delta potentials, Journal of Mathematical Physics **54**, 082107 (2013), with the permission of AIP Publishing.

In Figs. 8.3a and 8.3b, we illustrate the behaviour of Eq. (8.35) for a repulsive and an attractive potential, i.e., $U > 0$ and $U < 0$, respectively. In the first case, the particle experiences an ultrathin barrier; in the second case, there is only one bound state. The influence of the potential increases for values of $\mu \to 1$ and decreases for values of μ very close to the usual limit, i.e., $\mu \to 2$. This is connected with the asymptotic behaviour of the Green's function for $p \to \infty$, which is long-tailed for $\mu \to 1$ and, similar to the usual case, which is short-tailed for $\mu \to 2$. In addition, Eq. (8.35) represents an extension of the standard case to the fractional Schrödinger equation [279, 280].

To push the previous development further, we consider the fractional Schrödinger equation with the potential

$$V(x) = U_1\delta(x - l_1) + U_2\delta(x - l_2),$$

where U_1 and U_2 are constants, which is characterised by two localised terms, one at $x = l_1$ and the other at $x = l_2$. This double barrier structure may be viewed as the electronic analog of a Fabry–Perot interferometer. It may be also used to approximately describe several physical contexts such as the one related to resonant tunnelling in semiconductor quantum-well structures or in quantum transport, where the spatial separation of the barriers is large when compared with the individual barrier thickness [279]. By substituting this potential into Eq. (8.27), we obtain

$$i\hbar\frac{\partial}{\partial t}\Psi(x,t) = D_\mu\left(-\hbar^2\nabla^2\right)^{\frac{\mu}{2}}\Psi(x,t) + U_1\delta(x-l_1)\Psi(x,t) + U_2\delta(x-l_2)\Psi(x,t). \tag{8.39}$$

In order to solve this equation, we again use the Green's function approach and Laplace–Fourier transforms, as before. From Eq. (8.39), we obtain

$$\begin{aligned}\widetilde{\widetilde{G}}(p,x';s,t') &= \widetilde{\widetilde{G}}_f(p,s)e^{-i\frac{p}{\hbar}x'}e^{-st'} + U_1 e^{-i\frac{p}{\hbar}l_1}\widetilde{G}(l_1,x';s,t')\widetilde{\widetilde{\mathcal{G}}}_f(p,s)\\ &\quad + U_2 e^{-i\frac{p}{\hbar}l_2}\widetilde{G}(l_2,x';s,t')\widetilde{\widetilde{G}}_f(p,s).\end{aligned} \tag{8.40}$$

The inverse Fourier and Laplace transforms of Eq. (8.40) yield

$$\begin{aligned}G(x,x';t,t') &= G_{\rm f}(x-x',t)\theta(t-t') + U_1\int_0^t d\bar{t}G_{\rm f}(x-l_1,t-\bar{t})G(l_1,x';\bar{t},t')\\ &\quad + U_2\int_0^t d\bar{t}G_{\rm f}(x-l_2,t-\bar{t})G(l_2,x';\bar{t},t').\end{aligned} \tag{8.41}$$

The first term of Eq. (8.41) corresponds to the Green's function of the free case; the other terms account for the effect of the potential on the first term. In order to formally determine the Green's function, we have to find the functions $G(l_1,x';t,t')$ and $G(l_2,x';t,t')$. After some calculation, it is possible to show that they can be written, respectively, as

$$\begin{aligned}G(l_1,x';t,t') &= \theta(t-t')\int_0^t d\vartheta\Upsilon(t-\xi)\\ &\quad\times\left[G_{\rm f}(l_1-x',\xi) + U_2\int_0^\xi d\eta\Delta_1(\xi,\eta,\vartheta;x')\right]\end{aligned} \tag{8.42}$$

and

$$\begin{aligned}G(l_2,x';t,t') &= \theta(t-t')\int_0^t d\xi\Upsilon(t-\xi)\\ &\quad\times\left[G_{\rm f}(l_1-x',\xi) + U_1\int_0^\xi d\eta\Delta_2(\xi,\eta,\vartheta;x')\right],\end{aligned} \tag{8.43}$$

where

$$\Delta_1(\xi,\eta,\vartheta;x') = G_{\rm f}(l_1-l_2,\xi-\eta)G_{\rm f}(l_2-x',\eta) - G_{\rm f}(0,\vartheta-\eta)G_{\rm f}(l_1-x',\eta) \tag{8.44a}$$

and

$$\Delta_2(\xi,\eta,\vartheta;x') = G_{\rm f}(l_1-l_2,\xi-\eta)G_{\rm f}(l_1-x',\eta) - G_{\rm f}(0,\vartheta-\eta)G_{\rm f}(l_2-x',\eta). \tag{8.44b}$$

In Eqs. (8.42) and (8.43), we have introduced the quantities $\Upsilon(t)$ and $\Xi(t)$ defined, respectively, as

$$\Upsilon(t) = \Xi(t) + \sum_{n=1}^{\infty} (U_1 U_2)^n \int_0^t dt_n \Theta(t - t_n) \cdots \int_0^{t_2} dt_1 \Theta(t_2 - t_1) \int_0^{t_1} d\xi \Theta(t_1 - \xi) \Xi(\xi), \tag{8.45}$$

with

$$\Theta(t) = \int_0^t d\zeta \Lambda_{U_1}(t - \zeta) \Lambda_{U_2}(\zeta),$$

in which

$$\Lambda_{U_1,U_2} = G_f(|l_1 - l_2|, t) + \int_0^t d\xi \Delta_{U_1,U_2}(\xi) G_f(|l_1 - l_2|, t - \xi),$$

and

$$\Xi(t) = \delta(t) + \frac{U_2^2 t^{-\frac{1}{\mu}}}{U_2 - U_1} \mathcal{E} \mathrm{E}_{1-\frac{1}{\mu},1-\frac{1}{\mu}} \left(U_2 \mathcal{E} t^{1-\frac{1}{\mu}} \right) - \frac{U_1^2 t^{-\frac{1}{\mu}}}{U_2 - U_1} \mathcal{E} \mathrm{E}_{1-\frac{1}{\mu},1-\frac{1}{\mu}} \left(U_1 \mathcal{E} t^{1-\frac{1}{\mu}} \right). \tag{8.46}$$

The results are an extension, to the fractional case, of results already obtained [281], which are also characterised by a relaxation process different from the usual process.

Let us now analyse the changes produced on the solution when fractional time derivatives are incorporated in the above equations [282]. For simplicity, we again focus our analysis on Eq. (8.27), without loss of generality. The formal solution of this equation, when the usual time derivative is replaced with a fractional time derivative in the Caputo sense, is

$$\Psi(x,t) = \frac{1}{\Gamma(1-\gamma)} \int_0^t \frac{d\bar{t}}{(t-\bar{t})^\gamma} \int_{-\infty}^{\infty} dx' G_\gamma(x, x'; \bar{t}, t') \Phi(x'). \tag{8.47}$$

The Green's function is obtained by solving the following equation:

$$i\hbar \frac{\partial^\gamma}{\partial t^\gamma} G_\gamma(x, x'; t, t') - \mathcal{H} G_\gamma(x, x'; t, t') = i\hbar \delta(x - x') \delta(t - t'), \tag{8.48}$$

with $0 < \gamma < 1$, subject to the boundary conditions $G(\pm\infty, x'; t, t') = 0$, and the fractional time derivative defined as (see Section 2.3):

$$\frac{\partial^\gamma}{\partial t^\gamma} G_\gamma(x, x'; t, t') = \frac{1}{\Gamma(n-\gamma)} \int_0^t \frac{d\bar{t}}{(t-\bar{t})^{\gamma+n-1}} G_\gamma^{(n)}(x, x'; \bar{t}, t'), \tag{8.49}$$

where $n-1 < \gamma < n$ and $G_\gamma^{(n)}(x, x'; t, t') = \partial_t^n G_\gamma(x, x'; t, t')$. Employing the previous procedure of calculation, it is possible to show that the solution of Eq. (8.48) in the Fourier space is

$$\begin{aligned} G_\gamma(p, x'; t, t') &= \overline{G}_{f,\gamma}(p, t) e^{-i\frac{p}{\hbar}x'} \theta(t-t') \\ &\quad + U \int_0^t \overline{G}_{f,\gamma}(p, \bar{t}) \widetilde{G}_\gamma(0, x'; t-\bar{t}, t') d\bar{t}, \end{aligned} \tag{8.50}$$

with

$$\overline{G}_{f,\gamma}(p, t) = t^{\gamma-1} \mathrm{E}_{\gamma,\gamma}\left(-iD_\mu |p|^\mu t^\gamma / \hbar\right),$$

$$G_\gamma(0, x'; t, t') = G_{f,\gamma}(x', t)\theta(t-t') + \theta(t-t') \int_0^t G_{f,\gamma}(x', \eta) \Delta_{U,\gamma}(t-\eta) d\eta, \tag{8.51}$$

and

$$\Delta_{U,\gamma}(t) = U\mathcal{E} t^{(1-1/\mu)\gamma-1} \mathrm{E}_{\gamma-\gamma/\mu, \gamma-\gamma/\mu}\left(U\mathcal{E} t^{\gamma-\gamma/\mu}\right).$$

Notice again that the presence of the generalised Mittag-Leffer function in the solution is responsible for a power-law behaviour of $\overline{G}_{f,\gamma}(p, t)$ in the asymptotic limit of $|p| \to \infty$. In contrast, the preceding case was governed by a stretched exponential in the Fourier space, which changes the behaviour of the solutions. By performing the inverse Fourier transform of Eq. (8.50), we obtain

$$\begin{aligned} G_\gamma(x, x'; t, t') &= G_{f,\gamma}(x - x', t)\theta(t-t') \\ &\quad + U \int_0^t G_{f,\gamma}(x, \bar{t}) \widetilde{G}_\gamma(0, x'; t-\bar{t}, t') d\bar{t}, \end{aligned} \tag{8.52}$$

where

$$G_{f,\gamma}(x, t) = \frac{1}{\mu |x|} \mathrm{H}_{3,3}^{2,1}\left[\frac{|x|}{(D_\mu i t/\hbar)^{\frac{1}{\mu}}} \middle| \begin{matrix} (1,1) \left(\gamma, \frac{\gamma}{\mu}\right) \left(1, \frac{1}{2}\right) \\ \left(1, \frac{1}{\mu}\right) \left(1, \frac{1}{\mu}\right) \left(1, \frac{1}{2}\right) \end{matrix}\right]. \tag{8.53}$$

Figure 8.4a shows the solution given by Eq. (8.47) for $\gamma = 1/2$ and for different values of μ, and Fig. 8.4b illustrates the behaviour of the Green's function given

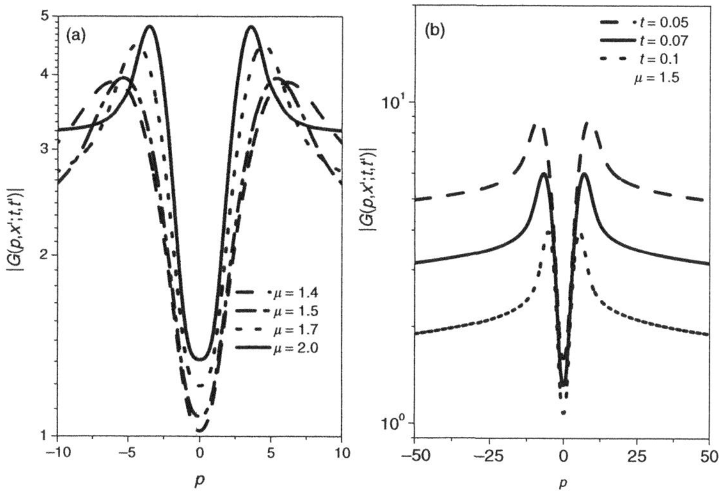

Figure 8.4 Behaviour of Eq. (8.50) for different values of μ (a) and t (b) by considering, for illustrative purposes, $U = 1$, $x' = 0$, $\hbar = 1$, and $D_\mu = 1$ [276]. Modified from E. K. Lenzi, H. V. Ribeiro, M. A. F. dos Santos, R. Rossato, and R. S. Mendes, Time dependent solutions for a fractional Schrödinger equation with delta potentials, Journal of Mathematical Physics **54**, 082107 (2013), with the permission of AIP Publishing.

by Eq. (8.50) for different times, when $\gamma = 1/2$ and $\mu = 1.5$. Note that, unlike previous results obtained for $\gamma = 1$ (see Fig. 8.3, where $|G(p, x'; t, t')| \to 1$ to $|p| \to \infty$), the case $\gamma \neq 1$ presents a different asymptotic behaviour for the Green's function. This is connected with the presence of the generalised Mittag-Leffler function in the solution of the free case, which introduces a nonexponential asymptotic behaviour for the solution.

In the next section, we use the continuous-time random walk (CTRW) approach as a guide to obtain consistent extensions of the Schrödinger equation incorporating nonlocal effects [274]. As an application, we solve the problem of a free particle in a half-space, obtaining the time-dependent solution for an arbitrary initial condition [273].

8.3 CTRW and the Fractional Schrödinger Equation

We first notice that diffusion equations can be obtained from the CTRW by performing a suitable choice of the probability density function (pdf) [41, 283]. This means that to describe a diffusive process subject, for example, to an absorbent or reflecting surface, it is possible to use a CTRW approach or a diffusion equation. In addition, memory effects or other characteristics which are not conveniently described by usual diffusion equations may be described by the CTRW approach if an appropriate choice of probability density function is done. Depending on the effects considered, we are led to diffusion-like equations which may have, for example, fractional time derivatives [284], spatial fractional derivatives [285], and fractional time derivatives of distributed order [286]. The boundary conditions have to be carefully considered when the CTRWs are formulated to investigate these phenomena. It should be mentioned that the spatial operator may not preserve its form depending on the boundary conditions employed [287]. Typical examples are the cases of a system restricted to a half-space with reflecting boundary conditions [288] or absorbing boundary conditions [289]. These situations may be associated with diffusion equations with a fractional operator which are different from the ones obtained when the system is not restricted. These examples suggest that the Schrödinger equation may have different representations in terms of the spatial operator, depending on the specific situation considered.

In short, the procedure followed here basically consists in obtaining diffusion-like equations from the CTRW approach. It is based on the operators present in these equations that modify the usual Schrödinger equation in order to incorporate the nonlocal operators.This avoids the cumbersome calculations to find the solutions which may arise in other formulations of the Schrödinger equation.

Let us present the method by means of the concrete problem of a particle in a half-space in the absence of potential, which has as an initial condition an arbitrary wave function $\overline{\Psi}(x)$ satisfying the boundary conditions $\Psi(0,t) = \Psi(\infty,0) = 0$. To obtain a Schrödinger-like equation in terms of the fractional operators, we first investigate the corresponding diffusion equation. More precisely, by using the CTRW approach, the mathematical structure of this diffusion equation based on a long-tailed distribution for the jumping probability density is considered. To do this, we use some developments to get a diffusion-like equation for the system in half-space with absorbing boundary conditions, i.e., $\rho(0,t) = \rho(\infty,t) = 0$ [288]. The spatial operator obtained from this analysis will be used to formulate the Schrödinger equation and, consequently, to avoid possible inconsistencies [274].

The probability $\rho(x,t)dx$ of finding a particle in $[x, x+dx]$ at instant t satisfies the generalised master equation [288],

$$\rho(x,t) = \delta(x) \int_t^{\infty} dt' \psi(t') + \int_0^{\infty} dx' \int_0^t dt' \rho(x',t') \Lambda(x,x') \psi(t-t'), \tag{8.54}$$

where $\psi(t)$ represents the waiting time distribution and $\Lambda(x,x')$ is the probability density that the particle jumping from x' arrives at $[x, x+dx]$. The second term of Eq. (8.54) involves a spatial integration over the interval $[0,\infty)$ since the particles are constrained to stay in the half-space. Suitable choices for $\psi(t')$ in Eq. (8.54) have to be made in order to fulfil the conditions required before. One of them concerns with a short-tailed waiting time distribution such as

$$\psi(t) = \frac{1}{\tau} e^{-t/\tau},$$

which implies handling a first-order time derivative, as in the case of the usual diffusion equation. The other choice to be made is about the jumping probability, which needs to incorporate a long-tailed behaviour and to satisfy the boundary condition imposed on $x = 0$. A typical mathematical behaviour of the probability jumping $\rho(x,t)$ that is able to satisfy the requirements of the system is obtained if (for the reflective case, see Ref. [288]):

$$\Lambda(x,x') \propto \left(\frac{1}{|x-x'|^{1+\alpha}} - \frac{1}{|x+x'|^{1+\alpha}} \right).$$

This expression naturally accomplishes the symmetries of the sine Fourier transform. A similar situation was worked out to investigate Riesz fractional derivative in the presence of an absorbing boundary [289]. By substituting these assumptions in Eq. (8.54) and performing some calculations, it is possible to show that the diffusion equation which emerges in this scenario can be written as

$$\frac{\partial \rho(x,t)}{\partial t} = K_\alpha \widetilde{\partial}_x^\alpha \rho(x,t) , \tag{8.55}$$

with

$$\widetilde{\partial}_x^\alpha \rho(x,t) = A_\alpha \frac{\partial^2}{\partial x^2} \int_0^\infty dx' \rho(x',t) \left(|x-x'|^{1-\alpha} - |x+x'|^{1-\alpha} \right), \tag{8.56}$$

where

$$A_\alpha = \sqrt{\frac{\pi}{2}} \frac{1}{\Gamma(2-\alpha) \sin(\pi\alpha/2)}.$$

Based on these results, a possible extension to the Schrödinger equation, which fulfils the conditions required for a particle in a half-space, is

$$i\hbar \frac{\partial \Psi(x,t)}{\partial t} = -D_\alpha \widetilde{\partial}_x^\alpha \Psi(x,t) , \tag{8.57}$$

where, as before, D_α is a constant such that for $\alpha = 2$ it reduces to the usual form, i.e., $D_{\alpha=2} = \hbar^2/(2m)$. Thus, Eq. (8.27) is the usual form of the Schrödinger equation when $\alpha = 2$, however now restricted to the interval $[0, \infty)$. The spatial fractional operator in Eq. (8.57) is different from that considered before [264, 266, 267, 274, 290–292] since the conditions required by the system restrict the movement of the particle to a half-space. The spatial operator may change its form when other conditions are imposed on the system.

Now, we solve Eq. (8.27) for an arbitrary initial condition such that $\Psi(0,t) = \Psi(\infty,0) = 0$. Applying the sine Fourier transform and its inverse to handle Eq. (8.57), we obtain

$$i\hbar\frac{\partial}{\partial t}\Psi(k,t) = D_\alpha k^\alpha \Psi(k,t) \tag{8.58}$$

because $\mathcal{F}_S\left\{\widetilde{\partial}_x^\alpha \Psi(x,t)\right\} = -k^\alpha \Psi(k,t)$ with k nonnegative. The solution of Eq. (8.58) is

$$\Psi(k,t) = \Psi(k,0)e^{-i(D_\alpha/\hbar)k^\alpha t}. \tag{8.59}$$

The inverse Fourier transform yields

$$\Psi(x,t) = \int_0^\infty dx' \overline{\Psi}(x')G(x,x',t), \tag{8.60}$$

with

$$G(x,x',t) = \frac{1}{\alpha|x-x'|}\mathrm{H}_{2,2}^{1,1}\left[z_- \middle| \begin{matrix}\left(1,\frac{1}{\alpha}\right)\left(1,\frac{1}{2}\right)\\ (1,1)\left(1,\frac{1}{2}\right)\end{matrix}\right] - \frac{1}{\alpha|x+x'|}\mathrm{H}_{2,2}^{1,1}\left[z_+ \middle| \begin{matrix}\left(1,\frac{1}{\alpha}\right)\left(1,\frac{1}{2}\right)\\ (1,1)\left(1,\frac{1}{2}\right)\end{matrix}\right], \tag{8.61}$$

where $\overline{D}_\alpha = iD_\alpha/\hbar$ and $z_\pm = x \pm x'/\left(\overline{D}_\alpha t\right)^{\frac{1}{\alpha}}$. This solution satisfies Eq. (8.27) and the boundary conditions required by the system, in contrast to the situations discussed in Ref. [274]. For $|x| \to \infty$, the asymptotic limit of Eq. (8.61) for $\alpha < 2$ is given by

$$G(x,x',t) \sim \frac{1}{\pi}\frac{\sin(\pi\alpha/2)\,\Gamma(1+\alpha)}{\left(\overline{D}_\alpha t\right)^{\frac{1}{\alpha}}}\left(\frac{1}{|z_-|^{1+\alpha}} - \frac{1}{|z_+|^{1+\alpha}}\right). \tag{8.62}$$

Figure 8.5 illustrates the behaviour of the solution (8.60) by considering the initial condition $\overline{\Psi}(x) = \sqrt{4/\sqrt{\pi}}xe^{-x^2/2}$ and typical values of α. The asymptotic behaviour exhibited by the solution is a power law when long times are considered.

Figure 8.6, in turn, illustrates the time evolution of the solution for $\alpha = 1$ for the same initial condition used in Fig. 8.5. It shows that the solution is initially dominated by the behaviour of the wave function used here (short tailed) as the initial condition, whereas for long times it is governed by the Green's function containing the dynamical aspects of the system. A result of this kind may be extended to situations with different spreading regimes. In these cases, the spatial fractional operators of distributed order in the Schrödinger equation will be obtained from the prescription used here.

The equation which emerges from this prescription satisfies the conditions imposed on the system (e.g., boundary conditions) and incorporates the nonlocal effects manifested by the presence of unusual differential operators. This can be

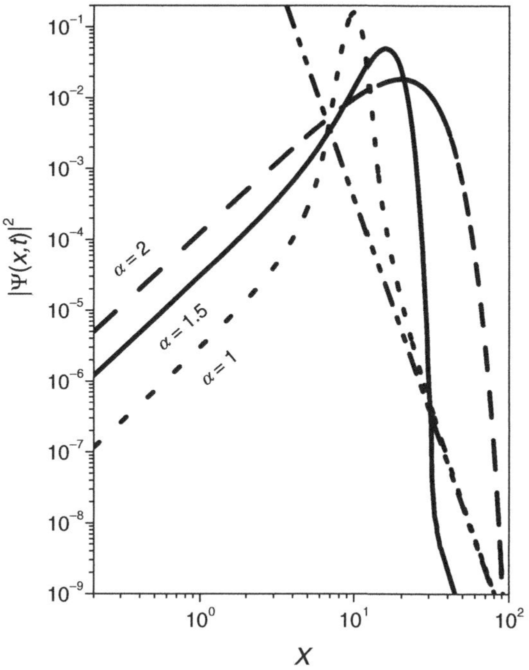

Figure 8.5 $|\Psi(x,t)|^2$ versus x for different values of α, $D_\alpha = 1$, and $t = 10$. A dashed-dotted line was incorporated to highlight the power-law behaviour of the solution in the asymptotic limit, i.e., $x \to \infty$, when $\alpha \neq 2$ [273]. Modified from E. K. Lenzi, H. V. Ribeiro, H. Mukai, and R. S. Mendes, Continuous-time random walk as a guide to fractional Schrödinger equation, Journal of Mathematical Physics **51**, 092102 (2010), with the permission of AIP Publishing.

seen in the following simple way. The usual spatial operator of the Schrödinger equation, written for a lattice containing essentially only first neighbours, may be put in the form:

$$\partial_x^2 \psi \sim \psi_{n+1} - 2\psi_n + \psi_{n-1}.$$

However, for those situations in which effective contributions of a large number of neighbouring terms have to be taken into account, the problem has to be formulated in terms of fractional derivatives. From the phenomenological point of view, an example is the specific heat of noncrystalline solids at very low temperature, which can be investigated preliminarily by considering a fractional Schrödinger equation, as we discuss in detail in Section 8.5. Other scenarios in which fractional extensions of the Schrödinger equation may be useful are characterised by self-similarity, memory, and non-Gaussian fluctuations, among others [266], when we consider the anomalous diffusion phenomenon.

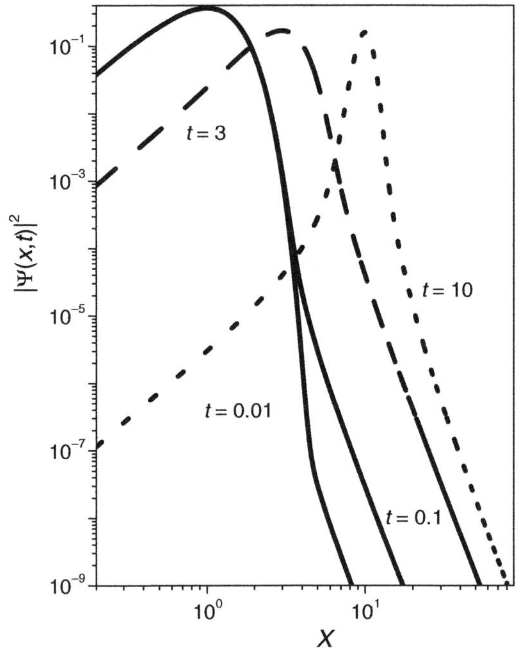

Figure 8.6 $|\Psi(x,t)|^2$ versus x, for $D_\alpha = 1$ and $\alpha = 1$ [273]. Modified from E. K. Lenzi, H. V. Ribeiro, H. Mukai, and R. S. Mendes, Continuous-time random walk as a guide to fractional Schrödinger equation, Journal of Mathematical Physics **51**, 092102 (2010), with the permission of AIP Publishing.

8.4 Memory and Nonlocal Effects

To investigate in more detail memory and nonlocal effects, we introduce the following time-dependent Schrödinger equation [293]:

$$\imath\hbar\int_0^t d\tau\,\gamma(t-\tau)\frac{\partial}{\partial\tau}\psi(x,\tau) = -\frac{\hbar^2}{2M}\frac{\partial^2}{\partial x^2}\psi(x,t) + V(x,t)\psi(x,t) + \int_0^t d\tau\int_{-\infty}^{\infty} d\xi\, U(x-\xi,t-\tau)\psi(\xi,\tau), \tag{8.63}$$

where $\gamma(t)$ is a memory kernel for which the assumption

$$\lim_{t\to\infty}\gamma(t) = \lim_{s\to 0} s\hat{\gamma}(s) = 0, \qquad \text{with} \qquad \hat{\gamma}(s) = \mathcal{L}\left[\gamma(t)\right],$$

is satisfied. The nonlocal term is represented as a convolution integral of the function $U(x,t)$ and wave function $\psi(x,t)$. In the absence of the potential and of the nonlocal term, i.e., for $V(x,t) = 0$ and $U(x,t) = 0$, the Schrödinger equation is connected with the diffusion-like equation:

$$\int_0^t d\tau \gamma(t-\tau)\frac{\partial}{\partial \tau}\psi(x,\tau) = D_\gamma \frac{\partial^2}{\partial x^2}\psi(x,t), \tag{8.64}$$

with diffusion coefficient D_γ. Equation (8.64) was obtained from an overdamped generalised Langevin equation for a free particle in the presence of a friction memory kernel $\gamma(t)$ [294]. Moreover, such an integral equation for the probability distribution function was obtained in the analysis of an anomalous diffusive process subordinated to normal diffusion under operational time, where the memory kernel is connected with the cumulative distribution function of waiting times; i.e., it represents the probability that the system will make no step before the time t [295].

In what follows, we consider different forms of the potential energy function and nonlocal term to obtain the wave function and probability distribution function.

8.4.1 Absence of the Nonlocal Term

Let us consider Eq. (8.63) in the absence of the nonlocal term. By using the separation ansatz $\Psi(x,t) = \Phi(x)T(t)$, we obtain the following equations:

$$\int_0^t d\tau \gamma(t-\tau)\frac{d}{d\tau}T(\tau) = -\imath\frac{\lambda}{\hbar}T(t) \tag{8.65a}$$

and

$$\frac{d^2}{dx^2}\Phi(x) + \frac{2M}{\hbar^2}\left[\lambda - V(x)\right]\Phi(x) = 0, \tag{8.65b}$$

where λ is the separation constant that corresponds to the energy. Equation (8.65b), for different forms of the potential energy function, is widely investigated in the literature. Application of the Laplace transform to Eq. (8.65a) yields for the time-dependent function $T(t)$ the following solution:

$$T_n(t) = T_n(0)\mathcal{L}^{-1}\left[\frac{\hat{\gamma}(s)}{s\hat{\gamma}(s) + \imath\frac{\lambda_n}{\hbar}}\right], \qquad \hat{\gamma}(s) = \mathcal{L}\left[\gamma(t)\right]. \tag{8.66}$$

The case $\gamma(t) = \delta(t)$ leads to the known result

$$T_n(t) = T_n(0)e^{-\imath\frac{\lambda_n}{\hbar}t},$$

obtained for the Schrödinger equation. If we consider a power-law memory kernel in the form

$$\gamma(t) = \frac{t^{-\mu}}{\Gamma(1-\mu)},$$

then we obtain the fractional Schrödinger equation [255]:

$${}_0^C\mathrm{D}_t^{\mu}\Psi(x,t) = \frac{\imath\hbar}{2M}\nabla^2\Psi(x,t) - \frac{\imath}{\hbar}V(x)\Psi(x,t),$$

where ${}_0^C\mathrm{D}_t^{\mu}f(t)$ is the Caputo fractional derivative, as defined in Section 2.3. The solution is thus

$$\begin{aligned} T_n(t) &= T_n(0)\mathrm{E}_{\mu}\left(-\imath\frac{\lambda_n}{\hbar}t^{\mu}\right) \\ &= T_n(0)\left[\mathrm{E}_{2\mu}\left(-\frac{\lambda_n^2}{\hbar^2}t^{2\mu}\right) - \imath\frac{\lambda_n}{\hbar}t^{\mu}\mathrm{E}_{2\mu,\mu+1}\left(-\frac{\lambda_n^2}{\hbar^2}t^{2\mu}\right)\right], \end{aligned} \tag{8.67}$$

where $\mathrm{E}_{\alpha}(z) = \mathrm{E}_{\alpha,1}(z)$ and $\mathrm{E}_{\alpha,\beta}(z)$ are, respectively, the one- and two-parameter Mittag-Leffler functions (cf. Section 1.2.3). The particular case $\mu = 1$ yields

$$T_n(t) = T_n(0)\left[\cos\left(\frac{\lambda_n}{\hbar}t\right) - \imath\sin\left(\frac{\lambda_n}{\hbar}t\right)\right] = T_n(0)e^{-\imath\frac{\lambda_n}{\hbar}t},$$

which is obtained by noting that, from Eqs. (1.121) and (1.122), we have $\mathrm{E}_{2,1}\left(-z^2\right) = \cos z$ and $\mathrm{E}_{2,2}\left(-z^2\right) = \sin z/z$. Thus, the probability distribution function is

$$|\psi_n(x,t)|^2 = |\Psi_n(x)|^2\left\{\left[\mathrm{E}_{2\mu}\left(-\frac{\lambda_n^2}{\hbar^2}t^{2\mu}\right)\right]^2 + \frac{\lambda_n^2}{\hbar^2}t^{2\mu}\left[\mathrm{E}_{2\mu,\mu+1}\left(-\frac{\lambda_n^2}{\hbar^2}t^{2\mu}\right)\right]^2\right\}. \tag{8.68}$$

In the limit of long times, it decays to zero as

$$|\psi_n(x,t)|^2 \simeq |\Psi_n(x)|^2\frac{\hbar^2}{\lambda_n^2}\frac{t^{-2\mu}}{\Gamma^2(1-\mu)}, \tag{8.69}$$

where we have used the asymptotic limit, i.e.,

$$\mathrm{E}_{\alpha,\beta}(z) \simeq -\frac{z^{-1}}{\Gamma(\beta-\alpha)}, \quad z \to \infty, \tag{8.70}$$

as can be easily checked with the help of Eq. (1.120).

Free Particle Solution

We consider first the zero potential energy case and use the Green's function approach to find the solution. The Fourier–Laplace transform applied to Eq. (8.63), when $\Psi(x,0) = \Phi(x) = \delta(x)$, yields the Green's function in the Fourier–Laplace space as:

$$\tilde{\hat{G}}(\kappa, s) = \frac{\hat{\gamma}(s)}{s\hat{\gamma}(s) + \imath \frac{\hbar}{2M}|\kappa|^2}, \tag{8.71}$$

where $\tilde{\hat{G}}(\kappa, s) = \mathcal{F}[\hat{G}(x,s)] = \mathcal{F}[\mathcal{L}[G(x,s)]]$ and $\tilde{\Psi}(\kappa, 0) = \mathcal{F}[\delta(x)] = 1$. Note that one can use the Fourier transform with $\kappa \rightarrow p/\hbar$ to obtain the function in the momentum space. After finding the Green's function, the wave function can be written as

$$\psi(x,t) = \int_{-\infty}^{\infty} d\xi \; G(x-\xi, t)\Phi(\xi), \tag{8.72}$$

where the Green's function can be considered as a propagator; from Eq. (8.71) one easily concludes that $\int_{-\infty}^{\infty} dx G(x,t) = 1$.

If $\gamma(t) = \delta(t)$, for the Green's function we easily obtain

$$G(x,t) = \frac{1}{\sqrt{4\pi \imath \frac{\hbar}{2M} t}} e^{-\frac{x^2}{4\imath \frac{\hbar}{2M} t}}, \tag{8.73}$$

from which, by exchanging $\imath\hbar/2M \rightarrow D$, we obtain the Gaussian probability distribution function for the usual diffusion equation.

If we consider instead a power-law memory kernel, in the form

$$\gamma(t) = \frac{t^{-\mu}}{\Gamma(1-\mu)}, \qquad 0 < \mu < 1, \tag{8.74}$$

by using relation (1.131), for the Green's function we find

$$\begin{aligned} \hat{G}(\kappa, t) &= \mathrm{E}_{\mu}\left(-\imath \frac{\hbar}{2M} t^{\mu} |\kappa|^2\right) \\ &= \mathrm{E}_{2\mu}\left(-\frac{\hbar^2}{4M^2} t^{2\mu}|\kappa|^4\right) - \imath \frac{\hbar}{2M} t^{\mu}|\kappa|^2 \mathrm{E}_{2\mu,\mu+1}\left(-\frac{\hbar^2}{4M^2} t^{2\mu}|\kappa|^4\right). \end{aligned} \tag{8.75}$$

Using Eqs. (1.132), (1.163a), and (1.182) we obtain

$$\begin{aligned} G(x,t) = \frac{1}{4|x|} \Bigg\{ & \mathrm{H}^{2,1}_{3,3}\left[\frac{|x|}{\sqrt{\frac{\hbar}{2M} t^{\mu}}} \middle| \begin{matrix} (1,\frac{1}{4}), (1,\frac{\mu}{2}), (1,\frac{1}{2}) \\ (1,1), (1,\frac{1}{4}), (1,\frac{1}{2}) \end{matrix}\right] \\ & - \imath \mathrm{H}^{2,1}_{3,3}\left[\frac{|x|}{\sqrt{\frac{\hbar}{2M} t^{\mu}}} \middle| \begin{matrix} (\frac{1}{2},\frac{1}{4}), (1,\frac{\mu}{2}), (1,\frac{1}{2}) \\ (1,1), (\frac{1}{2},\frac{1}{4}), (1,\frac{1}{2}) \end{matrix}\right] \Bigg\}, \end{aligned} \tag{8.76}$$

where we observe the presence of the H-function of Fox. We see that the Green's function can be separated into real and imaginary parts as follows:

$$G(x,t) = \mathcal{R}(x,t) - \imath \mathcal{I}(x,t). \tag{8.77}$$

The solution of this equation can be represented in terms of Wright function, introduced in Section 1.2.4, as [255]:

$$W(-z;-\beta,1-\beta)=\sum_{n=0}^{\infty}\frac{1}{\Gamma(-\beta n+1-\beta)}\frac{(-z)^n}{n!}, \tag{8.78}$$

where the inverse Fourier transform is performed directly on the function

$$\hat{G}(\kappa,t)=\mathrm{E}_{\mu}\left(-\imath\frac{\hbar}{2M}t^{\mu}|\kappa|^2\right).$$

Therefore, it follows that

$$\begin{aligned} G(x,t) &= \frac{1}{2|x|}\mathrm{H}_{1,1}^{1,0}\left[\frac{|x|}{\sqrt{\imath\frac{\hbar}{2M}t^{\mu}}}\,\middle|\,\begin{matrix}(1,\frac{\mu}{2})\\(1,1)\end{matrix}\right] \\ &= \frac{1}{2\sqrt{\imath\frac{\hbar}{2M}t^{\mu}}}\mathrm{H}_{1,1}^{1,0}\left[\frac{|x|}{\sqrt{\imath\frac{\hbar}{2M}t^{\mu}}}\,\middle|\,\begin{matrix}(1-\frac{\mu}{2},\frac{\mu}{2})\\(0,1)\end{matrix}\right] \\ &= \frac{1}{2\sqrt{\imath\frac{\hbar}{2M}t^{\mu}}}M_{\mu/2}\left(\frac{|x|}{\sqrt{\imath\frac{\hbar}{2M}t^{\mu}}}\right). \end{aligned} \tag{8.79}$$

Note that by exchanging again $\imath\hbar/2M \to D$, the solution (8.79) corresponds to the one obtained for fractional diffusion equation [40].

Let us now consider the distributed order Schrödinger equation in noninteger dimensions, which can be rewritten as [281]

$$\imath\hbar\int_0^1 d\bar{\gamma}\,\tau^{\bar{\gamma}-1}p(\bar{\gamma})\frac{\partial^{\bar{\gamma}}}{\partial\tau^{\bar{\gamma}}}\Psi(x,\tau)=\hat{H}\Psi(x,t), \tag{8.80}$$

where τ is the characteristic time, $p(\bar{\gamma})$ is a weight function, and the fractional derivative is that of Caputo. Taking into account its definition introduced in Section 2.3, Eq. (8.80) can be rewritten as

$$\imath\hbar\int_0^t d\tau\int_0^1 d\bar{\gamma}\,\tau^{\bar{\gamma}-1}p(\bar{\gamma})\frac{(t-\tau)^{-\bar{\gamma}}}{\Gamma(1-\bar{\gamma})}\frac{\partial}{\partial\tau}\Psi(x,\tau)=\hat{H}\Psi(x,t), \tag{8.81}$$

which has the same form as Eq. (8.63), where

$$\gamma(t)=\int_0^1 d\bar{\gamma}\,\tau^{\bar{\gamma}-1}p(\bar{\gamma})\frac{t^{-\bar{\gamma}}}{\Gamma(1-\bar{\gamma})}$$

is a distributed-order memory kernel. For such memory kernels, by applying the inverse Fourier transform to Eq. (8.71), it follows that

$$\hat{G}(x,s) = \frac{1}{s}\sqrt{\frac{s\hat{\gamma}(s)}{4\imath\frac{\hbar}{2M}}}\exp\left(-\sqrt{\frac{s\hat{\gamma}(s)}{\imath\frac{\hbar}{2M}}}|x|\right). \tag{8.82}$$

The Tauberian theorem [296] states that, for a slowly varying function $L(t)$, we have

$$\lim_{t\to\infty}\frac{L(at)}{L(t)} = 1, \qquad a > 0. \tag{8.83}$$

Thus, if

$$\hat{r}(s) \simeq s^{-\rho}L\left(\frac{1}{s}\right), \quad s\to 0, \quad \rho \geq 0, \tag{8.84}$$

then

$$r(t) = \mathcal{L}^{-1}\left[\hat{r}(s)\right](t) \simeq \frac{1}{\Gamma(\rho)}t^{\rho-1}L(t), \quad t\to\infty. \tag{8.85}$$

The theorem may now be applied to Eq. (8.82) in order to find the asymptotic properties of the Green's function. For the weight function $p(\bar{\gamma}) = 1$, we have

$$\hat{\gamma}(s) = \frac{s\tau - 1}{s\log s\tau},$$

in such a way that $\lim_{t\to\infty}\gamma(t) = \lim_{s\to 0}s\hat{\gamma}(s) = 0$. Thus, we may obtain the distributed-order Schrödinger equation in the form:

$$\imath\hbar\int_0^1 d\bar{\gamma}\,\tau^{\bar{\gamma}-1}{}_C D_t^{\bar{\gamma}}\Psi(x,t) = -\frac{\hbar^2}{2M}\Psi(x,t), \tag{8.86}$$

whose Green's function, in the limit $t\to\infty$, is

$$G(x,t) \simeq \frac{1}{\sqrt{4\imath\frac{\hbar}{2M}\log\frac{t}{\tau}}}\exp\left(-\frac{|x|}{\sqrt{\imath\frac{\hbar}{2M}\log\frac{t}{\tau}}}\right). \tag{8.87}$$

To proceed further, we consider a distributed-order memory kernel with the weight function $p(\bar{\gamma}) = \nu\bar{\gamma}^{\nu-1}$ such that

$$\hat{\gamma}(s) = \frac{\nu\gamma(\nu,-\log s\tau)}{s(-\log s\tau)^{\nu}} \simeq \frac{\nu\Gamma(\nu)}{s(-\log s\tau)^{\nu}}, \tag{8.88}$$

where $\gamma(\alpha,x)$ is the incomplete gamma function defined in Eq. (1.75). It has the form $\gamma(\nu,x) \simeq \Gamma(\nu)$ for large x (small s, thus large $-\log s\tau$). By applying the Tauberian theorem in the limit $t\to\infty$, for the Green's function we obtain

$$G(x,t) \simeq \sqrt{\frac{\Gamma(1+\nu)}{4\imath\frac{\hbar}{2M}\log^{\nu}\frac{t}{\tau}}}\exp\left[-\sqrt{\frac{\Gamma(1+\nu)}{\imath\frac{\hbar}{2M}\log^{\nu}\frac{t}{\tau}}}|x|\right]. \tag{8.89}$$

The case of distributed-order memory kernel with the weight function

$$p(\lambda) = \begin{cases} \frac{1}{\lambda_2-\lambda_1}, & \text{for} \quad 0 \le \lambda_1 < \lambda < \lambda_2 \le 1, \\ 0, & \text{otherwise,} \end{cases}$$

which means that

$$\gamma(t) = \frac{1}{\lambda_2-\lambda_1}\int_{\lambda_1}^{\lambda_2}\frac{t^{-\lambda}}{\Gamma(1-\lambda)}d\lambda$$

or

$$\hat{\gamma}(s) = \frac{1}{\lambda_2-\lambda_1}\frac{(s\tau)^{\lambda_2}-(s\tau)^{\lambda_1}}{s\log s\tau}$$

yields the following behaviour for the Green's function:

$$G(x,t) \simeq \sqrt{\frac{t^{-\lambda_1}}{4\imath\frac{\hbar}{2M}\log\left(\frac{t}{\tau}\right)^{\lambda_2-\lambda_1}}}\exp\left[-\sqrt{\frac{\left(\frac{t}{\tau}\right)^{-\lambda_1}}{\imath\frac{\hbar}{2M}\log\left(\frac{t}{\tau}\right)^{\lambda_2-\lambda_1}}}|x|\right]. \tag{8.90}$$

Such distributed-order memory kernels have been used in different contexts, in the distributed-order relaxation [36, 297, 298] and diffusion [101, 219, 299] equations, as well as in the Langevin equation with distributed-order friction memory kernel [300]. These works show that they are useful tools for modelling ultraslow relaxation and diffusion processes.

Delta Potential Energy Function

Additionally, we find the Green's function in the case of a Dirac delta potential energy function $V(x,t) = U\delta(x)$, where U is a constant. In the Fourier–Laplace space it follows that

$$\tilde{\hat{G}}(\kappa,s) = \frac{\hat{\gamma}(s)}{s\hat{\gamma}(s)+\imath\frac{\hbar}{2M}|\kappa|^2} - \frac{\frac{\imath U}{\hbar}}{s\hat{\gamma}(s)+\imath\frac{\hbar}{2M}|\kappa|^2}\hat{G}(x=0,s). \tag{8.91}$$

Applying the inverse Fourier transform and setting $x = 0$, we obtain

$$\hat{G}(x=0,s) = \frac{1}{2s}\frac{\frac{s\hat{\gamma}(s)}{\imath\frac{\hbar}{2M}}}{\sqrt{\frac{s\hat{\gamma}(s)}{\imath\frac{\hbar}{2M}}}+\frac{MU}{\hbar^2}}. \tag{8.92}$$

Thus, the Green's function becomes

$$\tilde{\hat{G}}(\kappa,s) = \frac{\hat{\gamma}(s)}{s\hat{\gamma}(s)+\imath\frac{\hbar}{2M}|\kappa|^2}\frac{\sqrt{s\hat{\gamma}(s)}}{\sqrt{s\hat{\gamma}(s)}+\frac{MU}{\hbar^2}\sqrt{\imath\frac{\hbar}{2M}}} = \tilde{\hat{P}}(\kappa,s)\hat{Q}(s), \tag{8.93}$$

from which we then find

$$\tilde{G}(\kappa,t) = \int_0^t d\tau \tilde{P}(\kappa,t-\tau)Q(\tau), \tag{8.94}$$

where

$$\tilde{P}(\kappa,t) = \mathcal{L}^{-1}\left[\hat{\tilde{P}}(\kappa,s)\right] \quad \text{and} \quad Q(t) = \mathcal{L}^{-1}\left[\hat{Q}(s)\right].$$

In the Laplace space, the Green's function reads:

$$\hat{G}(x,s) = \frac{1}{\sqrt{4\iota\frac{\hbar}{2M}}}\frac{\hat{\gamma}(s)}{\sqrt{s\hat{\gamma}(s)} + \frac{MU}{\hbar^2}\sqrt{\iota\frac{\hbar}{2M}}}e^{-\sqrt{\frac{s\hat{\gamma}(s)}{\iota\frac{\hbar}{2M}}}|x|}. \tag{8.95}$$

Therefore, by choosing a particular expression for the memory kernel $\gamma(t)$, by using either Eq. (8.94) or Eq. (8.95), we can obtain the Green's function $G(x,t)$.

8.4.2 Presence of Nonlocal Term

Let us find the formal solution of Eq. (8.63) for $V(x,t) = 0$. Using the Fourier–Laplace transform we obtain the Green's function as

$$\hat{\tilde{G}}(\kappa,s) = \frac{\hat{\gamma}(s)}{s\hat{\gamma}(s) + \frac{\iota}{\hbar}\hat{\tilde{U}}(\kappa,s) + \iota\frac{\hbar}{2M}|\kappa|^2}. \tag{8.96}$$

From now on, we consider some different expressions for the nonlocal term and construct the propagator $G(x,t)$. Some of these nonlocal terms have been already considered [258, 259].

First case: $U(x,t) = \zeta\delta(x)\delta(t)$

For this nonlocal term, the Green's function (8.96) becomes

$$\hat{\tilde{G}}(\kappa,s) = \frac{\hat{\gamma}(s)}{s\hat{\gamma}(s) + \iota\frac{\zeta}{\hbar} + \iota\frac{\hbar}{2M}|\kappa|^2}, \tag{8.97}$$

where $\hat{\tilde{U}}(\kappa,s) = \zeta$. This situation corresponds to the time-dependent Schrödinger equation with constant potential energy function ζ.

We first use the kernel $\gamma(t) = \delta(t)$, which leads to the ordinary time-dependent Schrödinger equation. For the Green's function, we find

$$G(x,t) = \mathcal{F}^{-1}\left[e^{-\left(\iota\frac{\hbar}{2M}|\kappa|^2 + \iota\frac{\zeta}{\hbar}\right)t}\right] = \frac{1}{\sqrt{4\pi\iota\frac{\hbar}{2M}t}}e^{\frac{\iota}{\hbar}\left(\frac{Mx^2}{2t} - \zeta t\right)}, \tag{8.98}$$

from which it follows that the wave function is

$$\psi(x,t) = \frac{1}{\sqrt{4\pi\imath\frac{\hbar}{2M}t}} e^{-\imath\frac{\zeta}{\hbar}t} \int_{-\infty}^{\infty} d\xi\, e^{\frac{\imath}{\hbar}\frac{M(x-\xi)^2}{2t}} \phi(\xi). \tag{8.99}$$

For a power-law memory kernel like (8.74), Eq. (8.63) becomes the time-dependent fractional Schrödinger equation with the fractional Caputo time derivative. The Green's function is now

$$\begin{aligned} G(x,t) &= \mathcal{F}^{-1}\left[\sum_{n=0}^{\infty}\left(-\imath\frac{\zeta}{\hbar}\right)^n t^{\mu n}\mathrm{E}_{\mu,\mu n+1}^{n+1}\left(-\imath\frac{\hbar}{2M}t^{\mu}|\kappa|^2\right)\right] \\ &= \frac{1}{2|x|}\sum_{n=0}^{\infty}\frac{(-\imath\frac{\zeta}{\hbar})^n}{n!}t^{\mu n}\mathrm{H}_{2,2}^{2,0}\left[\frac{|x|}{\sqrt{\imath\frac{\hbar}{2M}t^{\mu}}}\middle|\begin{matrix}(\mu n+1,\frac{\mu}{2}),(1,\frac{1}{2})\\(1,1),(n+1,\frac{1}{2})\end{matrix}\right], \end{aligned} \tag{8.100}$$

where $\mathrm{E}_{\alpha,\beta}^{\delta}(z)$ is the three-parameter Mittag-Leffler function. In the absence of the nonlocal term ($\zeta = 0$), the infinite series representation of the H-function of Fox reduces the solution to Eq. (8.79). The solution (8.100) can be separated into real and imaginary parts if we use the formula

$$\imath^n = \left(\cos\frac{\pi}{2} + \imath\sin\frac{\pi}{2}\right)^n = \cos\frac{n\pi}{2} + \imath\sin\frac{n\pi}{2},$$

before we use the inverse Fourier transform.

Second case: $U(x,t) = \eta\delta(x)\frac{t^{\lambda-1}}{\Gamma(\lambda)}$

If we use the Laplace transform $\hat{\tilde{U}}(\kappa,s) = \eta s^{-\lambda}$ in the Green's function, we obtain

$$\hat{\tilde{G}}(\kappa,s) = \frac{\hat{\gamma}(s)}{s\hat{\gamma}(s) + \imath\frac{\eta}{\hbar}s^{-\lambda} + \frac{\imath\hbar}{2M}|\kappa|^2} = \frac{s^{\lambda}\hat{\gamma}(s)}{s^{1+\lambda}\hat{\gamma}(s) + \imath\frac{\eta}{\hbar} + \frac{\imath\hbar}{2M}|\kappa|^2 s^{\lambda}}. \tag{8.101}$$

Now, we can find the wave function for particular forms of the memory kernel and nonlocal terms.

The case $\gamma(t) = \delta(t)$ yields

$$\begin{aligned} G(x,t) &= \mathcal{F}^{-1}\left[\sum_{n=0}^{\infty}\left(-\imath\frac{\eta}{\hbar}\right)^n t^{(\lambda+1)n}\mathrm{E}_{1,(\lambda+1)n+1}^{n+1}\left(-\imath\frac{\hbar}{2M}t|\kappa|^2\right)\right] \\ &= \frac{1}{2|x|}\sum_{n=0}^{\infty}\frac{(-\imath\frac{\eta}{\hbar})^n}{n!}t^{(\lambda+1)n}\mathrm{H}_{2,2}^{2,0}\left[\frac{|x|}{\sqrt{\imath\frac{\hbar}{2M}t}}\middle|\begin{matrix}((\lambda+1)n+1,\frac{1}{2}),(1,\frac{1}{2})\\(1,1),(n+1,\frac{1}{2})\end{matrix}\right]. \end{aligned} \tag{8.102}$$

For the power-law memory kernel (8.74), the Green's function becomes

$$G(x,t) = \mathcal{F}^{-1}\left[\sum_{n=0}^{\infty}\left(-\imath\frac{\eta}{\hbar}\right)^{n} t^{(\mu+\lambda)n}\mathrm{E}_{\mu,(\mu+\lambda)n+1}^{n+1}\left(-\imath\frac{\hbar}{2M}t^{\mu}|\kappa|^{2}\right)\right]$$
$$= \frac{1}{2|x|}\sum_{n=0}^{\infty}\frac{(-\imath\frac{\eta}{\hbar})^{n}}{n!}t^{(\mu+\lambda)n}\mathrm{H}_{2,2}^{2,0}\left[\frac{|x|}{\sqrt{\imath\frac{\hbar}{2M}t^{\mu}}}\,\middle|\,\begin{matrix}((\mu+\lambda)n+1,\frac{1}{2}),(1,\frac{1}{2})\\(1,1),(n+1,\frac{1}{2})\end{matrix}\right]. \tag{8.103}$$

Third case: $U(x,t) = \frac{\vartheta}{2\cos(\pi\lambda/2)\Gamma(-\lambda)}|x|^{-\lambda-1}\delta(t)$.

For this nonlocal term we find $\tilde{\hat{U}}(\kappa,s) = \vartheta|\kappa|^{\lambda}$. The Green's function is

$$\tilde{\hat{G}}(\kappa,s) = \frac{\hat{\gamma}(s)}{s\hat{\gamma}(s)+\imath\frac{\vartheta}{\hbar}|\kappa|^{\lambda}+\frac{\imath\hbar}{2M}|\kappa|^{2}}. \tag{8.104}$$

For $\gamma(t) = \delta(t)$, we obtain

$$G(x,t) = \mathcal{F}^{-1}\left[\sum_{n=0}^{\infty}\frac{(-\imath\frac{\vartheta}{\hbar})^{n}}{n!}t^{n}|\kappa|^{\lambda n}\mathrm{E}_{1,n+1}^{n+1}\left(-\imath\frac{\hbar}{2M}t|\kappa|^{2}\right)\right]$$
$$= \frac{1}{2|x|}\sum_{n=0}^{\infty}\frac{(-\imath\frac{\vartheta}{\hbar})^{n}}{n!}t^{n}\frac{1}{|x|^{\lambda n}}\mathrm{H}_{2,2}^{1,1}\left[\frac{|x|}{\sqrt{\imath\frac{\hbar}{2M}t}}\,\middle|\,\begin{matrix}(1,\frac{1}{2}),(\frac{\lambda n+2}{2},\frac{1}{2})\\(\lambda n+1,1),(\frac{\lambda n+2}{2},\frac{1}{2})\end{matrix}\right]. \tag{8.105}$$

The calculation of the Green's function for the power-law memory kernel is straightforward.

Fourth case: $U(x,t) = \varepsilon\sqrt{\frac{\varrho}{\pi}}e^{-\varrho x^{2}}\delta(t)$

For this nonlocal term, we have

$$\tilde{\hat{U}}(\kappa,s) = \varepsilon e^{-\frac{\kappa^{2}}{4\varrho}}.$$

The Green's function in the Fourier–Laplace space is

$$\tilde{\hat{G}}(\kappa,s) = \frac{\hat{\gamma}(s)}{s\hat{\gamma}(s)+\imath\frac{\varepsilon}{\hbar}e^{-\frac{\kappa^{2}}{4\varrho}}+\imath\frac{\hbar}{2M}|\kappa|^{2}}. \tag{8.106}$$

The simple form of the memory kernel $\gamma(t) = \delta(t)$ yields

$$G(x,t) = \mathcal{F}^{-1}\left[\sum_{n=0}^{\infty}\left(-\iota\frac{\varepsilon}{\hbar}\right)^n t^n \sum_{j=0}^{\infty}\frac{(-1)^j}{j!}\frac{n^j}{2^{2j}\varrho^j}\kappa^{2j}\mathrm{E}_{1,n+1}^{n+1}\left(-\iota\frac{\hbar}{2M}t|\kappa|^{\alpha}\right)\right]$$

$$= \frac{1}{2|x|}\sum_{n=0}^{\infty}\frac{(-\iota\frac{\varepsilon}{\hbar})^n t^n}{n!}\sum_{j=0}^{\infty}\frac{(-1)^j}{j!}\frac{n^j}{2^{2j}\varrho^j}\frac{1}{|x|^{2j}}$$

$$\times \mathrm{H}_{3,3}^{2,1}\left[\frac{|x|}{\left(\iota\frac{\hbar}{2M}t\right)^{1/2}}\,\middle|\,\begin{matrix}(1,\frac{1}{2}),(n+1,\frac{1}{2}),(j+1,\frac{1}{2})\\(2j+1,1),(n+1,\frac{1}{2}),(j+1,\frac{1}{2})\end{matrix}\right]. \quad (8.107)$$

Note that in the limit $\rho \to \infty$, from (8.107) it follows that

$$G(x,t) = \frac{1}{2|x|}\mathrm{H}_{1,1}^{1,0}\left[\frac{|x|}{\left(\iota\frac{\hbar}{2M}t\right)^{1/2}}\,\middle|\,\begin{matrix}(1,\frac{1}{2})\\(1,1)\end{matrix}\right]\sum_{n=0}^{\infty}\frac{(-\iota\frac{\varepsilon}{\hbar})^n t^n}{n!}$$

$$= \frac{1}{\sqrt{4\pi\iota\frac{\hbar}{2M}t}}e^{-\frac{x^2}{4\iota\frac{\hbar}{2M}t}}e^{-\frac{\iota\varepsilon}{\hbar}t}, \quad (8.108)$$

which corresponds to Eq. (8.98) for a nonlocal term of the form $U(x,t) = \varepsilon\delta(x)\delta(t)$, since

$$\lim_{\rho\to\infty}\varepsilon\sqrt{\frac{\rho}{\pi}}e^{-\rho x^2} = \varepsilon\delta(x),$$

because

$$\lim_{\eta\to 0+}\frac{1}{\sqrt{4\pi\eta}}e^{-\frac{x^2}{4\eta}} = \delta(x), \quad \rho \to \frac{1}{4\eta}.$$

We notice that, for different forms of the memory kernel and nonlocal terms, we can find exact wave functions by using the Green's function approach. The results have been represented in terms of Mittag-Leffler and the H-function of Fox and are valid for diffusion-like equations with memory and nonlocal terms if we simply perform the substitution

$$\iota\frac{\hbar}{2M} \to D.$$

8.5 Nonlocal Effects on the Energy Spectra

In this section, we consider a fractional model to describe a system whose low temperature spectrum of elementary excitations is composed of two gases of quasiparticles: the phonon gas and an ideal gas of (new) quasiparticles coming from a specific Hamiltonian. If the physical system behaves like this, we can show that it is possible to obtain a temperature dependence for the specific heat in good agreement with some well-known experimental data [301].

In the crystalline order, the atoms forming the solid suppress their individual displacements in favour of collective movements whose quantum manifestations are the phonons. In a crystal system, this order is enough to describe its thermal behaviour at very low temperature. If the system is metallic, another contribution, coming from the Exclusion Principle, has to be taken into account, but the global picture is found from the collective movements of the system. The case of a non-crystalline solid is surely more complicated, because, since the pioneering work of Zeller and Pohl, it is well known that the thermal properties of these materials are very different from the crystalline ones [302].

Some efforts have been made in the past to establish the physical nature of the elementary excitations present in these systems. By following an analogy to what happens in the physics of superfluid helium, a new contribution to the specific heat of a glassy system from the extra density of states produced by roton-like excitations has been proposed [303–306]. In particular, it was proposed that all liquids and glasses possess a new fundamental excitation [305]. This excitation should be a localised region of somewhat lower or higher density than the host matrix. These localised regions should have the property that they are free to propagate throughout the crystal, such that the glassy system should be composed of a "gas" of excitations. These excitations are supposed to explain most of the properties of the liquids and glasses.

To investigate a possible source for these "new" quasiparticles, we assume that the system is described by a Hamiltonian that can be written in the form

$$\widehat{\mathcal{H}} = \widehat{\mathcal{H}}_{\mathrm{F}} \oplus \widehat{\mathcal{H}}_{\mathrm{D}}, \tag{8.109}$$

where $\widehat{\mathcal{H}}_{\mathrm{F}}$ is a conventional Hamiltonian incorporating fractional derivatives in the kinetic energy term (as we discuss below) and $\widehat{\mathcal{H}}_{\mathrm{D}}$ is related to the presence of phonons in the system, also incorporating an excess term, typical of glasses. The form (8.109) can be justified if we admit the existence of two kinds of decoupled elementary excitations governing the behaviour of the system at very low temperature, as mentioned above. The partition function factorises and the specific heat is shown to be a sum of two contributions:

$$C(T,V,N) = C_{\mathrm{F}}(T,V,N) + C_{\mathrm{D}}(T,V,N).$$

For the fractional part, we can consider the following effective Hamiltonian:

$$\begin{aligned}\widehat{\mathcal{H}}_{\mathrm{F}} &= \frac{1}{2m_\alpha}\int d\mathbf{r}\psi^\dagger(\mathbf{r},t)\left(-\hbar^2\nabla^2\right)^{\alpha/2}\psi(\mathbf{r},t)\\ &\quad + \frac{1}{2}\int d\mathbf{r}\int d\mathbf{r}'\psi^\dagger(\mathbf{r},t)\psi^\dagger(\mathbf{r}',t)U(|\mathbf{r}-\mathbf{r}'|)\psi(\mathbf{r},t)\psi(\mathbf{r}',t),\end{aligned} \tag{8.110}$$

where $\psi^\dagger(\mathbf{r},\mathbf{t})$ and $\psi(\mathbf{r},\mathbf{t})$ are second-quantised operators, m_α is an effective constant, the last term is the interaction between the components of the system, and

$$\left(-\hbar^2\nabla^2\right)^{\alpha/2}\psi(\mathbf{r},t) \equiv \int \frac{d\mathbf{p}}{(2\pi\hbar)^3} e^{i\mathbf{p}/\hbar\cdot\mathbf{r}} |\mathbf{p}|^\alpha \psi(\mathbf{p},t)$$

is the quantum Reisz operator, which introduces a nonlocal character in the kinetic energy term (see Section 2.4). A remarkable characteristic of this procedure is that the solutions of the equation of motion (see below) remind us of a Lévy distribution, instead of a Gaussian distribution. Another direct consequence, via the Heisenberg equation for $\psi(\mathbf{r},\mathbf{t})$, verified from Eq. (8.110), concerns the dynamical equation for $\psi(\mathbf{r},\mathbf{t})$, which is actually given by

$$\begin{aligned} i\hbar\frac{\partial}{\partial t}\psi(\mathbf{r},t) &= \frac{1}{2m_\alpha}\left(-\hbar^2\nabla^2\right)^{\alpha/2}\psi(\mathbf{r},t) \\ &+ \int d\mathbf{r}' V(|\mathbf{r}-\mathbf{r}'|)\psi^\dagger(\mathbf{r}',t)\psi(\mathbf{r}',t)\psi(\mathbf{r},t). \end{aligned} \tag{8.111}$$

Equation (8.111) is a Schrödinger-like equation with fractional derivatives applied to the spatial variable, instead of the usual ones. This equation, without the interaction term but incorporating an external potential, was analysed in several scenarios, in a first quantised perspective [263, 266, 274]. In fact, it has been applied to analyse the energy spectra of a hydrogen-like atom, a fractional oscillator in the semiclassical approximation, the parity conservation law [266], quark–antiquark $q\bar{q}$ bound states treated within the non-relativistic potential picture [263], and the quantum scattering problem [274]. In particular, in this context for the free case (i.e., absence of interaction) with $\psi(\mathbf{r},0) = \delta(\mathbf{r})$, the solution of Eq. (8.111) is a Lévy-like distribution, i.e., it is

$$\psi(\mathbf{r},t) = \int \frac{d\mathbf{p}}{(2\pi\hbar)^3} e^{i\mathbf{p}\cdot\mathbf{r}/\hbar} e^{-i|\mathbf{p}|^\alpha/(2m_\alpha\hbar)t}. \tag{8.112}$$

The quantum statistics which emerge from the above scenario by using the thermal Green's function approach may be related to the dynamical aspects of the $\psi(\mathbf{r},t)$. We define the one-particle Green's function as [307]

$$G(1,1') = \frac{1}{i}\langle \mathbf{T}(\psi(1)\psi^\dagger(1'))\rangle, \tag{8.113}$$

where the thermodynamical averages, $\langle\cdots\rangle$, are evaluated by taking the grand canonical ensemble into account; $\mathbf{T}$ is the Dyson time-ordering operator; and 1 and $1'$ correspond to the variables $\mathbf{r}_1, t_1$ and $\mathbf{r}_{1'}, t_{1'}$, respectively. From this equation, we can define the correlation functions

$$G^{>}(1,1') = \frac{1}{i}\langle \psi(1)\psi^\dagger(1')\rangle, \tag{8.114a}$$

and

$$G^{<}(1,1') = \pm\frac{1}{i}\langle\psi^{\dagger}(1')\psi(1)\rangle, \tag{8.114b}$$

where $>$ and $<$ represent the Green's function to $t_1 > t_{1'}$, $G = G^{>}$ and $t_1 < t_{1'}$, $G = G^{<}$. The upper (lower) sign corresponds to the bosonic (fermionic) case and from Eq. (8.110) and Eq. (8.114a) it is possible to show that

$$G^{<}(1,1')|_{t_1=0} = \pm e^{\beta\mu}G^{>}(1,1')|_{t_1=-i\beta},$$

by using the cyclic invariance of the trace, $\mathrm{Tr}(\hat{A}\hat{B}) = \mathrm{Tr}(\hat{B}\hat{A})$. Similar to what is done in the nonfractional case, we may introduce the spectral function, $A(\mathbf{p},\omega)$, defined as

$$A(\mathbf{p},\omega) = G^{>}(\mathbf{p},\omega) \mp \mathbf{G}^{<}(\mathbf{p},\omega),$$

and express $G^{<}$ and $G^{>}$ as follows:

$$G^{>}(\mathbf{p},\omega) = (1 \pm f(\omega))A(\mathbf{p},\omega) \tag{8.115a}$$

and

$$G^{<}(\mathbf{p},\omega) = f(\omega)A(\mathbf{p},\omega), \tag{8.115b}$$

i.e., in terms of the spectral function, with $f(\omega) = 1/(e^{(\omega-\mu)} \pm 1)$. Using these Green's functions, we may obtain thermodynamical quantities such as the average of particle density with momentum $\mathbf{p}$ and energy ω, i.e., $\langle n(\mathbf{p},\omega)\rangle = G^{<}(\mathbf{p},\omega)$, the pressure $P(\beta,\mu) = \int_{-\infty}^{\mu} d\mu'\langle n(\mathbf{p},\omega)\rangle$, and the average of the energy

$$\langle H\rangle = V\int_{-\infty}^{\infty}\frac{d\omega}{2\pi}\int\frac{d^3\mathbf{p}}{(2\pi\hbar)^3}\frac{\omega + |\mathbf{p}|^{\alpha}/2m_{\alpha}}{2}G^{<}(\mathbf{p},\omega). \tag{8.116}$$

We now calculate the specific heat obtained from the Hamiltonian (8.109). As underlined before, it can be written as a sum of two independent contributions. In a simplified form [308], it reads

$$C = A_0T^{3/\alpha} + B_0T^3, \tag{8.117}$$

where A_0 and B_0 are temperature independent. The first contribution comes from the kinetic term present in (8.110). It is reduced to the contribution of an ideal Bose gas when the kinetic energy term is the usual one, i.e., for $\alpha = 2$. The second term is the usual Debye contribution.

In Fig. 8.7, the specific heat behaviour of three noncrystalline samples is shown for very low temperatures. The agreement between the predictions of this model is very good for SiO_2 ($C_F \approx T^n, n \approx 1.15$) and GeO_2 ($C_F \approx T^n, n \approx 0.98$), and quite

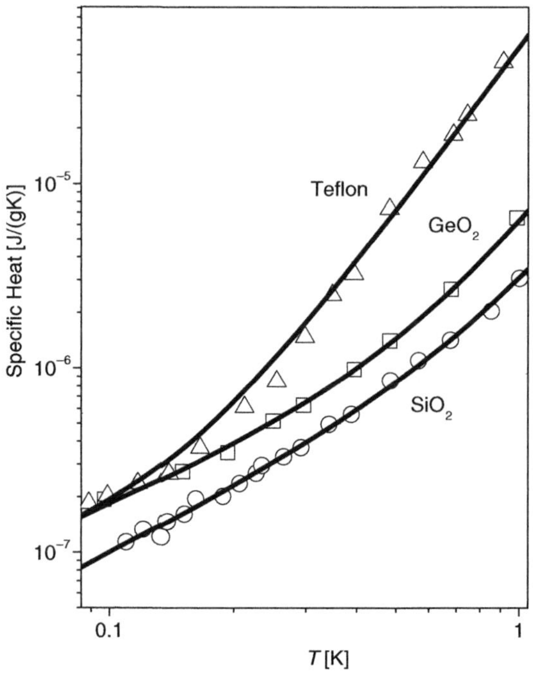

Figure 8.7 Specific heat for three noncrystalline samples versus T. The values of parameters A_0, B_0, and α for SiO_2 are 1.40×10^{-6}[J/(g K$^{1+3/\alpha}$)], 1.70×10^{-6}[J/(g K^4)], and 2.60, respectively. For GeO_2, we have that $A_0 = 1.74 \times 10^{-6}$[J/(g K$^{1+3/\alpha}$)], $B_0 = 4.63 \times 10^{-6}$[J/(g K^4)], and $\alpha = 3.05$. Finally, for Teflon, we have $A_0 = 6.45 \times 10^{-7}$[J/(g K$^{1+3/\alpha}$)], $B_0 = 5.50 \times 10^{-5}$[J/(g K^4)], and $\alpha = 4.5$ [301]. Modified from E. K. Lenzi, N. G. C. Astrath, R. Rossato, and L. R. Evangelista, Non-local effects on the thermal behaviour of non-crystalline solids, Brazilian Journal of Physics **39**, 507–510 (2009), with the permission of Springer.

satisfactory for Teflon [309]. The same good agreement is found with the measured values for ethanol [310], shown in Fig. 8.8. In this case, the contribution coming from the fractional Hamiltonian is of the form $C_{\mathrm{F}} \approx T^{1.1}$. Except for ethanol, the exponents lie between 1.0 and 1.5, approximately; i.e., the temperature behaviour of the specific heat is not linear at all.

For what concerns $\widehat{\mathcal{H}}_{\mathrm{D}}$ one can use an effective Hamiltonian [311]:

$$\widehat{\mathcal{H}}_{\mathrm{D}}(r,p,T) = p^2 + a\,r^2 - b\sqrt{T}\,r^3\,e^{-c\sqrt{T}\,r}, \tag{8.118}$$

where p and r are the rescaled momentum and position variables, respectively; a, b, and c are constants; and T is the absolute temperature. In terms of these rescaled variables, it was shown that the specific heat can be written as

$$C = E_0 T^3 \int_0^{r*} dr \int_0^{p*} dp p^2 r^2 \left[\frac{\widehat{\mathcal{H}}_{\mathrm{D}}(r,p,T)}{e^{\widehat{\mathcal{H}}_{\mathrm{D}}(r,p,T)} - 1}\right]^2 e^{\widehat{\mathcal{H}}_{\mathrm{D}}(r,p,T)}, \tag{8.119}$$

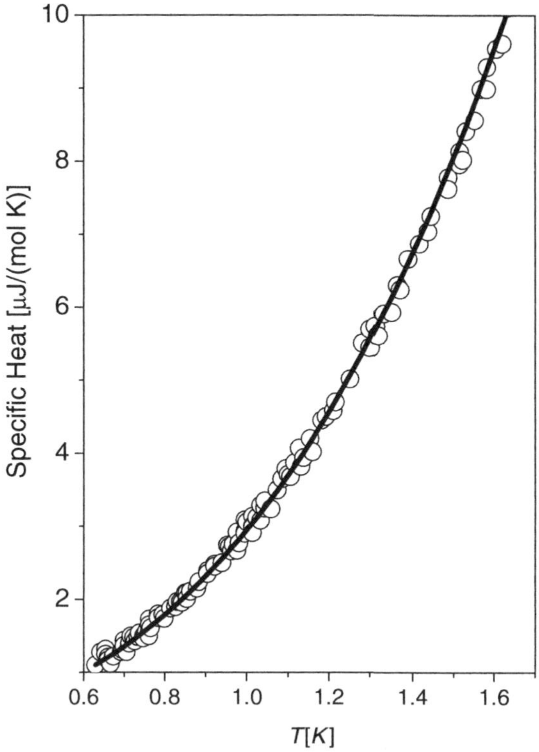

Figure 8.8 Specific heat of ethanol versus T. The values of parameters A_0, B_0, and α for SiO_2 are 1.05×10^{-3} [J/(mol $K^{1+3/\alpha}$)], 1.89×10^{-3}[J/(mol K^4)], and 2.70, respectively [301]. Modified from E. K. Lenzi, N. G. C. Astrath, R. Rossato, and L. R. Evangelista, Non-local effects on the thermal behaviour of non-crystalline solids, Brazilian Journal of Physics **39**, 507–510 (2009), with the permission of Springer.

where $E_0 = (4\pi k_B^2)^2(2m/\hbar^2)^{3/2}$, $r* = r_0/\sqrt{k_B T}$, and $p* = p_0/\sqrt{2mk_B T}$, with r_0 and p_0 being the cut-off values for the variables. Written in the form (8.118), it is evident that the anharmonic contribution may be negligible for very low temperature and also for room temperature.

The complete scenario could be then as follows. At very low temperature, the dynamics of the glassy system is governed by a Hamiltonian in the form (8.109), with $\widehat{\mathcal{H}}_F$ given by the kinetic term of (8.110), and $\widehat{\mathcal{H}}_D$ having a form similar to (8.118). At low temperature, i.e., near the temperature of the boson peak (between 5 and 50 K), the dynamics is still governed by (8.109), but the contribution coming from (8.110) is negligible, with the term represented by $\widehat{\mathcal{H}}_D$, with its anharmonic part, playing the dominant role. In this framework, the specific heat of the system can be well described by a Hamiltonian written in the general form (8.109), with each term contributing more significantly or not according to the range of temperatures considered. To put this approach on a firmer ground, it is necessary to justify the assumption of a decoupling between the different elementary excitations, which permits us to assume the form (8.109) and to justify also the physical basis of a fractional kinetic energy term. The gain with this kind of approach lies in the nonlocal character of the low temperature Hamiltonian, represented by fractional terms in the kinetic energy. This new element can be the source of a very rich spectral distribution of energies and can indicate a possible mechanism to explain the nonconventional thermal behaviour of glasses. If this picture holds true, each glass system will reach the more appropriate value of α to express the importance of nonlocal effects on its dynamics.

9

Anomalous Diffusion and Impedance Spectroscopy

This chapter describes some analytical results obtained by means of a pioneering application of fractional diffusion equations to the electrochemical impedance technique employed to investigate properties of condensed matter samples. The first part of the chapter focuses on some basic aspects of the impedance spectroscopy and the continuum Poisson–Nernst–Planck (PNP) model governing the behaviour of mobile charges. In this model, the fundamental equations to be solved are the continuity equations for the positive and negative charge carriers coupled with Poisson's equation for the electric potential across the sample. The diffusion equation is then rewritten in terms of fractional time derivatives and the predictions of this new model are analysed, emphasising the low frequency behaviour of the impedance by means of analytical solutions. The model is reformulated with the introduction of the fractional equations of distributed order for the bulk system. As a step further, the proposition of a new model – the so-called PNPA model, where "A" stands for anomalous – is built by extending the use of fractional derivatives to the boundary conditions, stated in terms of an integro-differential expression governing the interfacial behaviour. Some experimental data are invoked just to test the robustness of the model in treating interfacial effects in the low frequency domain.

9.1 Impedance Spectroscopy: Preliminaries

The electrochemical impedance technique is used to investigate electrical properties of liquid materials [312]. The sample is submitted to an ac voltage of small amplitude to assure that its response to the external signal is linear. Thus, the impedance, $Z(\omega)$, is measured as a function of the frequency $f = \omega/2\pi$ of the applied voltage, $V(t)$, with a typical amplitude V_0. In the low frequency region, of particular importance is the role of the mobile ions regarding the value of the measured impedance because they contribute to the electrical current [152].

In this frequency region, the theoretical analysis of the influence of the ions on the electrical impedance is usually performed by solving the continuity equations for the positive and negative ions and the equation of Poisson for the actual electric potential across the sample. This is the so-called PNP model.

The standard analysis predicts that the real part of the electrical impedance is frequency independent up to the relaxation frequency, whereas the imaginary part in dc limit diverges as $1/\omega$. However, the experimental data are not always in agreement with these predictions. For this reason, several models have been proposed to account for the observed effect of the ions on the electrical response of the cell [313, 314].

There seems to be no doubt that the adsorption phenomenon has a central role [152]. This should be one of the first and main mechanisms. But the adsorption alone is not able to account for the experimental features found in these systems. Therefore, the role of nonblocking electrodes has also been invoked to shed some light on the problem. In some sense, both situations should be analysed by considering the adsorption process in which memory effects could play some role [151, 315, 316]. In this case, the kinetic equation at the interface incorporates memory effects in the process by means of three different kernels, whose importance is associated with the predominance of chemisorption or physisorption in the adsorption–desorption processes [316]. The net result is that the density of adsorbed particles at the surface can present oscillations with time before reaching a stationary regime [151, 316].

Another possible interfacial mechanism could be connected with the accumulation of ions near the surface, in the absence of adsorption phenomena. This accumulation should be connected with the necessity to modify Fick's law and to solve a modified diffusion equation. In this regard, a fractional diffusion equation could be the appropriate theoretical framework to take into account this effect on the electrical impedance.

A series of pioneering works, dedicated to addressing the electrochemical impedance by means of fractional calculus [317–320], has shown that the impedance at high frequency behaves as

$$Z \propto (i\omega)^{-\gamma/2},$$

and, at low frequency, as

$$Z \propto (i\omega)^{-\gamma},$$

where γ is the fractional coefficient. These results are a first generalisation, using fractional calculus, proposed for the behaviour of Warburg impedance [321] (high frequency) and for the predictions of the constant-phase element (CPE) impedance model (low frequency). In these approaches, the effective electric field inside the

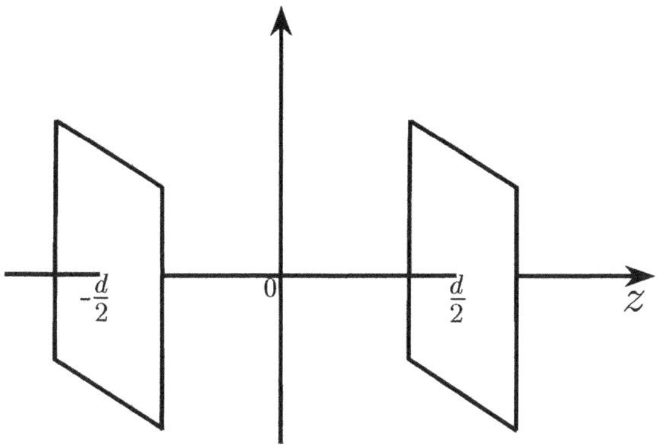

Figure 9.1 Geometry of the sample in the shape of a slab of thickness d. The surfaces located at $z = \pm d/2$ represent the electrodes.

sample is taken as the applied external field; i.e., no spatial distribution of charges is determined. This means that the influence of the ions on the spectroscopy impedance measurements is not taken into account in a complete manner. To achieve a more realistic description, a different strategy may be employed. It consists in solving the complete problem, i.e., taking into account the presence of a drift term in the diffusion equation coupled with Poisson's equation.

To establish the notation and the geometry representing a typical finite-length situation used in this chapter and in the next, we discuss a simple model to tackle the effect of the accumulated ions on the impedance spectroscopy response of the sample (see details in Ref. [152]). Consider an electrolytic cell in the shape of a slab of thickness d, limited by two flat surfaces placed in $z = \pm d/2$, where z is the axis, normal to the surfaces, of a Cartesian reference frame (see Fig. 9.1). This cell is initially filled with an isotropic liquid, inside which dimensionless ions – positive, with density $N_+(z,t)$, and negative, with density $N_-(z,t)$ – are dispersed, forming a homogeneous medium of dielectric constant ϵ (for simplicity, hereafter measured in units of ϵ_0). The surfaces are supposed by now to be nonadsorbing, and the monovalent ions, of charge q, are supposed to have the same mobility $\mu_+ = \mu_- = \mu$; before the application of an external field, the liquid is locally and globally neutral. When the external field is turned on, the liquid becomes locally charged but remains globally neutral.

For the investigations connected with impedance spectroscopy, we assume that the surfaces are prepared to work initially as blocking electrodes. In this case, for $V_0 = 0$, $N_+(z,t) = N_-(z,t) = N$, with N representing the equilibrium density of ions of positive and negative signs. If we neglect recombination of ions, we have

$$\int_{-d/2}^{d/2} N_+(z,t)dz = \int_{-d/2}^{d/2} N_-(z,t)dz = N\,d,$$

stating the *conservation of the number of particles*. If, on the other hand, $V_0 \neq 0$, $N_+(z,t) = N_-(-z,t)$. Since the amplitude of the external voltage is assumed to be small, the actual densities of ions differ only slightly from N and we can write

$$N_+(z,t) = N + n_+(z,t) \quad \text{and} \quad N_-(z,t) = N + n_-(z,t), \tag{9.1}$$

where $n_\pm(z,t)$ represent the bulk densities of ions due to the presence of the external field. From the hypothesis of global neutrality, we have

$$\int_{-d/2}^{d/2} n_+(z,t)dz = \int_{-d/2}^{d/2} n_-(z,t)dz = 0. \tag{9.2}$$

Under these assumptions, the problem of obtaining the current flowing through the electrolytic cell can be solved by considering the three fundamental equations of the problem. The first two are the continuity equations,

$$\frac{\partial}{\partial t}n_\pm(z,t) = -\frac{\partial}{\partial z}j_\pm(z,t), \tag{9.3}$$

in which the densities of currents for positive and negative ions are

$$j_\pm(z,t) = -D\left[\frac{\partial}{\partial z}n_\pm(z,t) \pm \frac{Nq}{k_B T}\frac{\partial}{\partial z}V(z,t)\right], \tag{9.4}$$

where k_B is the Boltzmann constant, T is the absolute temperature, and D is the diffusion coefficient. The third one is Poisson's equation:

$$\frac{\partial^2}{\partial z^2}V(z,t) = -\frac{q}{\epsilon}\left[n_+(z,t) - n_-(z,t)\right]. \tag{9.5}$$

Equation (9.4) holds in the limit $|n_\pm(z,t)| \ll N$, which may always be true if the amplitude of the external voltage is low, as assumed before. The concept of electrical impedance is meaningful only in this framework because the system behaves in a linear manner; indeed, Eqs. (9.3) and (9.5) are linear.

These equations may be solved with the imposed boundary conditions of *blocking electrodes* on $j_\pm(z,t)$, i.e.,

$$j_\pm\left(\pm\frac{d}{2},t\right) = 0, \tag{9.6}$$

and on the *difference of potential* $V(z,t)$:

$$V\left(\pm\frac{d}{2},t\right) = \pm\frac{V_0}{2}e^{i\omega t}. \tag{9.7}$$

If we can assume that

$$V(z,t) = \phi(z)e^{i\omega t}, \tag{9.8}$$

then the electric field in the cell may be easily obtained as

$$E(z,t) = -\frac{\partial V}{\partial z} = -\phi'(z)e^{i\omega t}, \tag{9.9}$$

where $\phi'(z) = d\phi(z)/dz$. To obtain the electric current flowing through the external circuit, we have to determine the total charge density at the interface, $Q_s(t)$. It is connected with the electric field at one interface as follows:

$$E\left(\frac{d}{2},t\right) = -\frac{Q_s(t)}{\epsilon}, \tag{9.10}$$

which, using (9.9), gives the total charge as

$$Q(t) = Q_s(t)\,S = \epsilon\phi'(d/2)Se^{i\omega t}, \tag{9.11}$$

in which S is the area of the electrode. The searched electrical current is then

$$I(t) = \frac{dQ(t)}{dt} = i\omega\epsilon\phi'(d/2)Se^{i\omega t}. \tag{9.12}$$

The electrical impedance of the cell may be determined as

$$Z(\omega) = \frac{V_0 e^{i\omega t}}{I(t)} = \frac{V_0}{i\omega\epsilon\phi'(d/2)S}. \tag{9.13}$$

From $Z(\omega)$ we obtain the real, $R = \mathrm{Re}\,Z$, and the imaginary, $X = \mathrm{Im}\,Z$, parts of the electrical impedance, which are the quantities experimentally detectable.

As a simple application of the formalism sketched above, we determine an explicit analytical expression for the electrical impedance of a cell filled with an isotropic fluid having the geometry and subject to the conditions presented before (see Fig. 9.1). In the steady state, we may write for the densities of positive and negative ions the following expression:

$$n_{\pm}(z,t) = \rho_{\pm}(z)e^{i\omega t}, \tag{9.14}$$

in such a way that Eq. (9.5) is rewritten as

$$\frac{d^2\phi(z)}{dz^2} = -\frac{q}{\varepsilon}\left[\rho_+(z) - \rho_-(z)\right], \tag{9.15}$$

if (9.8) is taken into account. The functions $\rho_{\pm}(z)$ in Eq. (9.15) are solutions of Eq. (9.3), which now assumes the form:

$$\frac{d^2\rho_{\pm}(z)}{dz^2} - \frac{1}{\lambda^2}\rho_{\pm}(z) + \frac{1}{2\lambda_{\mathrm{D}}^2}\rho_{\mp}(z) = 0, \tag{9.16}$$

where

$$\lambda = \lambda_D \sqrt{\frac{2}{1 + 2i(\omega/D)\lambda_D^2}}, \tag{9.17}$$

is an intrinsic length of the problem, and

$$\lambda_D = \sqrt{\frac{\varepsilon k_B T}{2Nq^2}} \tag{9.18}$$

is the Debye's screening length [322]. We look for solutions of Eq. (9.16) in the form

$$\rho_\pm(z) = C_\pm e^{\nu z}. \tag{9.19}$$

Substitution of (9.19) into Eq. (9.16) yields the following equation:

$$\begin{bmatrix} \nu^2 - \frac{1}{\lambda^2} & \frac{1}{2\lambda_D^2} \\ \frac{1}{2\lambda_D^2} & \nu^2 - \frac{1}{\lambda^2} \end{bmatrix} \begin{pmatrix} C_+ \\ C_- \end{pmatrix} = 0, \tag{9.20}$$

which admits solutions different from $C_+ = C_- = 0$ only if

$$\left(\nu^2 - \frac{1}{\lambda^2}\right) = \frac{1}{4\lambda_D^4}, \tag{9.21}$$

implying that

$$\frac{C_-}{C_+} = -2\lambda_D^2 \left(\nu^2 - \frac{1}{\lambda^2}\right).$$

Using now the condition $\rho_+(z) = \rho_-(-z)$, together with the condition from Eq. (9.2), namely,

$$\int_{-d/2}^{d/2} \rho_\pm(z)\, dz = 0, \tag{9.22}$$

we obtain the solutions

$$\rho_\pm(z) = \pm p_0 \sinh \beta z, \tag{9.23}$$

where

$$\beta = \frac{1}{\lambda_D}\sqrt{1 + i\frac{\omega}{D}\lambda_D^2}. \tag{9.24}$$

In Eq. (9.23), p_0 is an integration constant to be determined by the boundary conditions of blocking electrodes, Eqs. (9.6) and (9.7), the latter implying that

$$\phi\left(\pm\frac{d}{2}\right) = \pm\frac{V_0}{2}. \tag{9.25}$$

The electric potential may be obtained from Eq. (9.15), which assumes the form

$$\frac{d^2\phi(z)}{dz^2} = -2\frac{q}{\varepsilon} p_0 \sinh \beta z. \tag{9.26}$$

Since $\phi(z) = -\phi(-z)$, integration of (9.26) yields

$$\phi(z) = -\frac{2q}{\varepsilon\beta^2} p_0 \sinh \beta z + p_1 z, \tag{9.27}$$

where p_1 is another integration constant to be determined by the boundary conditions. Using previous results, the densities of currents defined by (9.4) are rewritten as

$$j_\pm(z) = \mp D\left[i\left(\frac{\omega}{D\beta}\right) p_0 \cosh \beta z + \frac{Nq}{k_B T} p_1\right]. \tag{9.28}$$

Using Eqs. (9.27) and (9.28), the boundary conditions (9.6) and (9.25) yield the system of equations

$$\begin{aligned} -\frac{2q}{\varepsilon\beta^2} p_0 \sinh\left(\beta\frac{d}{2}\right) + \frac{d}{2} p_1 &= \frac{V_0}{2} \\ i\left(\frac{\omega}{D\beta}\right) p_0 \cosh\left(\beta\frac{d}{2}\right) + \left(\frac{Nq}{k_B T}\right) p_1 &= 0. \end{aligned} \tag{9.29}$$

Solving (9.29) for p_0 and p_1, we obtain

$$p_0 = -\frac{Nq\beta}{2k_B T}\frac{V_0}{\Delta} \quad \text{and} \quad p_1 = \frac{i\omega}{2D}\cosh\left(\beta\frac{d}{2}\right)\frac{V_0}{\Delta}, \tag{9.30}$$

where

$$\Delta = \frac{1}{\lambda_D^2\beta}\sin\left(\beta\frac{d}{2}\right) + i\left(\frac{\omega d}{2D}\right)\cosh\left(\beta\frac{d}{2}\right).$$

Thus, the electrical problem is formally solved and the impedance, defined in Eq. (9.13), is analytically obtained as

$$Z = -i\frac{2}{\omega\varepsilon S\beta^2}\left[\frac{1}{\lambda_D^2\beta}\tanh\left(\frac{\beta d}{2}\right) + \frac{i\omega d}{2D}\right]. \tag{9.31}$$

When we consider the limit of true dielectric, for which $N = 0$ and $\lambda_D \to \infty$, the response of the sample is purely capacitive and we obtain

$$Z = \frac{1}{i\omega C}, \quad \text{where} \quad C = \varepsilon\frac{S}{d},$$

as expected for a parallel plate capacitor. In the low frequency limit, by rewriting $Z = R + iX$, it is possible to show that [152]

$$R(\omega \to 0) = \frac{\lambda_D^2 d}{\varepsilon D S}\left(1 - \frac{\lambda_D^4}{D^2}\omega^2\right) \tag{9.32}$$

and

$$X(\omega \to 0) = -2\frac{\lambda_D}{\varepsilon S \omega}\left(1 + \frac{\lambda_D^3 d}{2D^2}\omega^2\right). \tag{9.33}$$

In the limit of high frequency, we have

$$R(\omega \to \infty) = \frac{dD}{\varepsilon S \omega^2 \lambda_D^2}\left(1 - \frac{1}{d}\sqrt{\frac{2D}{\omega}}\right) \tag{9.34}$$

and

$$X(\omega \to \infty) = -\frac{d}{\varepsilon S \omega}\left(1 - \frac{D}{d\omega\lambda_D^2}\sqrt{\frac{2D}{\omega}}\right). \tag{9.35}$$

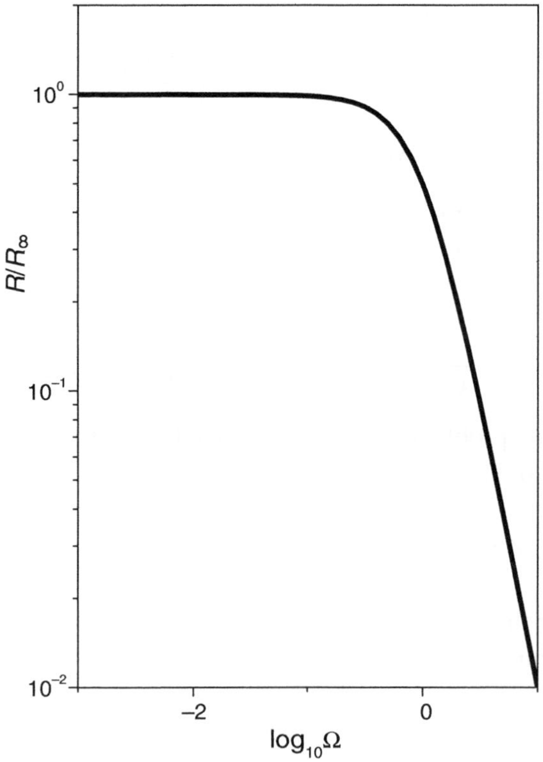

Figure 9.2 Real part of the impedance of the cell, R/R_∞ versus $\log_{10}\Omega$ for $M =$ 500, as predicted by Eq. (9.36).

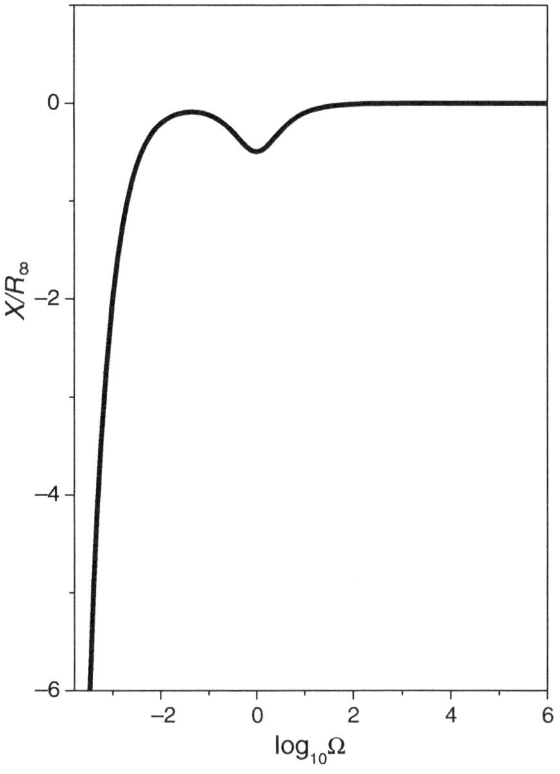

Figure 9.3 Imaginary part of the impedance of the cell, X/R_∞ versus $\log_{10}\Omega$ for $M = 500$, as predicted by Eq. (9.36).

The impedance defined by (9.31) may be put in a compact form as [323]

$$Z(\Omega) = R_\infty \frac{M\Omega\sqrt{1+i\Omega} - i\tanh\left(M\sqrt{1+i\Omega}\right)}{M\Omega(1+i\Omega)^{3/2}}, \tag{9.36}$$

where

$$R_\infty = \frac{\lambda_D^2}{\varepsilon D}\frac{d}{S}, \quad M = \frac{d}{2\lambda_D}, \quad \text{and} \quad \Omega = \frac{\omega}{\omega_D}, \tag{9.37}$$

with $\omega_D = D/\lambda_D^2$ being the Debye's circular frequency. For illustrative purposes, in Figs. 9.2 and 9.3 the general behaviour of the real and imaginary parts of the impedance, given by (9.36), are depicted for $M = d/2\lambda_D = 500$. There is a characteristic plateau in the low frequency limit of the real part; also characteristic is the presence of a minimum in the imaginary part of the impedance. These behaviours are analytically predicted by Eqs. (9.32)–(9.35).

9.2 The PNP Time Fractional Model

We present now an extension of the problem discussed in the previous section in terms of fractional equations [324]. A fractional time derivative will be incorporated in the diffusion equation which appears by combining Eq. (9.3) with Eq. (9.4). Following a procedure previously employed [317–320], which yielded useful results for the electrical impedance, we extend the time derivative present in Eq. (9.3) to the Riemann–Liouville fractional derivative by promoting the change

$$\frac{\partial n_{\pm}(z,t)}{\partial t} \longrightarrow \frac{\partial^{\gamma} n_{\pm}(z,t)}{\partial t^{\gamma}} \longrightarrow {}_{t_0}\mathrm{D}_t^{\gamma} n_{\pm}, (z,t),$$

i.e., by using the definition, Eq. (2.8),

$${}_{t_0}\mathrm{D}_t^{\gamma} n_{\pm}(z,t) = \frac{1}{\Gamma\,(m-\gamma)}\frac{d^m}{dt^m}\int_{t_0}^{t} d\bar{t}\,\frac{n_{\pm}(z,\bar{t}\,)}{\left(t-\bar{t}\right)^{\gamma+1-m}}, \tag{9.38}$$

where $m-1<\gamma<m$ and t_0 is related to the conditions initially imposed to the system (see Section 2.1). For the present analysis, we consider $t_0=-\infty$ to study the response of the system to the periodic applied potential defined above [46]. Substitution of Eq. (9.3) into Eq. (9.4), after taking into account the above conditions, yields

$${}_{-\infty}\mathrm{D}_t^{\gamma} n_{\pm}(z,t) = D_{\gamma}\frac{\partial^2}{\partial z^2} n_{\pm}(z,t) \pm \frac{NqD_{\gamma}}{k_BT}\frac{\partial^2}{\partial z^2}V(z,t), \tag{9.39}$$

where D_{γ} represents the diffusion coefficient and the second term results from the drift term with the assumption already made that $|n_{\pm}(z,t)| \ll N$.

When we formulate the mathematical problem in this way, the analysis may be done with a drift-diffusion term *and* simultaneously taking into account the equation of Poisson, as a consequence of the presence of ions. The spatial distribution of the electric field inside the sample can now be explicitly (and analytically) determined!

To go on with the investigation of the role of the ions regarding impedance spectroscopy by means of a fractional approach, it is necessary to obtain again the total current in the external circuit, taking into account the presence of the ions. This means that we have to solve Eq. (9.39) together with Eq. (9.5), considering the current density in the form (9.4), with D substituted with D_{γ}. As indicated before, we assume again that Eqs. (9.8) and (9.14) still hold. Thus, we obtain the following equations:

$$\rho_{\pm}''(z) - \frac{1}{\lambda_1^2}\rho_{\pm}(z) + \frac{1}{2\lambda_D^2}\rho_{\mp}(z) = 0, \tag{9.40}$$

with

$$\frac{1}{\lambda_1^2} = \frac{1}{2\lambda_D^2} + \frac{(i\omega)^\gamma}{D_\gamma}, \tag{9.41}$$

where λ_D is the Debye's screening length defined in Eq. (9.18). To obtain the above system of equations, we have used the result [46]

$$_{-\infty}\mathrm{D}_t^\gamma\left(e^{\lambda t}\right) = \lambda^\gamma e^{\lambda t}, \tag{9.42}$$

for $0 < \gamma \leq 1$ by replacing λ with $i\omega$.

The system of differential equations obtained for $\rho_+(z)$ and $\rho_-(z)$ may be solved by employing the same ansatz defined in Eq. (9.19), yielding a new system of equations:

$$\nu^2 C_\pm - \frac{1}{\lambda_1^2} C_\pm + \frac{1}{2\lambda_D^2} C_\mp = 0, \tag{9.43}$$

for which nontrivial solutions exist only if

$$\nu_{1,2} = \pm\beta = \pm\sqrt{\frac{1}{\lambda_D^2} + \frac{(i\omega)^\gamma}{D_\gamma}} \quad \text{and} \quad \nu_{3,4} = \pm\sqrt{\frac{(i\omega)^\gamma}{D_\gamma}}, \tag{9.44}$$

which become the characteristic exponents of the problem. The rest of the calculation is performed in a way very similar to the one developed in Section 9.1 and will be omitted to save space. Using the results obtained above, the impedance may be written in the form

$$Z = \frac{2}{i\omega\varepsilon S\beta^2}\left[\frac{1}{\lambda_D^2\beta}\tanh\left(\frac{\beta d}{2}\right) + \frac{(i\omega)^\gamma d}{2D_\gamma}\right]. \tag{9.45}$$

This formula is very similar to Eq. (9.31) deduced in the previous section. The important difference is again the fractional coefficient γ. The usual results are reobtained in the limit $\gamma \to 1$.

In Fig. 9.4, Re Z, as calculated by (9.45), is shown as a function of the frequency for typical values of material parameters in an electrolytic cell. The value $\gamma = 1.0$, shown by the dotted line, is depicted for comparative purposes because it represents the normal diffusive behaviour. The deviations from this normal behaviour are remarkable and more pronounced for decreasing values of γ.

In Fig. 9.5, Im Z, as calculated by (9.45), is also shown. In this case, the main differences are obtained for decreasing values of γ in the high frequency region.

As we have seen, the diffusion phenomena that occur near the interface are very complicated and involve several mechanisms. Since the surfaces are not homogeneous, the diffusion has different characteristics at different points in the sample. A generalisation of our approach according to the lines previously suggested [101, 219, 325, 326] can be useful to analyse the experimental data. More

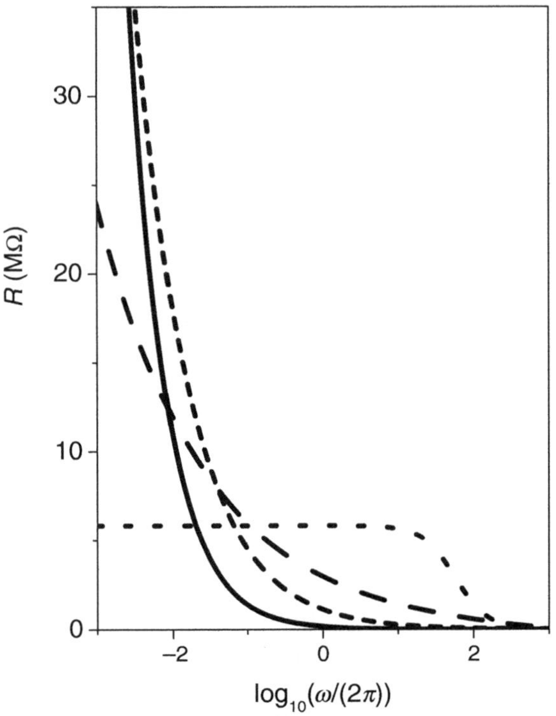

Figure 9.4 Real part of the impedance of the cell versus the frequency for the set of parameters: $\epsilon = 6.7\epsilon_0$ (where ϵ_0 is the dielectric permittivity of the free space), $N = 4 \times 10^{20}\,\mathrm{m}^{-3}$, $D \approx 8.2 \times 10^{-12}\,\mathrm{m}^2/\mathrm{s}$, $d = 25\,\mu\mathrm{m}$, and $S = 2 \times 10^{-4}\,\mathrm{m}^2$. With these values, $\lambda_D \approx 10^{-7}$ m. The curves were drawn for $\gamma = 0.1$ (solid line), $\gamma = 0.4$ (short dashed line), $\gamma = 0.7$ (dashed line), and $\gamma = 1.0$ (dotted line). Modified with permission from E. K. Lenzi, L. R. Evangelista, and G. Barbero, Fractional diffusion equation and impedance spectroscopy of electrolytic cells, Journal of Physical Chemistry B **113**, 11371–11374 (2009). Copyright 2009 American Chemical Society.

precisely, the previous results, obtained for the impedance in the framework of the fractional diffusion equation (9.39), may also be extended further by considering the fractional diffusion equations of distributed order. This will be discussed in more detail in the next section, but we present now a sketch of this approach to indicate its main lines.

A possible generalisation is to consider

$$\int_0^1 d\gamma p(\gamma)\, {}_{-\infty}\mathrm{D}_t^{\gamma}\,(\cdots)\,, \tag{9.46}$$

where $p(\gamma)$ is a distribution of γ with

$$\int_0^1 d\gamma p(\gamma) = 1, \tag{9.47}$$

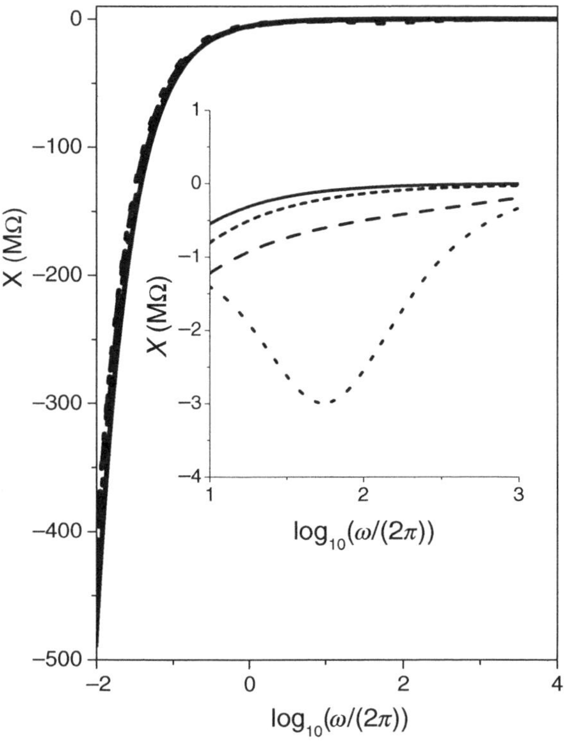

Figure 9.5 Imaginary part of the impedance of the cell versus the frequency for the same set of parameters and for the same values of γ shown in Fig. 9.4. The inset refers to high values of ω. Modified with permission from E. K. Lenzi, L. R. Evangelista, and G. Barbero, Fractional diffusion equation and impedance spectroscopy of electrolytic cells, Journal of Physical Chemistry B **113**, 11371–11374 (2009). Copyright 2009 American Chemical Society.

instead of the fractional operator

$$ {}_{-\infty}\mathrm{D}_t^{\gamma}\,(\cdots)\,. \tag{9.48}$$

For this case, the impedance is formally given by Eq. (9.45) with

$$(i\omega)^{\gamma} \to \int_0^1 d\gamma p(\gamma)(i\omega)^{\gamma}. \tag{9.49}$$

The same change needs to be performed in the quantity β in Eqs. (9.44) and (9.45).

A remarkable property of the fractional derivative of distributed order on the diffusion equation is the presence of different diffusive regimes [219]. This characteristic is evident in the quantities obtained from these equations, in particular the impedance. The physical insight we gain from this analysis can be better appreciated if we consider that $\gamma = 1$ represents the normal diffusion process, while $\gamma < 1$

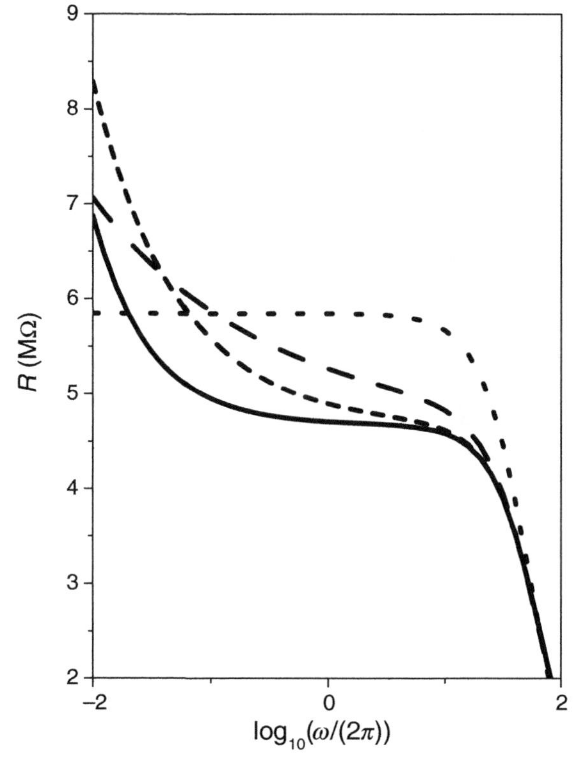

Figure 9.6 The real part of the impedance of the cell versus the frequency for the same set of parameters and for the same values of γ shown in Fig. 9.4. Modified with permission from E. K. Lenzi, L. R. Evangelista, and G. Barbero, Fractional diffusion equation and impedance spectroscopy of electrolytic cells, Journal of Physical Chemistry B **113**, 11371–11374 (2009). Copyright 2009 American Chemical Society.

represents a subdiffusive process. When two typical situations of this kind can be considered in a unique approach, we achieve a more complete phenomenological description of complex phenomena without having to deal with a large number of free parameters. A simple derivation will be discussed in Section 9.3.1.

For illustrative purposes, here we consider the case $p(\overline{\gamma}) = 1/5\delta(\gamma - 1/2) + 4/5\delta(\gamma - 1)$, which yields

$$\beta = \pm\sqrt{\frac{1}{\lambda_D^2} + \frac{1}{5}\frac{(i\omega)^\gamma}{D_\gamma} + \frac{4}{5}\frac{i\omega}{D_\gamma}}. \tag{9.50}$$

In Fig. 9.6, Re Z, calculated by (9.45) with the modifications presented above, is shown again for the same set of parameters as in Figs. 9.4 and 9.5. The existence of a plateau is evident for small values of γ (e.g., $\gamma = 0.1$ as shown in the figure).

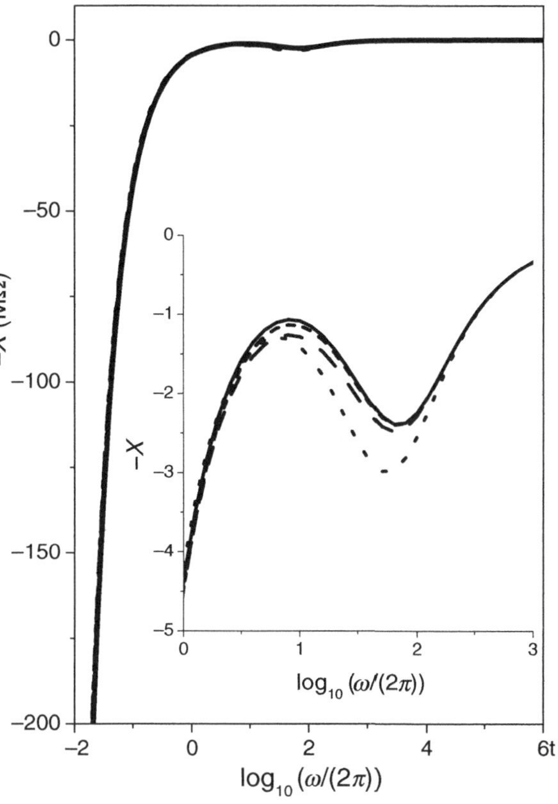

Figure 9.7 The imaginary part of the impedance of the cell versus the frequency for the same set of parameters and for the same values of γ shown in Fig. 9.6. The inset refers to high values of ω. Modified with permission from E. K. Lenzi, L. R. Evangelista, and G. Barbero, Fractional diffusion equation and impedance spectroscopy of electrolytic cells, Journal of Physical Chemistry B **113**, 11371–11374 (2009). Copyright 2009 American Chemical Society.

Moreover, the illustrative curves show that the low frequency behaviour of $\mathrm{Re}\, Z$ is strongly influenced by the anomalous diffusive regime, i.e., the values of $\gamma < 1$, whereas the high frequency behaviour is essentially the same as the one predicted by the normal diffusive behaviour $\gamma = 1$.

In Fig. 9.7, $\mathrm{Im}\, Z$ is shown as a function of the frequency. Again, the variations in the diffusive regime are sensible only in the case of high values of the frequency, as shown in the inset to this figure. The frequency dependence of $\mathrm{Re}\, Z$ and $\mathrm{Im}\, Z$ are similar to the ones experimentally observed in electrolytes and are usually interpreted in terms of the CPE impedance model [327–329].

To sum up, in this section we have presented a first complete, although simplified, approach to determining the impedance of an electrolytic cell when the phenomenon of anomalous diffusion is taken into account. The mathematical problem is complete in the following sense: the diffusion equation, with a drift term, is solved at the same time at which the influence of the ions present in the sample is considered by means of Poisson's equation. Furthermore, the diffusive process is considered in a general formalism represented first by the fractional diffusion equation for which anomalous diffusion (subdiffusive regime) can be found. This means that the diffusion of the ions follows a nonconventional behaviour in the sample. As a consequence, the trends of $\mathrm{Re}\,Z$ and $\mathrm{Im}\,Z$ as a function of the frequency are strongly dependent on the fractional index, i.e., on the kind of diffusive regime taken into account. The long-lasting memory behaviour, which in some sense arises in this context, may be a consequence of the conservation of the number of particles imposed on the system to correctly account for the interfacial behaviour [210].

9.3 Anomalous Diffusion and Memory Effects

In the preceding section, the contribution of the mobile ions to the electrical impedance of an electrolytic cell limited by perfect blocking electrodes was determined by considering the role of the anomalous diffusion process and memory effects. Fractional diffusion equations have been considered there as an appropriate theoretical approach to take into account the ionic redistribution effect on the impedance. The basic idea guiding this procedure is the connection usually found between fractional diffusion equations and anomalous diffusion processes.

In this section, we shall present the general expression for the electrical impedance that results from a complete mathematical model based on the fractional diffusion equation of distributed order. We show that, in the appropriate limits, this expression reduces to the one obtained for the usual diffusion [330, 331] as well as to the one corresponding to a single anomalous diffusive regime [324, 332]. Subsequently, we analyse the physical meaning and the consistency of these new expressions incorporating anomalous diffusion processes into the description of the effect of ions on the electrical impedance. This procedure may be interpreted as a new tool to face the enormous complexity of impedance spectroscopy data and is built by following a general but nevertheless simple mathematical formulation.

9.3.1 Superposition of Two Diffusive Regimes

We consider the same system as before, i.e., a slab of thickness d limited by two flat surfaces (electrodes) placed perpendicular to the z-axis of a Cartesian reference system, at the positions $z = \pm d/2$, as shown in Fig. 9.1.

For example, an elementary derivation for an alternative expression containing a superposition of two diffusive regimes (one of which corresponds to the usual diffusion, i.e., $\gamma = 1$) can be performed as a particular case of a more general approach dealing with fractional diffusion equations of distributed order, to be discussed in Section 9.3.2. For simplicity and just in order to illustrate the procedure, we consider only the diffusion equation for positive ions (a similar equation holds for negative ions) always assuming that $|n_\pm(z,t)| \ll N$, which may be written as

$$\frac{\partial n_+(z,t)}{\partial t} = D\frac{\partial}{\partial z}\left[\frac{\partial n_+(z,t)}{\partial z} + \frac{qN}{k_BT}\frac{\partial V}{\partial z}\right]. \tag{9.51}$$

Using Eqs. (9.8) and (9.14), we obtain

$$(i\omega)\rho_+(z) = D\left[\rho_+''(z) + \frac{qN}{k_BT}\phi''(z)\right]. \tag{9.52}$$

Now, we assume that the diffusion of ions also presents a fractional component and rewrite Eq. (9.51) as a superposition involving a normal diffusive regime and one single anomalous regime characterised by a fractional coefficient γ, in the form:

$$A\frac{\partial n_+}{\partial t} + B\frac{\partial^\gamma n_+}{\partial t^\gamma} = D\frac{\partial}{\partial z}\left[\frac{\partial n_+}{\partial z} + \frac{qN}{k_BT}\frac{\partial V}{\partial z}\right], \tag{9.53}$$

where A should be dimensionless, while the dimension of B is $t^{\gamma-1}$. In the linear limit, again using Eqs. (9.8) and (9.14), from Eq. (9.53), we get

$$\left[A(i\omega) + B(i\omega)^\gamma\right]\rho_+(z) = D\left[\rho_+''(z) + \frac{qN}{k_BT}\phi''(z)\right]. \tag{9.54}$$

The equation of Poisson (9.15) does not change. Equation (9.54) coincides with (9.52) if D is changed as

$$D \longrightarrow D_e = \frac{D}{A + B(i\omega)^{\gamma-1}}. \tag{9.55}$$

In this case, the impedance of the cell again has the same form as (9.31), namely

$$Z_e = -i\frac{2}{\omega\varepsilon\beta_e^2 S}\left[\frac{1}{\lambda_D^2\beta_e}\tanh\left(\beta_e\frac{d}{2}\right) + i\frac{\omega d}{2D_e}\right], \tag{9.56}$$

but β has changed to

$$\beta_e = \frac{1}{\lambda_D}\sqrt{1 + i\frac{\omega}{D_e}\lambda_D^2}. \tag{9.57}$$

To explore the behaviour of the impedance in this context, in Fig. 9.8, we show the frequency behaviour of the real part of Z, as predicted by Eq. (9.56) for different values of the fractional coefficient γ. Regarding the imaginary part, the model predicts that the slopes of the curve in the low frequency region and in the high frequency region are the same (see Fig. 9.9).

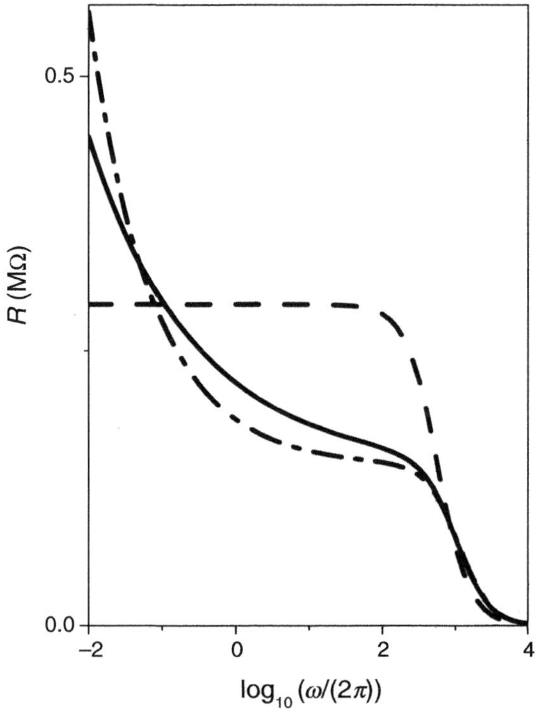

Figure 9.8 The real part of the impedance predicted by Eq. (9.56) for different values of γ: $\gamma = 0.5$ (dashed-dotted line), $\gamma = 0.7$ (solid line), and $\gamma = 1$ (dashed line). The curves were drawn for $S = 2 \times 10^{-4}\,\mathrm{m}^2$, $\varepsilon = 6.7\,\varepsilon_0$, $d = 25 \times 10^{-6}\,\mathrm{m}$, $\lambda = 1.076 \times 10^{-7}\,\mathrm{m}$, and $D = 8.2 \times 10^{-11}\,\mathrm{m}^2/\mathrm{s}$. Modified from L. R. Evangelista, E. K. Lenzi, G. Barbero, and J. R. Macdonald, Anomalous diffusion and memory effects on the impedance spectroscopy for finite-length situations, Journal of Physics: Condensed Matter **23**, 485005 (2011), with the permission of IOP Publishing.

Two limiting situations are considered now. The first is the case $\gamma = 1$ for which $B = 0$ and $A = 1$. In this limit, we recover Eq. (9.31). The other limit, which corresponds to a pure fractional case [41], can be obtained by using $A = 0$, and, consequently:

$$D_e = \frac{D}{B(i\omega)^{\gamma-1}} \quad \text{and} \quad \beta_e = \frac{1}{\lambda_D}\sqrt{1 + B\,\frac{(i\omega)^{\gamma}\,\lambda_D^2}{D}}.$$

In this case, the expression for Z, given by Eq. (9.56), becomes:

$$Z_e = -i\,\frac{2}{\omega\varepsilon\beta_e^2 S}\left\{\frac{1}{\lambda_D^2\beta_e}\tanh(\beta_e d/2) + B\,\frac{(i\omega)^{\gamma} d}{2D}\right\}, \tag{9.58}$$

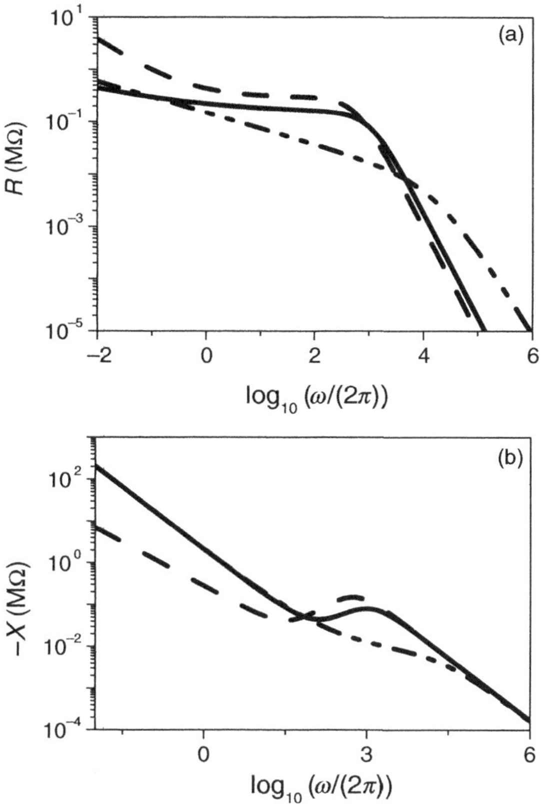

Figure 9.9 The real (a) and imaginary (b) parts of the impedance as predicted by Eqs. (9.56) (solid line) for $A = 1/2$ with $B = 1/2$ (9.59) (dashed-dotted line) for $B = 1$, and the model proposed in Ref. [332] (dashed line). The curves were drawn for $\gamma = 0.7$, $\tau = 1.411 \times 10^{-3}$ s, and $R_\infty = 2.973 \times 10^5$ Ω. The other parameters are the same as in Fig. 9.8. Modified from L. R. Evangelista, E. K. Lenzi, G. Barbero, and J. R. Macdonald, Anomalous diffusion and memory effects on the impedance spectroscopy for finite-length situations, Journal of Physics: Condensed Matter **23**, 485005 (2011), with the permission of IOP Publishing.

which can be rewritten in the form:

$$Z_c = R_c \left[\frac{B(i\omega\tau_c)^{\gamma-1}}{q_c^2} + \frac{\tanh(M\, q_c)}{M\ (i\omega\tau_c)\ q_c^3} \right], \tag{9.59}$$

where $R_c = d\tau_c/(\varepsilon S)$, $M = d/(2\lambda_D)$, $q_c = \sqrt{1 + B(i\omega\tau_c)^\gamma}$, and $\tau_c = (\lambda_D^2/D)^{1/\gamma}$.

Even a pure fractional expression like the one represented by Eq. (9.59), which, as discussed above, is obtained as a particular case of Eq. (9.53) when $A = 0$, is not rich enough to face the great complexity represented by the experimental data [333]. Other possibilities, including a model that combines fractional time derivatives with ordinary ones along the lines discussed for Eq. (9.53), might lead to physically satisfactory results and might possibly be needed in some situations to best explain experimental data. This generalisation is presented and discussed in the next section.

9.3.2 Distributed Order: General Case

We shall focus our attention on fractional time diffusion equations of distributed orders in general [286] and investigate the diffusive regimes which can be manifested by them. These equations may be formally written, for example, as

$$\int_0^1 d\gamma p(\gamma) \frac{\partial^\gamma}{\partial t^\gamma} n_\pm(z,t) = -\frac{\partial}{\partial z} j_\pm(z,t), \tag{9.60}$$

where $p(\gamma)$ is a distribution function of γ defined in Eq. (9.47) and the operator considered is that of Caputo (see Section 2.3), here implemented as follows:

$$\begin{aligned}\frac{\partial^\gamma}{\partial t^\gamma} n_\pm(z,t) &\equiv {}_{t_0}^{C}\mathrm{D}_t^\gamma \{n_\pm(z,t)\} \\ &= \frac{1}{\Gamma(k-\gamma)} \int_{t_0}^{t} d\bar{t} \frac{n_\pm^{(k)}(z,\bar{t})}{(t-\bar{t})^{1-\gamma+k}},\end{aligned} \tag{9.61}$$

with $0 < k \le 1$ and $n_\pm^{(k)}(z,\bar{t})$ representing the kth derivative. As before, we consider $t_0 \to -\infty$ to study the response of the system to the periodic applied potential [46].

Note that Eq. (9.60) exhibits the presence of a fractional time operator of distributed order which, depending on the choice of $p(\gamma)$, can account for several diffusive regimes of the ions in the system. Simple cases can be reobtained by a suitable choice of $p(\gamma)$. Some examples are the usual one, obtained for $p(\overline{\gamma}) = A\delta(\overline{\gamma}-1)$; the pure fractional case, obtained for $p(\overline{\gamma}) = B\delta(\overline{\gamma}-\gamma)$; and the case represented by Eq. (9.53), considered in the previous section, which is obtained by assuming that $p(\overline{\gamma}) = A\delta(\overline{\gamma}-1) + B\delta(\overline{\gamma}-\gamma)$. In the preceding expressions, A and B play the role of characteristic times, as we discuss later. Since we are considering only mobile positive and negative ions of a single type, with unit valence numbers, the distributed order approach considered here is such that each type of ions can show normal diffusion in one interval of time (or frequency) and an anomalous diffusion in another time interval. The order of these derivatives is consequently distributed according to the function $p(\gamma)$, which works as the weight factor for each regime (order).

Thus, the general expression for the impedance is [324]

$$Z = -i\frac{2}{\omega\varepsilon\beta^2 S}\left\{\frac{1}{\lambda_D^2\beta}\tanh\left(\frac{\beta d}{2}\right) + \frac{d}{2D}F(i\omega)\right\}, \tag{9.62}$$

where, now,

$$\beta = \frac{1}{\lambda_D}\sqrt{1 + F(i\omega)\,\frac{\lambda_D^2}{D}}, \tag{9.63}$$

with

$$F(i\omega) = \int_0^1 d\gamma p(\gamma)(i\omega)^{\gamma}. \tag{9.64}$$

The presence of $F(i\omega)$ in Eqs. (9.62) and (9.63) is responsible for the incorporation of an arbitrary number of diffusive regimes in the description of the diffusion of ions through the sample. It is noteworthy that the general expression for the impedance, Eq. (9.62), has exactly the same functional form of Eqs. (9.31), (9.56), and (9.58), which can be taken hereafter as its particular cases.

It is still possible to obtain again the same functional form for the impedance if a memory term is present in Eq. (9.51). Consider, for instance, the particular case in which $A = 1$ and $B = 0$, i.e., when the fractional time derivative is absent. To account for this memory effect, in this situation, Eq. (9.51) could be rewritten in the form

$$\frac{\partial n_+(z,t)}{\partial t} = D\frac{\partial}{\partial z}\left\{\frac{\partial n_+(z,t)}{\partial z} + \frac{qN}{k_B T}\frac{\partial V}{\partial z}\right\} + \int_{-\infty}^{t} dt'\alpha(t-t')n_+(z,t'), \tag{9.65}$$

where $\alpha(t)$ is the kernel function which quantifies the effect of $n_+(z,t')$ on $n_+(z,t)$. A similar equation is valid for the negative species $n_-(z,t)$. By performing the calculations as in the preceding section, we obtain for the electrical impedance the same expression as (9.56), with β_e given again by (9.57), but with a different "effective" diffusion coefficient; i.e., Eq. (9.55) now becomes

$$D_e = \frac{D}{1 + \overline{\alpha}(i\omega)/i\omega}, \tag{9.66}$$

where $\overline{\alpha}(i\omega)$ is the inverse transform of $\alpha(t)$ in the last term in the right hand side of (9.65).

This result illustrates the noteworthy fact that the presence of a general term accounting for some kind of memory effect implies a redefinition of the diffusion coefficient. This redefinition is similar to the one that occurs when fractional time derivatives are considered, as illustrated before in the case of Eq. (9.51). Since memory effects can be relevant to explain the diffusion of ions in actual samples, both formalisms presented in this section may eventually be helpful in discussing specific experimental data. In addition, the presence of the kernel $\alpha(t)$ introduces some freedom in the description of the frequency dependence of Z. This freedom in choosing the function $\alpha(t)$ can be useful in a phenomenological perspective of fitting complex experimental data. However, the renormalisation of D illustrated in (9.55) and (9.66) affects more sensibly the second term of (9.56) or (9.62) when the low frequency limit is considered. This seems to indicate that the first term always dictates the low frequency behaviour of the real and imaginary parts of Z_e.

Even in the general case of many diffusive regimes (i.e., several different values of γ), the system is governed by linear differential equations. For this reason, one expects that for the real and imaginary parts of Z, the relations of Kramers–Kronig (KK) hold. This is easily demonstrated for the case represented by Eq. (9.56). Since the general form of the electrical impedance is preserved, all the properties satisfied by Eq. (9.56) are satisfied also by Eq. (9.62) whenever the functions $F(i\omega)$ and $\alpha(i\omega)/i\omega$ are analytical functions of ω. For simple superpositions formed from the distribution $p(\gamma)$ considered here or by a suitable choice for $\alpha(i\omega)$, this analyticity is, in principle, always preserved [334].

In summary, by using a superposition of anomalous as well as usual diffusive regimes, general expressions for the electrical impedance can be analytically obtained. These expressions can be formally obtained by using the so-called fractional diffusion equations of distributed order that combine diffusion equations whose derivatives in the time domain are of arbitrary order in a simple mathematical manner. This combination of fractional time derivatives with ordinary ones may lead to physically satisfactory results and, as discussed before, may possibly be needed in some situations to best explain experimental data. Anyway, until now we have considered these extensions using boundary conditions of blocking electrodes. As we shall demonstrate in the next sections, a description like this is suitable to tackle bulk effects, but is not accurate enough to consider surface or interfacial phenomena, which play a crucial role in the impedance spectroscopy in general.

Indeed, further modifications in the boundary conditions are required to incorporate effects which were not conveniently described so far in association with fractional diffusion equations. Other possibilities to be explored could be the adoption of Chang–Jaffé boundary conditions to take into account specific adsorption at the electrodes [330, 335] or the consideration of the adsorption–desorption process as being governed by a typical balance equation characterising a chemical reaction of first kind (Langmuir's approximation), when the conservation of the number of particles is imposed [151, 210, 316].

9.4 Anomalous Interfacial Conditions

In this section, we face the mathematical problem of solving again a fractional diffusion equation of distributed order [101, 102, 219] for the mobile ions coupled to Poisson's equation for the electrical potential in the bulk, but now subjected to boundary conditions that are stated by means of an integro-differential expression. These boundary conditions embody, in particular, the usual kinetic equation for describing the adsorption–desorption process at the electrodes [152], but is expressed in terms of a temporal kernel that can be judiciously chosen to cover scenarios which are not suitably described within the usual framework of blocking electrodes.

In short, the procedure amounts to solving the fractional diffusion equation, for the bulk densities of ions $n_{\pm}(z,t)$ given by Eq. (9.60), with the current densities given again by Eq. (9.4). As pointed out before, the value of $\gamma \neq 1$ opens a richer scenario in which different diffusive regimes may occur depending on the distribution $p(\overline{\gamma})$ of the fractional time derivative of distributed order.

Of special interest may be the case worked out in Section 9.3.1, for which we can consider

$$p(\overline{\gamma}) = \tau\delta(\overline{\gamma} - 1) + \tau_{\gamma}\delta(\overline{\gamma} - \gamma), \tag{9.67}$$

which corresponds to a situation with two different regimes, one for long times and other one for short times [102]. An important remark in order here refers to the use of the Caputo fractional derivative instead of the Riemann–Liouville one [324]. In the preceding section, the fractional derivative was incorporated after linearising the system of partial differential equations with usual time derivatives. Here, in order to face the complete problem of the electrical response from the beginning and to avoid cumbersome calculations – which may appear due to the definition of the Riemann–Liouville fractional derivative – we use the Caputo fractional time derivative, Eq. (9.61). The results we obtain will reduce to the ones previously found in Section 9.2 in the appropriate limits.

The solutions for Eq. (9.60) will be subjected to the boundary conditions

$$j_{\pm}(d/2,t) = \int_{-\infty}^{t} d\overline{t}\,\overline{\kappa}(t-\overline{t})\frac{d}{d\overline{t}}n_{\pm}\left(d/2,\overline{t}\right) \tag{9.68a}$$

and

$$j_{\pm}(-d/2,t) = -\int_{-\infty}^{t} d\overline{t}\,\overline{\kappa}(t-\overline{t})\frac{d}{d\overline{t}}n_{\pm}\left(d/2,\overline{t}\right), \tag{9.68b}$$

where the right term in both expressions can be related to an adsorption–desorption process. If we take the expression

$$\overline{\kappa}(t) = \kappa e^{-t/\tau}, \tag{9.69}$$

where κ is a constant, then the description regards the adsorption–desorption processes at the surfaces governed by a kinetic equation that corresponds to the Langmuir approximation [152]. In the time domain, the kinetic equation may be written as

$$j_{\pm}(\pm\, d/2,t) = \frac{d\sigma(t)}{dt} = \kappa n_{\pm}(\pm d/2,t) - \frac{1}{\tau}\sigma(t), \tag{9.70}$$

where $\sigma(t)$ is the surface density of adsorbed particles. We can always consider other expressions for $\overline{\kappa}(t)$ in order to incorporate memory effects and, consequently, non-Debye relaxation processes in the analysis [173].

The electrical potential profile across the sample is determined again by Poisson's equation (9.5), which involves the difference between the densities of charged particles. These equations give the dynamics of the system and satisfy a balance equation in the form:

$$\frac{\partial}{\partial t}\int_{-d/2}^{d/2} dz\,\Psi_+(z,t) + \int_{-\infty}^{t} d\bar{t}\,\overline{\kappa}(t-\bar{t})\frac{d}{d\bar{t}}\left[\Psi_+\left(\frac{d}{2},\bar{t}\right) + \Psi_+\left(-\frac{d}{2},\bar{t}\right)\right] = 0, \qquad (9.71)$$

with

$$\Psi_\pm(z,t) = N_+(z,t) \pm N_-(z,t),$$

where $N_\pm(z,t)$ are the ones defined in Eq. (9.1). For the conditions employed here, we have

$$\int_{-d/2}^{d/2} \Psi_+(z,t)dz = \text{constant}.$$

A solution for the previous equations and, consequently, an expression for the electrical impedance may be found in the linear approximation by assuming also that $N_\pm(z,t) = N$ for $t = 0$. In addition, we also consider solutions in the form of Eq. (9.14) to analyse the impedance when the electrolytic cell is subjected to the time-dependent potential $V(z,t)$ in the form (9.8) and to the boundary conditions (9.7). Substitution of these quantities in Eqs. (9.5), (9.60), and (9.68b) yields a set of four coupled equations which may be decoupled by introducing the functions

$$\psi_+(z) = \rho_+(z) + \rho_-(z) \quad \text{and} \quad \psi_-(z) = \rho_+(z) - \rho_-(z). \qquad (9.72)$$

The first two equations are

$$\frac{d^2}{dz^2}\psi_\pm(z) = \alpha_\pm^2\psi_\pm(z), \qquad (9.73)$$

where

$$\alpha_-^2 = \frac{F(i\omega)}{D} + \frac{1}{\lambda_D^2} \quad \text{and} \quad \alpha_+^2 = \frac{F(i\omega)}{D}.$$

The other two equations are

$$D\frac{d}{dz}\psi_-(z) + \frac{2qD}{k_BT}N\frac{d}{dz}\phi(z) = \mp i\omega\,\overline{\kappa}(i\omega)\psi_-(z) \qquad (9.74a)$$

and

$$D\frac{d}{dz}\psi_+(z) = \mp i\omega\,\overline{\kappa}(i\omega)\psi_+(z), \qquad (9.74b)$$

at $z = \pm d/2$, with

$$\overline{\kappa}(i\omega) = e^{-i\omega t} \int_{-\infty}^{t} d\bar{t}\, \overline{\kappa}(t - \bar{t}) e^{i\omega\bar{t}}. \tag{9.75}$$

The solutions of Eq. (9.73) are of the form:

$$\psi_{\pm}(z) = C_{\pm,1} e^{\alpha_{\pm} z} + C_{\pm,2} e^{-\alpha_{\pm} z}, \tag{9.76}$$

where $C_{\pm,1}$ and $C_{\pm,2}$ may be determined by the boundary conditions and the symmetry of the potential: $V(z,t) = -V(-z,t)$, which implies $C_{-,1} = -C_{-,2}$ and, consequently,

$$\psi_{-}(z) = 2C_{-,1} \sinh(\alpha_{-} z), \tag{9.77a}$$

$$\phi(z) = -\frac{2q}{\varepsilon\alpha^2} C_{-,1} \sinh(\alpha_{-} z) + \overline{C} z. \tag{9.77b}$$

The constants $C_{-,1}$ and $\overline{C}$ are determined by solving the system of equations

$$\mathcal{E} C_{-,1} + \frac{qND\alpha_{-}\overline{C}}{k_B T \cosh(\alpha_{-} d/2)} = 0, \tag{9.78a}$$

and

$$-\frac{2q}{\varepsilon\alpha_{-}^2} C_{-,1} \sinh\left(\alpha_{-}\frac{d}{2}\right) + \frac{d}{2}\overline{C} = \frac{V_0}{2}, \tag{9.78b}$$

with

$$\mathcal{E} = F(i\omega) + i\alpha_{-}\omega\overline{\kappa}(i\omega) \tanh\left(\alpha_{-}\frac{d}{2}\right).$$

These equations were obtained from the corresponding boundary condition for $\psi_{-}(z)$, i.e., from Eq. (9.74a) and the condition imposed on the potential, for example, in $z = d/2$. Following the procedure employed in Ref. [152] and using the previous equations, we may obtain the current and the admittance, $Y = I/V$, of the sample (cell). For the calculations performed here, the admittance is given by

$$Y = 2iq\frac{\omega S}{\alpha_{-} V_0} \left\{ \frac{\varepsilon}{2q}\alpha_{-}\overline{C} - \left[\cosh\left(\alpha_{-}\frac{d}{2}\right) + \alpha_{-}\overline{\kappa}(i\omega) \sinh\left(\alpha_{-}\frac{d}{2}\right)\right] C_{-,1} \right\}. \tag{9.79}$$

The impedance of the cell, defined by $Z = 1/Y$, can be determined from the previous equation by means of simple calculation. It is given by

$$Z = \frac{2}{i\omega\varepsilon S\alpha_{-}^2} \frac{\tanh(\alpha_{-} d/2)/(\lambda_D^2 \alpha_{-}) + d\mathcal{E}/(2D)}{1 + \overline{\kappa}(i\omega)\left(1 + i\omega\lambda_D^2/D\right) \tanh(\alpha_{-} d/2)/(\lambda_D^2 \alpha_{-})}. \tag{9.80}$$

The presence of the kernel $\overline{\kappa}(t)$, transformed by means of Eq. (9.75), gives to the electrical impedance (9.80) a very general profile. As a matter of fact, for $p(\gamma) = \delta(\gamma - 1)$, with $\overline{\kappa}(t) = \kappa e^{-t/\tau}$, the case worked out in [152], in which adsorption-desorption phenomena are incorporated in the analysis by means of a kinetic balance equation at the surfaces, is reobtained. Moreover, for $\tau(\gamma) = \delta(\gamma - 1)$, with $\overline{\kappa}(t) = 0$, the usual form of the electrical impedance obtained in the situation of blocking electrodes, Eq. (9.31), is reobtained.

Figure 9.10a illustrates the behaviour of the real part of the impedance, i.e., $\text{Re}\, Z = R = \text{Re}\,(1/Y)$, by illustrating some choices for $\overline{\kappa}(i\omega)$, with $\gamma = 1$, in order to show the influence of the boundary conditions on the dynamics of the system. This figure considers $S = 2 \times 10^{-4}\text{m}^2$, $\varepsilon = 6.7\varepsilon_0$ (with $\varepsilon_0 = 8.85 \times 10^{-12}$) F/m, $D = 8.2 \times 10^{-11}\,\text{m}^2\text{s}^{-1}$, $\lambda \approx 3.04 \times 10^{-7}$ m, and $d = 25 \times 10^{-6}$ m [152]. The solid line is the case $\overline{\kappa}(i\omega) = 0$, which corresponds to the usual situation characterised by perfect blocking electrodes, i.e., $\kappa = 0$ and $\tau = 0$. The dashed line is the

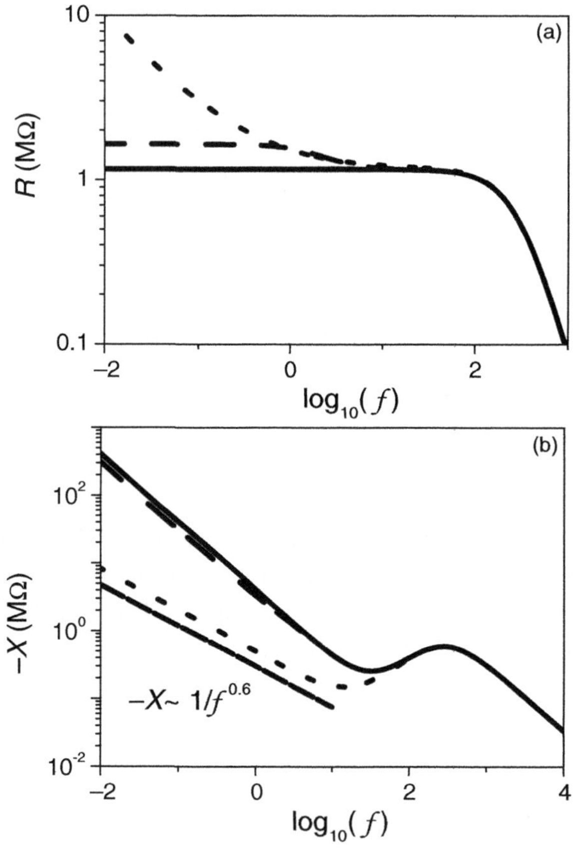

Figure 9.10 Real (a) and imaginary (b) parts of the electrical impedance of the cell versus the frequency of the applied voltage, $f = \omega/2\pi$, for different $\overline{\kappa}(i\omega)$ [336]. Modified from P. A. Santoro, J. L. de Paula, E. K. Lenzi, and L. R. Evangelista, Anomalous diffusion governed by a fractional diffusion equation and the electrical response of an electrolytic cell, Journal of Chemical Physics **135**, 114704 (2011), with the permission of AIP Publishing.

case represented by (9.81), which may be connected with an adsorption–desorption process with $\kappa = 10^{-6}$ m/s and $\tau = 0.1$ s. The dotted line considers

$$\overline{\kappa}(i\omega) = \frac{\kappa\tau}{(i\omega\tau)^{\vartheta}}, \tag{9.81}$$

with $\kappa = 10^{-3}$ m/s, $\tau = 10^{-4}$ s, and $\vartheta = 0.35$. This case shows that the effect of the boundary condition is to produce an anomalous diffusion. This anomalous behaviour is also manifested in Fig. 9.10b, which shows the imaginary part of the impedance, i.e., $\text{Im}\, Z = X = \text{Im}\,(1/Y)$. A straight line was incorporated in Fig. 9.10 to highlight the anomalous behaviour of the imaginary part at low frequency.

In Fig. 9.11, we consider the influence of the fractional diffusion equation on the impedance for the distribution, Eq. (9.67), which, as pointed out before, corresponds to a diffusive process characterized by two different regimes, the usual

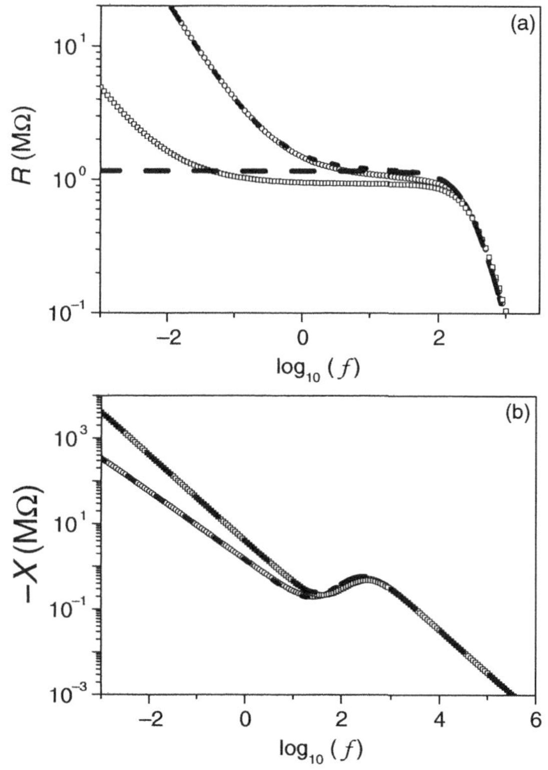

Figure 9.11 Real and imaginary parts of the electrical impedance of the cell versus the frequency of the applied voltage, $f = \omega/2\pi$, for different values of γ and ϑ [336]. Modified from P. A. Santoro, J. L. de Paula, E. K. Lenzi, and L. R. Evangelista, Anomalous diffusion governed by a fractional diffusion equation and the electrical response of an electrolytic cell, Journal of Chemical Physics **135**, 114704 (2011), with the permission of AIP Publishing.

and other dominated by the anomalous diffusion. We also consider in this analysis the kernel defined by Eq. (9.81) in order to investigate the effect of the bulk and the surface on the dynamics of the ions. The dashed line is the usual situation, i.e., $\kappa = 0$ and $\gamma = 1$. The squares correspond to the fractional diffusion with $\kappa = 0$, $\tau_1 = 0.8$, $\tau_\gamma = 0.2\,\mathrm{s}^{\gamma-1}$, and $\gamma = 0.25$. The circles are the fractional diffusion case with $\kappa = 5 \times 10^{-6}$ m/s, $\tau = 0.1\,\mathrm{s}^{-1}$, and $\vartheta = 0.25$. The dotted line corresponds to $\gamma \neq 1$ with the values of κ, τ, and ϑ equal to those of the circles. For this case, the effects of the boundary conditions are pronounced.

In Fig. 9.12, the same scenario with Eq. (9.81) is found. The bulk effects, i.e., the fractional diffusion equation, govern the real part of the impedance in the low frequency limit, whereas the effects of the boundary condition are verified at intermediate frequency with the presence of a second plateau. The imaginary part

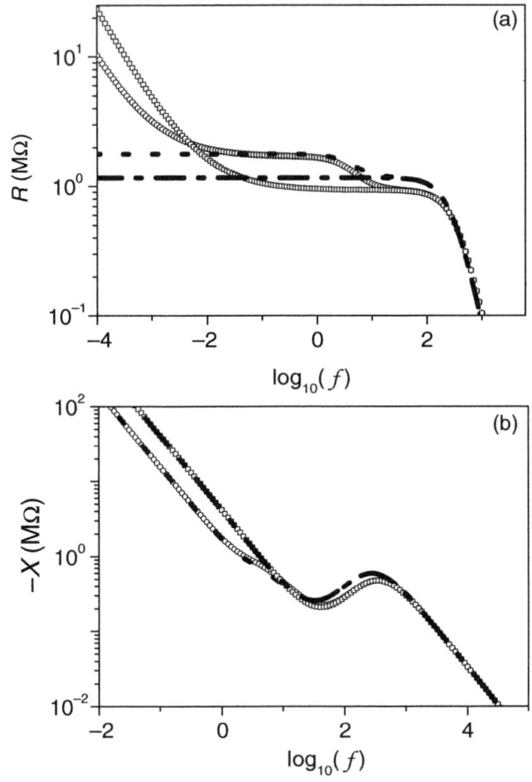

Figure 9.12 Real and imaginary parts of the electrical impedance of the cell versus the frequency of the applied voltage, $f = \omega/2\pi$, for different values of γ as in Fig. 9.11 [336]. Modified from P. A. Santoro, J. L. de Paula, E. K. Lenzi, and L. R. Evangelista, Anomalous diffusion governed by a fractional diffusion equation and the electrical response of an electrolytic cell, Journal of Chemical Physics **135**, 114704 (2011), with the permission of AIP Publishing.

of the impedance has the asymptotic behaviour governed by the usual limits; only the intermediate frequency range manifests the effects of the fractional derivative and the boundary conditions.

The boundary conditions considered here generalise the ones usually employed to investigate these systems and establish on analytical grounds a more powerful treatment of the electrical response of the system to the action of an external field. Along these lines, the very important situations presented in Refs. [152, 324] can be also reobtained from this formalism in the appropriate limits without any further problem. Other generalisations [332, 337] are still possible and can be considered in the same framework.

The results show that in the low frequency limit the behaviour of the real part of the impedance is more influenced by the bulk diffusion equation, while the effect of the processes governed by boundary conditions, i.e., by the surfaces, depends on $\overline{\kappa}(i\omega)$. If $\overline{\kappa}(i\omega) \propto 1/(i\omega\tau)^{\vartheta}$, then a strong influence on the electrical response of the system at low frequency, in contrast to what happens if $\overline{\kappa}(i\omega) \propto 1/(1+i\omega\tau)$, is found. These features may be related to the diffusive regimes of the ions in the sample which have a direct influence on the electrical response of the system.

Actually, the dynamical aspects of the motion of ions may provide an important bridge between the experimental impedance measurements and the theoretical description [337]. Experimental evidence of the unusual behaviour of the electrical response (which may be associated with anomalous diffusion) is observed, for example, in the dielectric dispersion of water [338] and in disordered solids [337].

A good example of anomalous behaviour may be found by analysing experimental data of complex fluids like liquid crystals [142]. To face the high complexity of these experimental data, instead considering the boundary conditions of Eq. (9.68b), we tackle a still more general set of boundary conditions in the form:

$$j_{\pm}(d/2,t) = \int_{-\infty}^{t} d\bar{t}\,\kappa(t-\bar{t})\frac{d^{\eta}}{d\bar{t}^{\eta}}n_{\pm}\big(d/2,\bar{t}\big) \tag{9.82a}$$

and

$$j_{\pm}(-d/2,t) = -\int_{-\infty}^{t} d\bar{t}\,\kappa(t-\bar{t})\frac{d^{\eta}}{d\bar{t}^{\eta}}n_{\pm}\big(-d/2,\bar{t}\big)\,. \tag{9.82b}$$

These conditions contain a fractional operator that, depending on the sign of η, may represent a fractional time derivative (if $\eta > 0$) or a fractional integral (if $\eta < 0$). For $0 < \eta \leq 1$, the term on the right side can be associated with an adsorption–desorption process. Indeed, for the specific case of a kernel like (9.81) with $\eta = 1$, we reobtain the adsorption–desorption processes at the surfaces governed by a kinetic equation that corresponds to the Langmuir approximation mentioned before.

Equations (9.82a) and (9.82b) embody, also as a particular case, the Chang–Jaffé boundary conditions [330, 335] and, consequently, can be related to the situation characterised by Ohmic electrodes [339].

The fundamental equations governing the dynamics of the system can be analytically solved [336], giving the impedance of the cell, in the form

$$Z = \frac{2}{i\omega\varepsilon S\beta^2} \frac{\tanh(\beta d/2)/(\lambda_D^2\beta) + d\mathcal{E}/(2D)}{1 + (i\omega)^{\eta-1}\kappa(i\omega)\left(1 + i\omega\lambda_D^2/D\right)\tanh(\beta d/2)/(\lambda_D^2\beta)}, \tag{9.83}$$

where, now,

$$\mathcal{E} = F(i\omega) + \beta(i\omega)^\eta\kappa(i\omega)\tanh\left(\beta\frac{d}{2}\right),$$

in which the transformed kernel is similar to the one defined in Eq. (9.75), i.e.

$$\kappa(i\omega) = e^{-i\omega t}\int_0^\infty dt' k(t-t')e^{i\omega t'}. \tag{9.84}$$

The asymptotic behaviour of Eq. (9.83) in the low frequency limit depends on the functional forms of $F(i\omega)$ and $\kappa(i\omega)$. To investigate this behaviour in detail, we consider that $F(i\omega) \sim (i\omega)^\delta$, with $0 < \delta \le 1$, and $\kappa(i\omega) \sim 1/(i\omega)^\nu$, with $0 < \nu < 1$, when $\omega \to 0$. These forms of $F(i\omega)$ and $\kappa(i\omega)$ will be also employed to analyse experimental data. Using them, we obtain the approximated result

$$Z \approx \frac{\lambda}{i\omega\varepsilon S} \frac{2 + \lambda_D(d - \lambda_D)F(i\omega)/D + (i\omega)^\eta d\kappa(i\omega)/D}{1 + (i\omega)^{\eta-1}\kappa(i\omega)/\lambda_D}, \tag{9.85}$$

in the limit $\omega \to 0$.

Let us now briefly analyse the experimental data. We focus on three representative samples, whose frequency behaviour of Z was found to be unusual, requiring a more powerful tool to explain it. The first sample was made by ITO electrodes, corresponding to a spin coating of 10% wt solution of LQ1800 (ITO10) and two samples of gold electrodes, corresponding to 2 and 20% wt solution of LQ1800 (respectively hereafter called AU2 and AU20), whose preparation was described elsewhere [142]. As a starting point, we have considered the possibility of two different regimes for the diffusion of ions through the sample in order to reproduce the behaviour of the experimental data which, in the low frequency limit, has a frequency dependence on the impedance, in contrast to the usual case, which is verified in the high frequency limit. This procedure was suggested by the existence of the polymide layers coated onto the electrodes which are in contact with the insulating medium of interest.

To mathematically account for two diffusive regimes, one of them being the usual and the other one anomalous, characterised by a fractional coefficient γ, it is convenient to assume, for instance, the superposition:

$$F(i\omega) = A(i\omega) + (1 - A)\,(i\omega)^{\gamma}, \tag{9.86}$$

where A is a discrete controlling parameter which assumes only two values; i.e., $A = 1$ corresponds to a pure usual diffusion ($\gamma = 1$) whereas $A = 0$ to a pure fractional, anomalous, regime ($\gamma \neq 1$). The experimental data obtained for the real part of impedance exhibit a plateau at high frequency and a power-law behaviour at low frequency; this kind of superposition is a suitable tool to analyse these data in a phenomenological perspective, because it represents a kind of mixing or interpolation between different diffusive regimes in different ranges of frequency. The plateau may be connected with an usual diffusive process and the power-law behaviour with an anomalous one.

Notice, however, that the presence of fractional derivatives in the diffusion equation does not change the asymptotic behaviour of the imaginary part of the impedance at low frequency [336, 340]. In this regard, the surface effects represented by the boundary conditions, i.e., Eq. (9.68b), actually play a prominent role and deserve a more detailed analysis. The frequency dependence of the imaginary part of the impedance is influenced by the boundary conditions, and, depending on them, different trends can be obtained [336]. This unusual behaviour of the imaginary part of the impedance may also be connected indeed to an anomalous diffusion process, because surfaces effects may induce anomalous diffusion [210].

To investigate these surface effects in more detail, we consider the composed kernel

$$\kappa(i\omega) = \kappa\tau\left[\frac{1}{(i\omega\tau)^{\vartheta_1}} + \frac{\kappa_1}{\kappa}\frac{1}{(i\omega\tau)^{\vartheta_2}}\right]. \tag{9.87}$$

This $\kappa(i\omega)$, chosen to allow for a phenomenological description of the surface effects on the ions, accounts for the existence of two processes occurring at the electrodes with a characteristic time, τ. Of these processes, one is delocalised along a characteristic surface thickness, $\kappa\tau$, and another one along the thickness $\kappa_1\tau$. The parameters ϑ_1, ϑ_2 are connected with effects produced by the surfaces, and γ with effects produced by the bulk on the dynamics of the mobile ions.

In Fig. 9.13, the frequency dependence of the real (R) and imaginary (X) parts of the electrical impedance are shown for the sample AU2 for the case in which the ion diffusion in the bulk was treated as usual, i.e., for $\gamma = 1$. The kernel corresponding to the surface part was such that the characteristic surface lengths were $\kappa\tau = \kappa_1\tau = 1.2\,\mu\text{m}$. These lengths have to be compared with the thickness

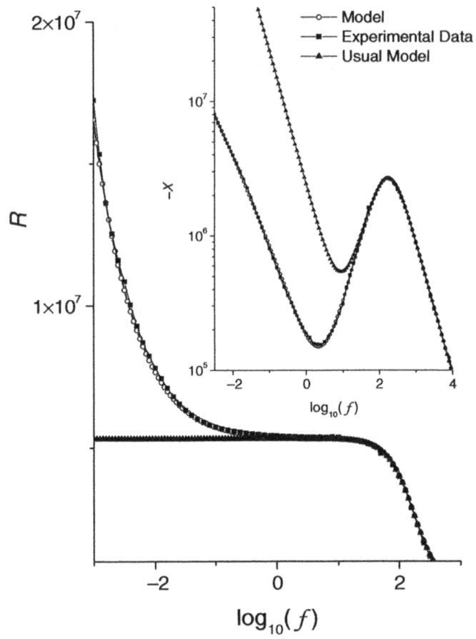

Figure 9.13 Real (R) and imaginary (X) parts of the electrical impedance of the cell versus the frequency of the applied voltage, $f = \omega/2\pi$. The figure was drawn for parameters (in SI units): $S = 10^{-4}$, $d = 36 \times 10^{-6}$, $\varepsilon = 7\varepsilon_0$, $D = 4.7 \times 10^{-12}$, $\kappa = 7.5 \times 10^{-7}$, $\tau = 1.6$, $\lambda_D = 6.574 \times 10^{-8}$, $\vartheta_1 = 0.45$, $\vartheta_2 = \vartheta_1/4$, $\kappa_1 = \kappa$, $\eta = 1$, and $\gamma = 1$ ($A = 1$). Note that the results predicted by the model described in previous section (open squares) and the usual model (triangles) are obtained from Eq. (9.83), with $\kappa = 0$ and $\gamma = 1$. The experimental data (closed squares) correspond to the sample AU2 (see the text) [142]. Modified with permission from F. Ciuchi, A. Mazzulla, N. Scaramuzza, E. K. Lenzi, and L. R. Evangelista, Fractional diffusion equation and the electrical impedance: Experimental evidence in liquid-crystalline cells, Journal of Physical Chemistry C **116**, 8773–8777 (2012). Copyright 2012 American Chemical Society.

of the sample, $d = 36\,\mu\text{m}$, and the Debye's screening length $\lambda_D \approx 0.065\,\mu\text{m}$. For this sample, the surface effects are very pronounced and delocalised along a mesoscopic length corresponding to more than 3% of the thickness of the sample. For comparative purposes, in the same figure is shown the usual behaviour, i.e., the usual diffusion equation governing the diffusion of bulk ions together with the blocking electrodes boundary conditions. This usual solution is characterised by a plateau in the low frequency region, which is absent in the experimental data. Thus, to explain the low frequency dependence of the electrical impedance we have used a simple kernel in which intervene two different power-law behaviours in the form

$$\frac{\kappa(i\omega)}{\kappa\tau} \approx \frac{1}{(i\omega\tau)^{0.45}} + \frac{1}{(i\omega\tau)^{0.11}}. \tag{9.88}$$

The same quantities are exhibited for the sample ITO10 in Fig. 9.14. For this sample, there is only one characteristic surface length, namely, $\kappa\tau = 1.2 \times 10^{-3}\,\mu\text{m}$

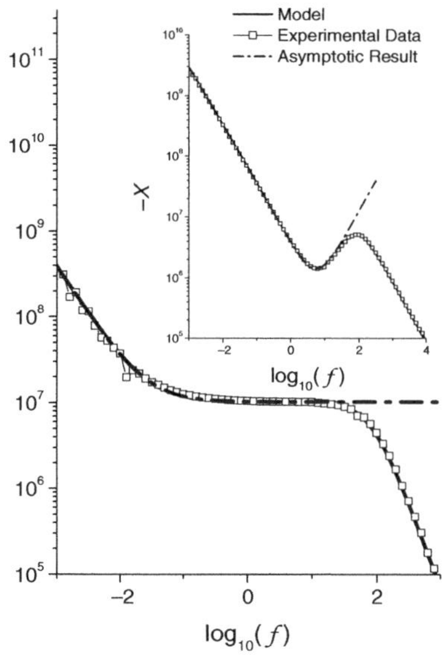

Figure 9.14 Real (R) and imaginary (X) parts of the electrical impedance of the cell versus the frequency of the applied voltage, $f = \omega/2\pi$. The parameters (in SI units) are $S = 10^{-4}$, $d = 37 \times 10^{-6}$, $\varepsilon = 7.8\varepsilon_0$, $D = 4 \times 10^{-12}$, $\kappa = 1.2 \times 10^{-6}$, $\tau = 10^{-3}$, $\lambda_D = 8.778 \times 10^{-8}$, $\vartheta_1 = 0.285$, $\kappa_1 = 0$, $\eta = 1$, and $\gamma = 1$ ($A = 1$). The experimental data (open squares) correspond to the sample ITO10 and the dashed-dotted line represents the results obtained from Eq. (9.85). Similar to what happens in Fig. 9.13, the solid line is obtained by means of Eq. (9.83) with the previous values for the parameters [142]. Modified with permission from F. Ciuchi, A. Mazzulla, N. Scaramuzza, E. K. Lenzi, and L. R. Evangelista, Fractional diffusion equation and the electrical impedance: Experimental evidence in liquid-crystalline cells, Journal of Physical Chemistry C **116**, 8773–8777 (2012). Copyright 2012 American Chemical Society.

($\kappa_1\tau = 0$) to be compared with the thickness of the sample, $d = 37\ \mu$m, and the Debye's screening length, $\lambda_D = 87.78 \times 10^{-3}\ \mu$m. We observe that the surface effects are still delocalised and are also important in determining the anomalous behaviour of the impedance. The low frequency behaviour is essentially governed by a kernel (surface) having the form $\kappa(i\omega)/\kappa\tau \approx 1/(i\omega\tau)^{0.285}$, but the diffusion of ions in the bulk is the normal one, i.e., $\gamma = 1$.

A still more complex behaviour was found in the AU20 sample, shown in Fig. 9.15. Now, the best agreement was obtained by invoking the anomalous process for the bulk, governed by a fractional diffusion equation with $\gamma = 0.16$, and a largely delocalised surface influence, characterised by the length $\kappa\tau = \kappa_1\tau \approx 10^{-3}\ \mu$m, while the thickness of the sample is $d = 36\ \mu$m and the Debye's screening length is $\lambda_D \approx 0.10337\ \mu$m.

In Fig. 9.16 the imaginary (X) versus the real part (R) of the electrical impedance is shown for the systems presented in Figs. 9.13, 9.14, and 9.15. The values of the

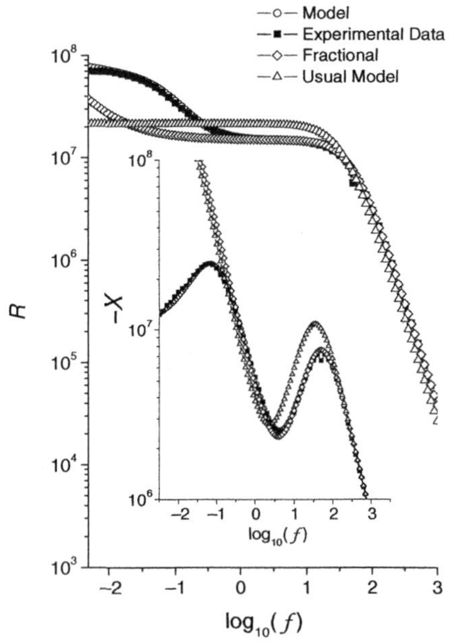

Figure 9.15 Real (R) and imaginary (X) parts of the electrical impedance of the cell versus the frequency of the applied voltage, $f = \omega/2\pi$. The parameters (in SI units) are $S = 10^{-4}$, $d = 36\times10^{-6}$, $\varepsilon = 8.5\varepsilon_0$, $D = 1.05\times10^{-12}$, $\kappa = 1.1\times10^{-7}$, $\tau = \times10^{-2}$, $\lambda_D = 1.0337 \times 10^{-7}$, $\vartheta_1 = 0.864$, $\vartheta_2 = 0.2$ $\kappa_1 = \kappa$, $\eta = 1$, and $\gamma = 0.16$. The experimental data correspond to the sample AU20. The circles correspond to results obtained from Eq. (9.83) with the previous values for the parameters. The fractional (diamond) and usual (triangle) models are obtained from Eq. (9.83) by considering $\kappa = 0$, with $\gamma \neq 1$, and $\kappa = 0$, with $\gamma = 1$ [142]. Modified with permission from F. Ciuchi, A. Mazzulla, N. Scaramuzza, E. K. Lenzi, and L. R. Evangelista, Fractional diffusion equation and the electrical impedance: Experimental evidence in liquid-crystalline cells, Journal of Physical Chemistry C **116**, 8773–8777 (2012). Copyright 2012 American Chemical Society.

parameters in agreement with the experimental data have shown that the imaginary part of the impedance is more influenced by the surface (i.e. the boundary conditions) at low frequency and by the bulk at high frequency. In the low frequency limit the behaviour of the real part of the impedance is more influenced by the bulk diffusion equation than the imaginary part is. Anyway, the surface conditions also play an important role in the behaviour of the real part of the impedance.

In summary, these results may be considered as evidence that the diffusion process of the ions in these electrolytic cells is anomalous and the mechanisms of these processes may be connected with the surface effects which induce an anomalous process on the bulk. These conclusions are reinforced by a “systematic” analysis of the experimental data in the framework of the anomalous Poisson–Nernst–Planck model proposed here and can be summarised as follows [341].

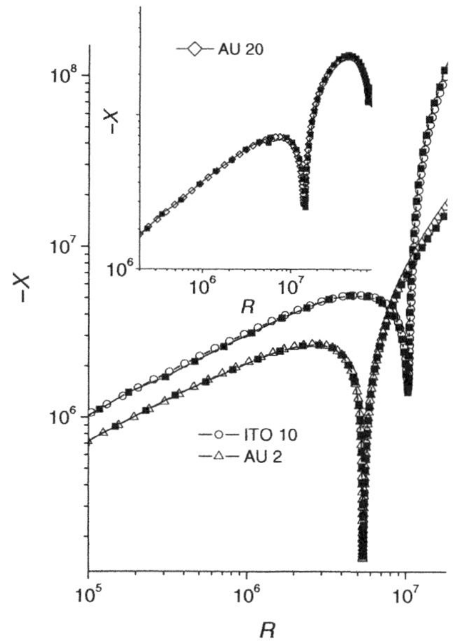

Figure 9.16 The imaginary versus the real part of the electrical impedance for the samples we are analysing: AU2 (bottom), ITO10 (middle), and AU20 (top). The closed squares correspond to the experimental data and the diamonds, circles, and triangles represent Eq. (9.83) with the parameters values used in Figs. 9.13, 9.14, and 9.15 [142]. Modified with permission from F. Ciuchi, A. Mazzulla, N. Scaramuzza, E. K. Lenzi, and L. R. Evangelista, Fractional diffusion equation and the electrical impedance: Experimental evidence in liquid-crystalline cells, Journal of Physical Chemistry C **116**, 8773–8777 (2012). Copyright 2012 American Chemical Society.

While in general it is not necessary to account for anomalous diffusion regimes occurring in the bulk, near to the interfaces the data may require a more involved theoretical interpretation. To account for the very rich behaviour of the electrical impedance response of these complex systems it is perhaps necessary and surely useful to consider anomalous response behaviour at the interfaces. This does not imply that the bulk medium in itself presents an anomalous diffusive behaviour but that the resulting response may be interpreted as phenomenologically anomalous. This sounds plausible because the overall response is the result of an interplay between bulk and surface effects. These surface effects are very complex and connected also with some peculiarities of the electrodes or the interfaces characterising the majority of real systems. Thus, a useful conceptual framework to analyse these data can be built by formulating the boundary conditions in such a way that they embody anomalous as well as conventional processes.

10

The Poisson–Nernst–Planck Anomalous Models

This chapter presents the pathway towards the construction of Poisson–Nernst–Planck anomalous (PNPA) models proposed to connect the anomalous diffusion phenomenon with the impedance spectroscopy. The first part is dedicated to analysing the conceptual links between a PNPA model and equivalent electrical circuits containing constant-phase elements (CPEs) in the low frequency domain. It is demonstrated, on analytical grounds, that the effect of a CPE in an equivalent electrical circuit may be represented by an appropriate term added to the boundary conditions of PNP or PNPA models. The second part recalls the fundamental equations of the PNPA models, including also reaction terms and Ohmic boundary conditions to account for more complex bulk and interfacial behaviour. It is shown that the formulation based on the fractional diffusion equations establishes on general theoretical grounds a connection between the PNPA models with an entire framework of continuum models and equivalent circuits with CPEs to analyse impedance data.

10.1 PNPA Models and Equivalent Circuits

As we have seen in Chapter 9, the continuum models frequently used to analyse the data are essentially based on diffusion-like equations for the ions, satisfying the Poisson's equation requirement for the electric potential (Poisson–Nernst–Planck or PNP model), or on equivalent electrical circuits [312]. There are various distributed circuit elements that can be incorporated into equivalent circuits [342]. However, a careful analysis is necessary before reaching to general conclusions about the data, since the incorrect choice of the equivalent circuit can lead to deceptive conclusions about the process that occurs in the sample [343].

Even more powerful and useful general models, such as ordinary (PNP) or anomalous diffusion (PNPA) ones, are not free from ambiguities [344]. On one hand, the PNPA models aim at incorporating behaviours that may not be well described in terms of usual diffusive PNP models. On the other hand, an important extension used in the framework of equivalent circuits is the CPE, whose presence can be connected with the necessity of describing unusual effects in many solid electrode–electrolyte interfaces. For instance, it has been pointed out that simple elements cannot describe frequency dispersion often found in the solid electrode–electrolyte interfacial region [345]. This behaviour can be related to surface disorder and roughness [346–349], electrode porosity [350], and electrode geometry [351]. The CPE in equivalent circuits is usually applied to describe a capacitance that exhibits frequency dispersion connected with these situations [352].

Having in mind the importance of these two approaches to analyse the experimental data, in this section we establish, in the low frequency limit, a connection between the predictions of PNPA models and the ones coming from equivalent circuits with CPE models in the context of the impedance spectroscopy response of an electrolytic cell. This point was worked out before from the perspective of a combination of trapping with diffusion transport and recombination [318, 319, 353]. Actually, it will be shown here that the low frequency behaviour obtained by means of CPE models may be very similar to the ones from PNPA models, if the boundary conditions are suitably represented by integro-differential boundary conditions accounting for unusual diffusive processes, as done in Chapter 9.

The discussion starts by establishing a connection between the general expression for the impedance, Eq. (9.83), and an equivalent circuit containing a CPE, in the low frequency region. In this limit, Eq. (9.83) becomes

$$Z_{\mathrm{PNPA}} \approx \frac{2\lambda_{\mathrm{D}}^2}{\varepsilon S} \frac{1}{i\omega\left[\lambda_{\mathrm{D}} + \kappa(i\omega)\right]} + \frac{\lambda_{\mathrm{D}}^2 d}{\varepsilon S D}, \tag{10.1}$$

for $\gamma = \eta = 1$. The values $\gamma = 1$ for the fractional coefficient and $A = 1$ in Eq. (9.86) were chosen to simplify the analysis and to connect the bulk effects to a simple association between resistive and capacitive elements, as illustrated in Fig. 10.1. Indeed, to consider $\gamma \neq 1$ introduces the additional term

$$\frac{d\lambda_{\mathrm{D}}}{2D}(1 - A)\Lambda(i\omega), \quad \text{with} \quad \Lambda(i\omega) = (i\omega)^{\gamma} - i\omega,$$

in the numerator of the first term of the right-hand side of Eq. (10.1), which does not give a relevant contribution to the impedance response in the low frequency limit. This feature was expected, for nonblocking boundary conditions, since in this limit the electrical response is essentially governed by surface terms [336, 341]. These terms, in turn, are responsible for an anomalous diffusion behaviour [173, 210]. This means that, depending on the physical system, to incorporate only fractional time derivatives in the bulk equation may not be sufficient to obtain a consistent behaviour for the electrical response in order to reproduce the experimental data in

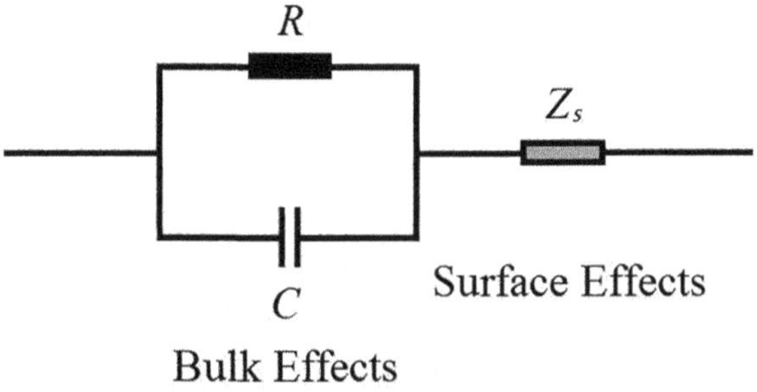

Figure 10.1 Illustration of a circuit in which the first part is a parallel association between a resistive (R) and a capacitive (C) element. The second part (Z_S) of the circuit is an arbitrary element or association of elements connected with the surface effects [354]. Modified with permission from E. K. Lenzi, J. L. de Paula, F. R. G. B. Silva, and L. R. Evangelista, A connection between anomalous Poisson–Nernst–Planck model and equivalent circuits with constant phase elements, Journal of Physical Chemistry C **117**, 23685–23690 (2013). Copyright 2013 American Chemical Society.

this limit. Anyway, they may play an important role in investigating the electrical response of a system when intermediate frequency values are considered.

The surface effects are expected to be connected with the second part of the circuit illustrated in Fig. 10.1, i.e., Z_S, which represents an arbitrary element or an association of elements. For this reason, an important issue is to know how $\kappa(i\omega)$ is connected with Z_S and, therefore, what the element is that appears when the connection is established. From the behaviour of Z, information about the surface of the electrode (roughness) can be obtained [347, 348, 355]. Analysing this quantity in the low frequency limit, we deduce that $Z \sim 1/(i\omega)^\delta$, with $0 < \delta < 1$. This behaviour is closely related to the roughness of the interface and, consequently, is connected with the fractal dimension of the interface, with δ approaching unity as the surface is made infinitely smooth. From the previous developments, depending on the boundary conditions that account for the surface effects, by means of a suitable choice of the kernel in Eq. (9.83), this behaviour of the impedance can be analytically obtained. This shows that the shape of the electrodes (for instance, roughness) may be also used as a guide to make a suitable choice of the kernel appearing in Eq. (9.83) to interpret experimental data. The discussion regarding roughness and the one-dimensional approach presented here concern the roughness of the electrodes at microscopic scales, as opposed to macroscopic scales (two- or three-dimensional cases) of an electrolytic cell where the exponent δ may incorporate stretches or other irregularities at microscopic scale [349, 356–362].

A comparison between $Z_{\rm PNPA}$ and the impedance obtained from the circuit of Fig. 10.1 in the low frequency limit,

$$Z_C = \frac{R}{1 + i\omega RC} + Z_S, \tag{10.2}$$

yields

$$Z_S \approx \frac{2\lambda_D^2}{\varepsilon S} \frac{1}{i\omega\left[\lambda_D + \kappa(i\omega)\right]}, \tag{10.3}$$

where $R = \lambda_D^2 d/(\varepsilon SD)$ is fixed in order to connect a bulk effect with the first part of the circuit. Equation (10.3) provides a connection between the surface effects represented by $\kappa(i\omega)$ and the circuit element or association, Z_S. Thus, for each $\kappa(i\omega)$ it is possible to search a simple circuit or an association of circuit elements with the same or equivalent behaviour of the impedance, when the low frequency limit is considered. A typical situation is the one characterised by perfectly blocking electrodes, obtained when $\kappa(i\omega) = 0$, which corresponds to a capacitive element. Other expressions for $\kappa(i\omega)$ account for physical processes connected with different surface effects and, therefore, to different elements contributing to Z_S.

Specifically, a relation between the CPE and the boundary conditions used in the PNPA model can be established at this point. To do this, it is useful to rewrite Z_S as

$$\frac{1}{Z_S} \approx \frac{\varepsilon S}{2\lambda_D} i\omega + \frac{\varepsilon S}{2\lambda_D^2} i\omega \overline{\kappa}_\zeta(i\omega) \tag{10.4}$$

and to assume $\kappa(i\omega) = \kappa\tau/(i\omega\tau)^\zeta$ [142, 336] which, in turn, implies a parallel association between a capacitor and a CPE. Indeed, Z_S can be identified with the following association

$$\frac{1}{Z_S} \approx \underbrace{\frac{\varepsilon S}{2\lambda_D} i\omega}_{1/Z_1} + \underbrace{\frac{\varepsilon S}{2\lambda_D} \frac{\kappa}{\lambda_D} (i\omega\tau)^{1-\zeta}}_{1/Z_2}, \tag{10.5}$$

which represents the association, exhibited in Fig. 10.2, between a capacitive element, Z_1, and a CPE, Z_2, where

$$Z_1 = \frac{1}{i\omega C_1}, \qquad \text{with} \qquad C_1 = \frac{\varepsilon S}{(2\lambda_D)},$$

and

$$Z_2 = \frac{1}{(i\omega)^{1-\zeta} C_2}, \qquad \text{with} \qquad C_2 = C_1 \frac{\kappa\tau^{1-\zeta}}{\lambda_D}.$$

Another possibility is to assume

$$\kappa(i\omega) = \frac{\kappa_{a,1}\tau_1}{(i\omega\tau_1)^{\zeta_1}} + \frac{\kappa_{a,2}\tau_2}{(i\omega\tau_2)^{\zeta_2}}, \tag{10.6}$$

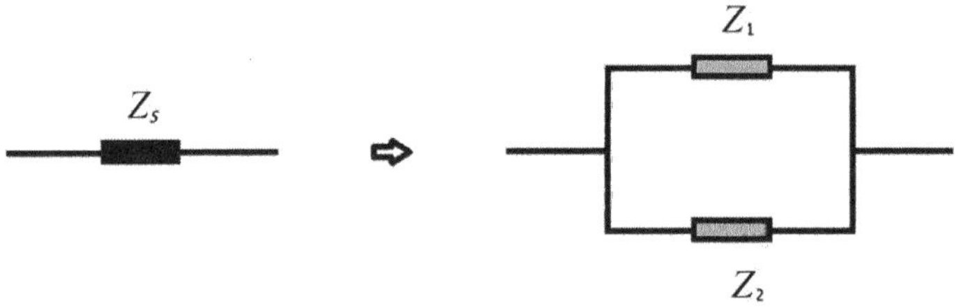

Figure 10.2 Circuit elements forming Z_S necessary to establish the connection with the PNPA model in the low frequency limit when $\kappa(i\omega)$ is given by Eq. (10.6) [354]. Modified with permission from E. K. Lenzi, J. L. de Paula, F. R. G. B. Silva, and L. R. Evangelista, A connection between anomalous Poisson–Nernst–Planck model and equivalent circuits with constant phase elements, Journal of Physical Chemistry C **117**, 23685–23690 (2013). Copyright 2013 American Chemical Society.

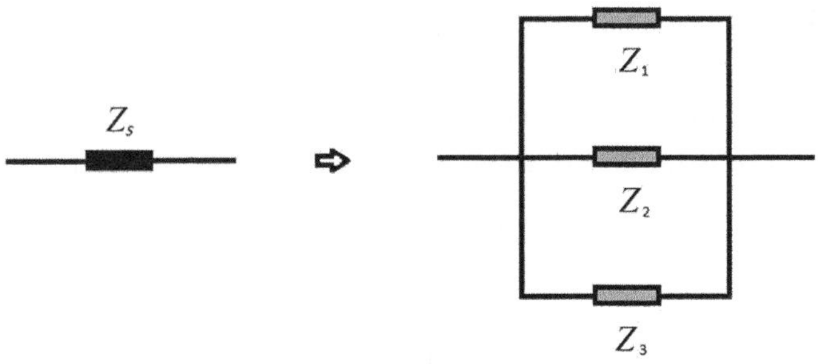

Figure 10.3 Circuit elements forming Z_S necessary to establish the connection with the PNPA model in the low frequency limit when $\kappa(i\omega) = \kappa_{a,1}\tau_1/(i\omega\tau_1)^{\zeta_1} + \kappa_{a,2}\tau_2/(i\omega\tau_2)^{\zeta_2}$. Modified with permission from E. K. Lenzi, J. L. de Paula, F. R. G. B. Silva, and L. R. Evangelista, A connection between anomalous Poisson–Nernst–Planck model and equivalent circuits with constant phase elements, Journal of Physical Chemistry C **117**, 23685–23690 (2013). Copyright 2013 American Chemical Society.

which implies

$$\frac{1}{Z_S} \approx \underbrace{\frac{\varepsilon S}{2\lambda_D} i\omega}_{1/Z_1} + \underbrace{\frac{\varepsilon S}{2\lambda_D}\frac{\kappa_{a,1}\tau_1}{\lambda_D}(i\omega\tau_1)^{1-\zeta_1}}_{1/Z_2} + \underbrace{\frac{\varepsilon S}{2\lambda_D}\frac{\kappa_{a,2}\tau_2}{\lambda_D}(i\omega\tau_2)^{1-\zeta_2}}_{1/Z_3} \tag{10.7}$$

and represents the association illustrated in Fig. 10.3. In Fig. 10.3, we have a capacitive element (Z_1) and two CPEs (Z_2 and Z_3), i.e.,

$$Z_1 = \frac{1}{i\omega C_1}, \quad \text{with} \quad C_1 = \varepsilon \frac{S}{2\lambda_D},$$

$$Z_2 = \frac{1}{(i\omega)^{1-\zeta_1} C_2}, \quad \text{with} \quad C_2 = \frac{\kappa_{a,1}\tau_1^{1-\zeta_1}}{\lambda_D C_1},$$

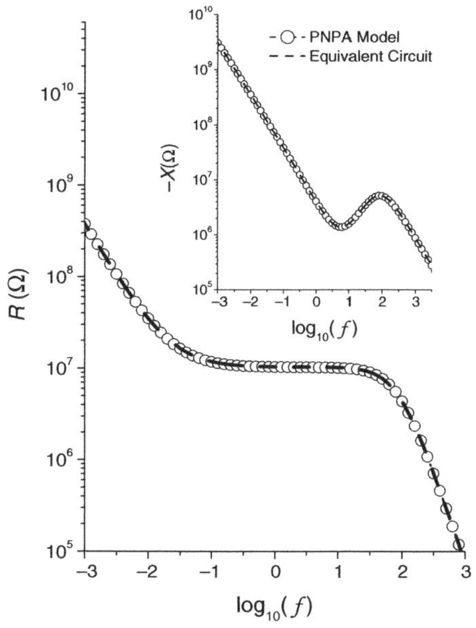

Figure 10.4 The real and the imaginary parts of the impedance for the PNPA model and the equivalent circuit arising when Eq. (10.3) is used. The circles represent the PNPA model and the dashed line is the equivalent circuit obtained from the connection established by Eq. (10.3) [354]. Modified with permission from E. K. Lenzi, J. L. de Paula, F. R. G. B. Silva, and L. R. Evangelista, A connection between anomalous Poisson–Nernst–Planck model and equivalent circuits with constant phase elements, Journal of Physical Chemistry C **117**, 23685–23690 (2013). Copyright 2013 American Chemical Society.

and

$$Z_3 = \frac{1}{(i\omega)^{1-\zeta_2} C_3}, \qquad \text{with} \qquad C_3 = \frac{\kappa_{a,2}\tau_2^{1-\zeta_2}}{\lambda_D C_1}.$$

Note that in order to consider $\eta \neq 1$ in the preceding calculations it is enough to replace $1 - \zeta_{1(2)}$ with $\eta - \zeta_{1(2)}$ (for $\eta - \zeta_{1(2)} > 0$).

Figure 10.4 illustrates the results for the PNPA model and the equivalent circuit which emerges from the connection established by Eq. (10.4). In this figure, $\kappa(i\omega) = \kappa\tau/(i\omega\tau)^{\zeta}$ and the values of the parameters are given in SI units: $\kappa = 10^{-6}$ m, $\tau = 10^{-3}$ s, $d = 37 \times 10^{-6}$ m, $\zeta = 0.287$, $\lambda_D = 8.6 \times 10^{-8}$ m $D = 4 \times 10^{-12}$ m^2/s, $S = 10^{-4}$ m^2, and $\varepsilon = 7.5\varepsilon_0$. For the case illustrated here, a good agreement between the PNPA model and the equivalent circuit is obtained when Eq. (10.3) is used.

The validity of the connection proposed in this section may be checked using another experimental scenario. The result in Fig. 10.5, illustrates the models discussed here and the experimental data of an electrolytic cell of salt ($CdCl_2H_2O$)

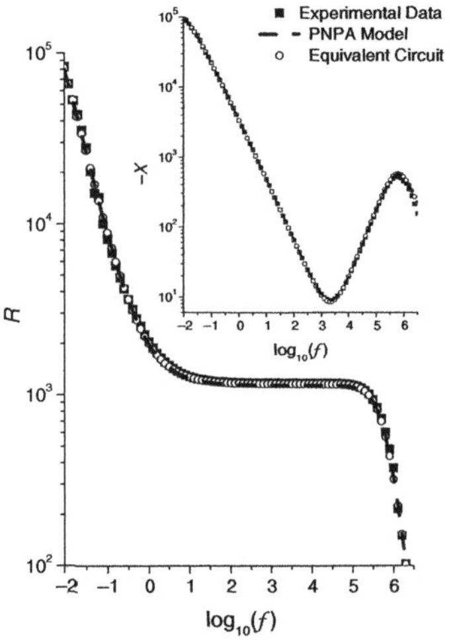

Figure 10.5 Comparison of the experimental data with the predictions of the model proposed here for the real, R, and imaginary, X, parts of the impedance. A good agreement between the experimental data and the predictions is obtained for the parameters: $S = 3.14 \times 10^{-4}$ m^2, $\epsilon = 80.03\epsilon_0$, $\gamma = 0.95$, $A = 0.98$, $B = 0.02$, $D = 3.05 \times 10^{-9}$ m$^2/s$, $d = 10^{-3}$ m, $\kappa_{a,1} = 8.67 \times 10^{-5}$ m/s, $\kappa_{a,2} = 6.24 \times 10^{-7}$ m/s, $\lambda_D = 6.24 \times 10^{-8}$ m, $\tau = 1.64 \times 10^{-3}$ s, $\zeta_1 = 0.158$, and $\zeta_2 = 0.899$ [354]. Modified with permission from E. K. Lenzi, J. L. de Paula, F. R. G. B. Silva, and L. R. Evangelista, A connection between anomalous Poisson–Nernst–Planck model and equivalent circuits with constant phase elements, Journal of Physical Chemistry C **117**, 23685–23690 (2013). Copyright 2013 American Chemical Society.

dissolved in Milli-Q deionised water (details about the experimental procedure can be found in Ref. [363]). The good agreement, obtained in the context of Fig. 10.4, which compares only the models, is also verified for the frequency range of Fig. 10.5 for the experimental data.

A connection between the PNPA models and an entire framework of equivalent circuits with CPE may be thus established on general theoretical grounds along the lines discussed before. The connection may be made analytically by a careful analysis of the low frequency limit, where the surface effects play an important role in the electrical response of an electrolytic cell. In this limit, the expression of the impedance obtained from the PNPA model and that obtained from an equivalent circuit with an arbitrary component Z_S may be compared. This comparison allows us to build a connection between $\kappa(i\omega)$ and Z_S, i.e, to propose an equivalence between these two approaches, which is very clear in the limit of low frequency but may also be valid in a broader frequency range, as illustrated in Fig. 10.4.

On the other hand, in the high frequency limit these approaches may lead to different results depending on the form of Z_S. Anyway, the present analysis permits us to conclude, on analytical grounds, that the effect of a CPE in an equivalent circuit may always be represented by an appropriate term in the boundary condition of PNP or PNPA models. This analysis is thus helpful in shedding some light on the possible meaning of a frequency-domain CPE, in terms of a condition formulated in the time domain at the electrodes. It offers at least two conceptual routes to face the complex richness of the impedance spectroscopy data and is entirely based on anomalous diffusion behaviour, both in the bulk and in the surface, as we shall discuss in more detail in the next section.

10.2 PNPA Models: A Framework

In this section, the Poisson–Nernst–Planck (PNP) diffusional model for the immittance or impedance spectroscopy response of an electrolytic cell in a finite-length situation is extended to a general framework in which anomalous diffusion is taken into account. In this new formalism, the bulk behaviour of the mobile charges is governed by a fractional diffusion equation in the presence of reaction terms. The solutions have to satisfy a general boundary condition embodying, in a single expression, most of the surface effects commonly encountered in experimental situations. Among these effects, we specifically consider the charge transfer process from an electrolytic cell to the external circuit and the adsorption–desorption phenomenon at the interfaces.

PNPA Models: Fundamental Equations

To establish a general and unified framework to investigate formal aspects of the immittance spectroscopy response, we recall the set of equations that represents the mathematical statement of a general PNPA diffusive model. The first of these equations is Eq. (9.60), i.e., the fractional diffusion equation of distributed order with a reaction term, in the form

$$\int_0^1 d\gamma\, p(\gamma)\frac{\partial^\gamma}{\partial t^\gamma} n_\alpha(z,t) = -\frac{\partial}{\partial z} j_\alpha(z,t) - \int_{-\infty}^{t} \varsigma(t-t') n_\alpha(z,t')\, dt', \tag{10.8}$$

where the bulk densities are $n_\alpha(z,t)$, with $\alpha = +$ for positive and $\alpha = -$ for negative ions. The drift-diffusion current densities are defined as

$$j_\alpha(z,t) = -D_\alpha \frac{\partial}{\partial z} n_\alpha(z,t) \mp \frac{qD_\alpha}{k_B T} n_\alpha(z,t)\frac{\partial}{\partial z} V(z,t). \tag{10.9}$$

Notice that this expression reduces to Eq. (9.4) only in the ac small-sign amplitude that requires $|n_\alpha(z,t)| \ll N$. The other is the equation of Poisson, Eq. (9.5), rewritten here to summarise the mathematical tool as follows:

$$\frac{\partial^2}{\partial z^2}V(z,t) = -\frac{q}{\epsilon}\left[n_+(z,t) - n_-(z,t)\right]. \tag{10.10}$$

To complete the set, we use boundary conditions stated in terms of integro-differential expressions as [364]

$$j_\alpha(z,t)\big|_{z=\pm\frac{d}{2}} = \pm\, k_{\alpha,e}E\,(z,t)\big|_{z=\pm\frac{d}{2}}$$
$$\pm \int_0^1 d\vartheta\; p(\vartheta) \int_{-\infty}^{t} d\bar{t}\mathcal{K}_a(t-\bar{t},\vartheta)\frac{\partial^\vartheta}{\partial \bar{t}^\vartheta} n_\alpha\big(z,\bar{t}\big)\Bigg|_{z=\pm\frac{d}{2}}\;, \tag{10.11}$$

where $k_{\alpha,e}$ represent the parameters of the Ohmic model, measured, e.g. in 1/(V m s) in the SI system [323]. The first term simply states that the ionic current density on the electrode is proportional to the surface electric field. The second term contains a temporal kernel $\mathcal{K}_a(t,\vartheta)$ convoluted with a fractional time derivative of the bulk density of ions, calculated on the electrodes. This boundary condition is built here to embody a large number of processes involving the surface (electrodes) and the bulk (sample) such as the charge transfer and the adsorption–desorption phenomenon at the electrode surfaces. It has, as particular cases, many other physical situations considered in different approaches [152, 323, 336, 363] (completely or partially blocking, Ohmic, and transparent electrodes; adsorption–desorption processes; and many others).

Built in this general form, Eq. (10.11) not only groups different known cases but also permits one to interpolate several contexts, eventually playing relevant roles in the description of the electrical response of a given system. Moreover, regarding the second term of the boundary condition, it may formally be obtained, as well as Eq. (10.8), in the context of the continuous-time random walk [41, 98, 130, 365] if reactive boundary conditions were considered, similar to the developments performed in Ref. [164, 185]. A particular choice of the kernel, as done in Refs. [141, 142, 363], may be dictated by the physical process manifested by the system to be considered. In this way as well, the set of Eqs. (10.8)–(10.10) together with the boundary conditions Eq. (10.11) constitute a new mathematical statement for the usual PNP and anomalous PNPA diffusional models, embodying a quite large class of particular situations considered before, but also pointing towards applications to novel scenarios in the field of electrochemical impedance and anomalous diffusion behaviour.

10.2.1 Analytical Solutions

An analytical solution to the previous equations and, consequently, an expression for the electrical impedance embodying all the situations mentioned before can be found in the linear approximation (small ac signal limit). We consider this approximation here to solve Eq. (10.8) for two diffusive regimes, as follows:

$$A\frac{\partial}{\partial t}n_\alpha(z,t) + B\frac{\partial^\gamma}{\partial t^\gamma}n_\alpha(z,t) = -\frac{\partial}{\partial z}j_\alpha(z,t) - \int_{-\infty}^{t} \zeta(t-t')n_\alpha(z,t')dt', \tag{10.12}$$

where γ, which is usually defined in the range $0 < \gamma < 1$, is now extended to the interval $0 < \gamma < 2$, in order to cover sub- and superdiffusive situations. In Eq. (10.12), the reaction term can be connected with reaction diffusion processes and anomalous diffusion [340, 366, 367]. Similar terms have also been used to build neural models of pattern formation where nonlocal effects play an important role [368, 369]. Typical situations connected with the previous equation may present different diffusive regimes such as the ones discussed in Refs. [101, 102] or the ones presented in Refs. [370–373], characterised by a finite phase velocity. The fractional operator considered here is Caputo's and the drift-diffusion current densities reduce to (9.4), with the diffusion coefficient for the mobile ions (assumed hereafter to be the same for positive and negative ones) denoted as D_α.

The procedure employed here is very similar to the one employed in Section 9.4. Using Eqs. (9.7), (9.8), and (9.14), the set of Eqs. (10.8), (10.10) and (10.12) becomes a new set of coupled equations which may be simplified further by using the auxiliary functions introduced in Eq. (9.72). The first two equations become

$$\frac{d^2}{dz^2}\psi_\pm(z) = \nu_\pm^2\psi_\pm(z), \tag{10.13}$$

where

$$\nu_-^2 = \frac{\Lambda\,(i\omega)}{D + 1/\lambda_{\rm D}^2} \quad \text{and} \quad \nu_+^2 = \frac{\Lambda\,(i\omega)}{D},$$

in which $\Lambda\,(i\omega) = A(i\omega) + B(i\omega)^\gamma + \zeta(i\omega)$.

The kernel in Eq. (10.12) is expressed as usual, i.e., as Eq. (9.75):

$$\zeta(i\omega) = e^{-i\omega t}\int_{-\infty}^{t} dt'\zeta(t-t')e^{i\omega t'}. \tag{10.14}$$

The other two equations connected with the boundary conditions become

$$\left.\left[D\frac{d}{dz}\psi_-(z) + \frac{2qN}{k_BT}D\frac{d}{dz}\phi(z)\right]\right|_{z=\pm\frac{d}{2}} = \pm\,\bar{k}_e\frac{d}{dx}\phi(z)\Big|_{z=\pm\frac{d}{2}} \mp\,\Upsilon_a(i\omega)\psi_-(z)\big|_{z=\pm\frac{d}{2}}, \tag{10.15}$$

and

$$D\frac{d}{dz}\psi_+(z)\bigg|_{z=\pm\frac{d}{2}} = \mp\,\Upsilon_a(i\omega)\psi_+(z)\big|_{z=\pm\frac{d}{2}}\,, \tag{10.16}$$

with

$$\Upsilon_a(i\omega) = e^{-i\omega t}\int_0^1 d\vartheta\, p(\vartheta)\,(i\omega)^{\vartheta}\int_{-\infty}^{t} d\bar{t}\,\mathcal{K}_a(t-\bar{t})e^{i\omega\bar{t}}, \tag{10.17}$$

and $\overline{k}_e = k_{+,e} - k_{-,e}$. The solutions of Eqs. (10.15) and (10.16) are $\psi_\pm(z) = C_{\pm,1}e^{\nu_\pm z} + C_{\pm,2}e^{-\nu_\pm z}$. Using the boundary conditions and invoking the symmetry of the potential, i.e., $V(z,t) = -V(-z,t)$, which implies $C_{-,1} = -C_{-,2}$, we obtain

$$\psi_-(z) = 2C_{-,1}\sinh(\nu_- z)\,, \tag{10.18a}$$

and

$$\phi(z) = -\frac{2q}{\varepsilon\nu_-^2}C_{-,1}\sinh(\nu_- z) + \overline{C}z, \tag{10.18b}$$

with $\overline{C}$ being determined by means of the boundary conditions, similar to $C_{\pm,1}$ and $C_{\pm,2}$. After some calculation, it is possible to show that the impedance, for the case discussed here, is given by the very general analytical expression

$$Z = \frac{2}{S\varepsilon\nu_-\Delta(i\omega)}\left[\frac{1}{\lambda_D^2\nu_-}\tanh\left(\nu_-\frac{d}{2}\right) + \frac{d}{2D}\mathcal{E}(i\omega)\right], \tag{10.19}$$

in which

$$\Delta(i\omega) = \frac{1}{\nu_-}\left\{\frac{1}{\lambda_D^2} + \frac{\chi\left[\Lambda(i\omega)+\omega_e\right]}{D}\right\}(i\omega+\omega_e) + \left\{\frac{1}{\lambda_D^2} + \frac{\chi\left[i\omega+\omega_e\right]}{D}\right\}\Upsilon_a(i\omega)\tanh\left(\nu_-\frac{d}{2}\right),$$

with

$$\mathcal{E}(i\omega) = \chi\left[\Lambda(i\omega)+\omega_e\right] + \chi\nu_-\Upsilon_a(i\omega)\tanh\left(\nu_-\frac{d}{2}\right),$$

$$\chi = \frac{1}{1-\lambda_D^2\omega_e/D},$$

and $\omega_e = \overline{k}_e q/\varepsilon$. Equation (10.19) represents a further extension of previous results [152, 324, 336, 363] to an even more general context, with the surface effects described by the boundary condition given by Eq. (10.11). Indeed, all the cases discussed in these works can hereafter be considered as particular cases of the present approach.

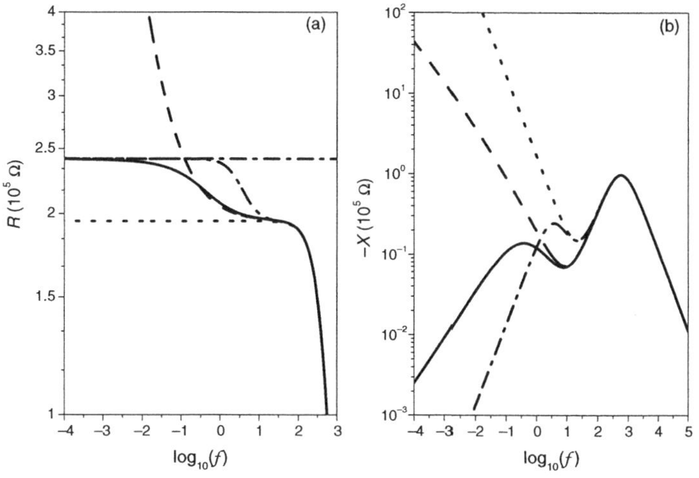

Figure 10.6 (a) The behaviour of the real, $R = \mathrm{Re}\,Z$, and (b) the imaginary, $X = \mathrm{Im}\,Z$, parts of the impedance versus the frequency in the absence of the reaction term. The solid line corresponds to the case $\kappa_{a,1} \neq \kappa_{a,2} \neq 0$, and $\overline{k}_e \neq 0$. The dashed line is the case $\kappa_{a,1} \neq \kappa_{a,2} \neq 0$ with $\overline{k}_e = 0$. The dotted line is the case $\kappa_{a,1} = \kappa_{a,2} = 0$ and $\overline{k}_e = 0$. The dashed-dotted line is characterised by $\kappa_{a,1} = \kappa_{a,2} = 0$ and $\overline{k}_e \neq 0$. For simplicity, we consider the following values for the parameters: $S = 2 \times 10^{-3}\,\mathrm{m}^2$, $\varepsilon = 80\varepsilon_0$ ($\varepsilon_0 = 8.85 \times 10^{-12}\,\mathrm{C}^2/(\mathrm{Nm}^2)$), $\gamma = 1, A = 1, B = 0, D = 2 \times 10^{-9}\,\mathrm{m}^2/\mathrm{s}$, $d = 10^{-3}\,\mathrm{m}$, $\kappa_{a,1} = 10^{-5}\,\mathrm{m/s}$, $\kappa_{a,2} = 1.95 \times 10^{-6}\,\mathrm{m/s}$, $\overline{k}_e = 10^{11}\,(\mathrm{msV})^{-1}$, $\lambda_D = 7.437 \times 10^{-7}\,\mathrm{m}$, $\tau = 0.01\,\mathrm{s}$, $\eta_1 = 0.6$, and $\eta_2 = 0.31$ [364]. Modified from E. K. Lenzi, M. K. Lenzi, F. R. G. B. Silva, G. Gonçalves, R. Rossato, R. S. Zola, and L. R. Evangelista, A framework to investigate the immittance responses for finite length-situations: Fractional diffusion equation, reaction term, and boundary conditions, Journal of Electroanalytical Chemistry, **712**, 82–88 (2014). Copyright (2014), with permission from Elsevier.

10.2.2 Charge Transfer and Adsorption Processes

Figure 10.6 shows the behaviour of the real and imaginary parts of Eq. (10.19) in absence of the reaction term, but considering the presence of charge transfer, i.e., $\overline{k}_e \neq 0$, and an adsorption process characterised by [364]

$$\Upsilon_a(i\omega) = \kappa_{a,1}\tau(i\omega\tau)^{1-\eta_1} + \kappa_{a,2}\tau(i\omega\tau)^{1-\eta_2}. \tag{10.20}$$

As before, this expression for $\Upsilon_a(i\omega)$ essentially describes the presence of a double adsorbing layer, each layer characterised by an effective thickness ($\kappa_{a,1}\tau$ and $\kappa_{a,2}\tau$) and a frequency dependence (represented by the exponents η_1 and η_2) to account for the different interactions in these layers. The dashed line is obtained for $\overline{k}_e = 0$. The dotted line is the standard model of blocking electrodes, i.e., $\overline{k}_e = 0$ and $\kappa_{a,1} = \kappa_{a,2} = 0$. The solid line combines the situation discussed for $\mathcal{K}_a(i\omega)$ with $\overline{k}_e \neq 0$, and the dashed-dotted line is the case $\kappa_{a,1} = \kappa_{a,2} = 0$, with $\overline{k}_e \neq 0$.

The results of Fig. 10.6a show that in the low frequency domain the effect of charge transfer governs the electrical response of the system, even in the presence of an adsorption–desorption process. This feature is also present in the imaginary part of the impedance, as illustrated in Fig. 10.6b. As a consequence, we can notice the presence of a second plateau in the real part of the impedance, in the low frequency domain, while the imaginary part of the impedance decreases and goes to zero in this limit. Thus, the influence of the processes governed by $\Upsilon_a(i\omega)$ in the form of Eq. (10.20), when the charge transfer is present, i.e., $\overline{k}_e \neq 0$, is absent in the low frequency domain since $\Upsilon_a(i\omega) \to 0$, when $\omega \to 0$.

This can be verified for the adsorption–desorption processes characterised by

$$\Upsilon_a(i\omega) = \kappa_a \tau (i\omega\tau)^{1-\delta},$$

with $\omega \to 0$ and $0 < \delta < 1$. This point is illustrated by the results shown in Fig. 10.6 and by the asymptotic (analytical) results obtained from Eq. (10.19). Specifically, for the real part of the impedance, we have

$$\mathrm{Re}\, Z \sim \frac{2\lambda_{\mathrm{D}}}{\overline{k}_e q S}\left[1 + \frac{\overline{k}_e q}{D\varepsilon}\lambda_{\mathrm{D}}^2\left(\frac{d}{2\lambda_{\mathrm{D}}} - 1\right)\right], \tag{10.21}$$

which only involves the constants connected with the bulk and with the charge transfer.

In Fig. 10.7, the Nyquist plot of the cases shown in Fig. 10.6 is presented. Notice that the main difference, as discussed in Fig. 10.6, is found in the low frequency limit. The first semicircle can be connected with the bulk effects, and the second part of Fig. 10.7 can be due to the effects present in the low frequency limit, i.e., the surface effects. The cases characterised by $\overline{k}_e \neq 0$ present an additional semicircle which is not present when $\mathcal{K}_a(i\omega) \sim 1/(i\omega\tau)^{\delta}$ or $\mathcal{K}_a(i\omega) = 0$, with $\overline{k}_e = 0$. These considerations may be useful when we investigate experimental data and can be used to establish a relation between Eq. (10.19) and an approach involving equivalent circuits.

Many physical situations [141, 363, 374, 375] exhibit, in the low frequency limit, an asymptotic behaviour governed by $Z \sim 1/(i\omega)^{\sigma}$, with $0 < \sigma < 1$, which is the behaviour essentially described by Eq. (10.19), for $\omega \to 0$, when $\mathcal{K}_a(i\omega) \sim 1/(i\omega\tau)^{\delta}$, for $\overline{k}_e = 0$. This behaviour is absent if the usual PNP diffusional model is considered, and has been also connected with the roughness of the surface [356, 358–360].

The adsorption process may have a pronounced influence when $\overline{k}_e \neq 0$, in the low frequency limit, if the second term of Eq. (10.11) has the following asymptotic behaviour: $\Upsilon_a(i\omega) \sim \kappa\tau$. Actually, a typical situation can be found if we use the generalised Chang–Jaffé boundary conditions accounting for specific

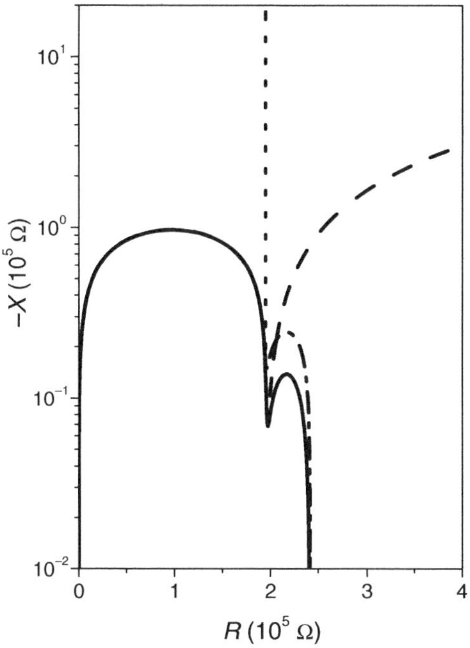

Figure 10.7 $-X$ (imaginary part) versus R (real part) for the cases discussed in Fig. 10.6 in order to show the differences in the low frequency limit. The same legend as in Fig. 10.6 applies here [364]. Modified from E. K. Lenzi, M. K. Lenzi, F. R. G. B. Silva, G. Gonçalves, R. Rossato, R. S. Zola, and L. R. Evangelista, A framework to investigate the immittance responses for finite length-situations: Fractional diffusion equation, reaction term, and boundary conditions, Journal of Electroanalytical Chemistry, **712**, 82–88 (2014). Copyright (2014), with permission from Elsevier.

ion adsorption at the interfaces [376]. The original Chang–Jaffé electrode-reaction boundary conditions [377] were introduced by Friauf [378] to investigate partial-blocking effects. Subsequently, extended Chang–Jaffé boundary conditions have been considered [379] and were also generalised to include specific ion adsorption, a few years later [335, 380]. These boundary conditions can be implemented in the present framework by incorporating $\kappa_{\rm cj}\tau$ into $\Upsilon_a(i\omega)$, where $\kappa_{\rm cj}$ is a single Chang–Jaffé parameter. In this scenario, we obtain

$$\mathrm{Re}\, Z \sim \frac{\left(2D + d\kappa_{\rm cj}\tau\right)\varepsilon\lambda_{\rm D}^2 + 2\bar{k}_e q\lambda_{\rm D}^3\left[d/\left(2\lambda_{\rm D}\right) - 1\right]}{S\bar{k}_e Dq\varepsilon\left[1 + \varepsilon\kappa_{\rm cj}\tau/\left(\bar{k}_e q\lambda_{\rm D}\right)\right]}. \tag{10.22}$$

Figure 10.8 illustrates the behaviour of the impedance for

$$\Upsilon_a(i\omega) = \kappa_{\rm cj}\tau + \kappa_{a,1}\tau_1(i\omega\tau_1)^{1-\eta_1} + \kappa_{a,2}\tau_2(i\omega\tau_2)^{1-\eta_2},$$

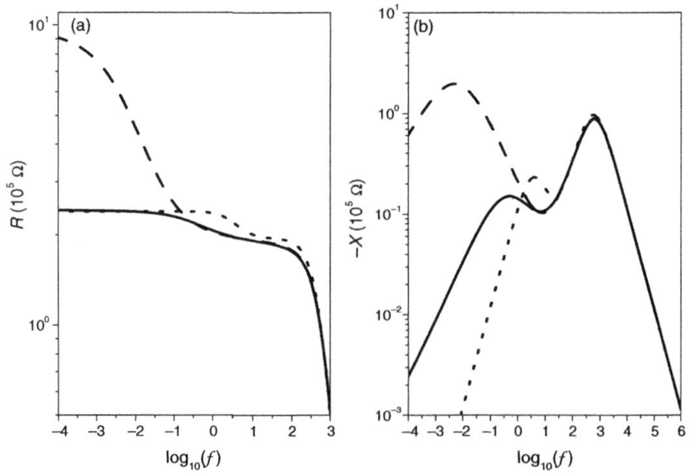

Figure 10.8 (a) Trends of the real, $\mathrm{R} = \mathrm{Re}\,Z$, and (b) the imaginary, $\mathrm{X} = \mathrm{Im}\,Z$, parts of the impedance versus frequency in absence of the reaction term. The solid line corresponds to the case $\gamma = 0.7$, $\kappa_{a,1} \neq \kappa_{a,2} \neq 0$, $\kappa_{\mathrm{cj}} = 0$, and $\overline{k}_e \neq 0$. The dashed line is the case $\gamma = 0.8$, $\kappa_{a,1} \neq \kappa_{a,2} \neq 0$, $\kappa_{\mathrm{cj}} \neq 0$ with $k_e = 0$. The dotted line is the case $\gamma = 0.9$, $\kappa_{a,1} = \kappa_{a,2} = 0$, $\kappa_{\mathrm{cj}} \neq 0$, and $\overline{k}_e \neq 0$. The figures were drawn for the following values of the parameters: $S = 2 \times 10^{-3}\,\mathrm{m}^2$, $\varepsilon = 80\varepsilon_0$, $D = 2 \times 10^{-9}\,\mathrm{m}^2/\mathrm{s}$, $d = 10^{-3}\,\mathrm{m}$, $\kappa_{a,1} = 10^{-5}\,\mathrm{m/s}$, $\kappa_{a,2} = 1.95 \times 10^{-6}\,\mathrm{m/s}$, $\overline{k}_e = 10^{11}\,(\mathrm{msV})^{-1}$, $\lambda_{\mathrm{D}} = 7.43 \times 10^{-8}\,\mathrm{m}$, $\kappa_{\mathrm{cj}} = 10^{-4}\,\mathrm{m/s}$, $\tau = 0.01\,\mathrm{s}$, $A = 0.8$, $\eta_1 = 0.6$, and $\eta_2 = 0.31$ [364]. Modified from E. K. Lenzi, M. K. Lenzi, F. R. G. B. Silva, G. Gonçalves, R. Rossato, R. S. Zola, and L. R. Evangelista, A framework to investigate the immittance responses for finite length-situations: Fractional diffusion equation, reaction term, and boundary conditions, Journal of Electroanalytical Chemistry, **712**, 82–88 (2014). Copyright (2014), with permission from Elsevier.

with $\zeta(i\omega) = 0$ and $\overline{k}_e \neq 0$. Notice that, in the low frequency limit, the behaviour of the real and imaginary parts of the impedance is governed by the Chang–Jaffé condition and the charge transfer. Figure 10.9 illustrates the effects produced on the electrical response by the reaction term, showing that it influences the behaviour of the electrical response in the low frequency limit.

Finally, in order to illustrate how the formalism presented here works when applied to a specific experimental context, the theoretical predictions for Z are compared with the experimental data. Both the experimental data and the theoretical predictions are presented in Fig. 10.10. Note that a suitable choice of the boundary conditions, to describe the experimental data, can be made by analysing the Nyquist plot of the experimental data and comparing it with the predictions shown in Fig. 10.7. The behaviour of the imaginary part of the impedance also plays an important role in the form of the boundary condition. In the low frequency limit,

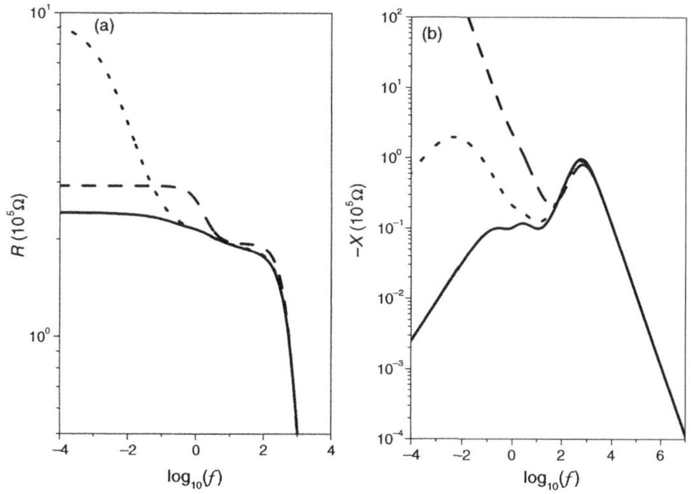

Figure 10.9 R = Re Z (a) and the imaginary, X = ImZ (b) parts of the impedance versus frequency in the presence of the reaction term, chosen to be given by $\zeta(i\omega) = 5i\omega/(10 + i\omega)$, which can be formally connected with the process of generation and recombination of ions [101, 348]. The solid line corresponds to the case $\gamma = 0.7$, $\kappa_{a,1} \neq \kappa_{a,2} \neq 0$, $\kappa_{\mathrm{cj}} = 0$, and $\overline{k}_e \neq 0$. The dotted line is the case $\gamma = 0.8$, $\kappa_{a,1} \neq \kappa_{a,2} \neq 0$, $\kappa_{\mathrm{cj}} \neq 0$ with $k_e = 0$. The dashed line is the case $\gamma = 1$, $\kappa_{a,1} = \kappa_{a,2} \neq 0$, $\kappa_{\mathrm{cj}} \neq 0$, and $\overline{k}_e = 0$. The other parameters are the same as in Fig. 10.8 [364]. Modified from E. K. Lenzi, M. K. Lenzi, F. R. G. B. Silva, G. Gonçalves, R. Rossato, R. S. Zola, and L. R. Evangelista, A framework to investigate the immittance responses for finite length-situations: Fractional diffusion equation, reaction term, and boundary conditions, Journal of Electroanalytical Chemistry, **712**, 82–88 (2014). Copyright (2014), with permission from Elsevier.

it can be directly connected with the surface effects and, consequently, with the statement of the boundary conditions [354].

These comparisons suggest the choice of Eq. (10.20), with $\overline{k}_e = 0$. This accounts, in the range of frequency considered in Fig. 10.10, for an adsorption–desorption-like process occurring at the electrodes limiting the electrolytic solution, with $\kappa_{a,1}\tau$ and $\kappa_{a,2}\tau$ being two effective thicknesses. They indeed account for the spatial extensions of two different layers near the electrode surface, in which the effective interaction changes behaviour. More precisely, the behaviour of the interaction in each of these layers is governed, in the frequency domain, by the exponents η_1 and η_2, which, in turn, indicate how these layers interplay to build an effective diffusive layer in the neighbourhood of the electrodes, in a phenomenological perspective. In this regard, we notice that the behaviour of the impedance in the low frequency limit resembles the one for semiconductor materials obtained from the trapping mechanisms, showing an anomalous diffusion [352, 353].

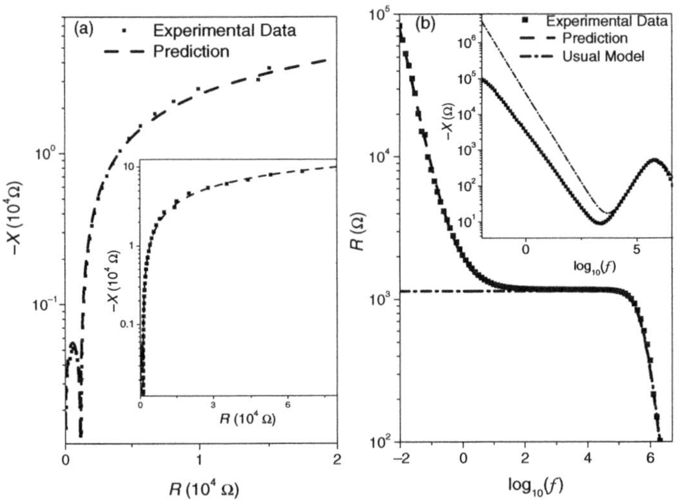

Figure 10.10 Comparison of the experimental data with the predictions of the model presented here for the real, $R = \mathrm{Re}Z$ (a), and imaginary, $X = \mathrm{Im}Z$ (b), parts of the impedance. A good agreement between the experimental data and the predictions was obtained for a relatively small set of parameters (see the text) [364]. Modified from E. K. Lenzi, M. K. Lenzi, F. R. G. B. Silva, G. Gonçalves, R. Rossato, R. S. Zola, and L. R. Evangelista, A framework to investigate the immittance responses for finite length-situations: Fractional diffusion equation, reaction term, and boundary conditions, Journal of Electroanalytical Chemistry, **712**, 82–88 (2014). Copyright (2014), with permission from Elsevier.

The results show that, in the limit of low frequency, the surface effects may govern the dynamical behaviour of the mobile charges. They reinforce the idea that the general framework presented here may be helpful to face real problems connected with the impedance spectroscopy response of electrolytic cells. In this new formalism, it is also possible to describe relevant and complex scenarios because, besides containing as particular cases many other useful models already proposed, it also accounts, in a synthetic and mathematically unified way, for the possible presence of different regimes (usual or anomalous) in the diffusion of mobile charges. In addition, the present framework can be connected with other formalisms such as the ones based on equivalent circuits with CPEs. This connection was analytically demonstrated in Section 10.1 by means of a suitable choice of kernels in the boundary conditions.

To sum up, the models proposed here contain, as particular cases that depend on the specific situation considered, equivalent circuits with CPE. A formulation of this kind provides a simple interpretation of these constant-phase elements in terms of a continuum PNPA description, i.e., an approach that uses fractional calculus with general boundary conditions expressed also in terms of fractional derivatives if needed.

References

[1] E. Butkov, *Mathematical physics* (Addison-Wesley, Reading, Mass., 1973).

[2] A. A. Kilbas, H. M. Srivastava, and J. J. Trujillo, *Theory and applications of fractional differential equations* (Elsevier, Amsterdam, 2006).

[3] M. Abramowitz and I. A. Stegun, *Handbook of mathematical functions* (National Bureau of Standards, Washington, D.C., 1972).

[4] V. K. Tuan and M. Saigo, Convolution of Hankel transform and its application to an integral involving Bessel functions of first kind, Internat. J. Math. Math. Sci. **18**, 545–550 (1995). See also L. Britvina, Generalized convolutions for the Hankel transform and related integral operators, Math. Nachr. **280**, 962–970 (2007).

[5] R. Piessens, *The Hankel transform*, in *The transforms and applications handbook*, 2nd edition. Edited by Alexander D. Poularikas (CRC Press, Boca Raton, 2000).

[6] G. Fikioris, *Mellin transform method for integral evaluation-introduction and applications to electromagnetics* (Morgan & Claypool, San Rafael, 2007).

[7] J. Bertrand, P. Bertrand, and J. P. Orvalez, *The Mellin transform*, in *The transforms and applications handbook*, 2nd edition. Edited by Alexander D. Poularikas (CRC Press, Boca Raton, 2000).

[8] H. Hochstadt, *The functions of mathematical physics* (Wiley-Interscience, New York, 1971).

[9] H. Cramér, *Mathematical methods of statistics* (Princeton University Press, Princeton, 1946).

[10] G. N. Watson, *A treatise on the theory of Bessel functions* (Cambridge University Press, Cambridge, 1944).

[11] L. Schwartz, *Mathematical for the physical sciences* (Addison-Wesley, Reading, Mass., 1966).

[12] P. M. Morse and H. Feshbach, *Methods of theoretical physics* (McGraw-Hill, New York, 1953), Vol. I.

[13] I. S. Gradshtein and I. M. Ryzhik, *Table of integrals, series, and products*, 7th edition (Elsevier Inc, Academic Press Inc., Amsterdam, 2007).

[14] H. W. Wyld, *Mathematical methods for physics* (Perseus Books, Reading, Mass., 1976).

[15] F. Bowman, *Introduction to Bessel functions* (Courier Corporation, Chelmsford, Mass., 2012).

[16] M. G. Mittag-Leffler, Sur la nouvelle fonction $\mathrm{E}_\alpha(x)$, Comptes Rendus Acad. Sci. Paris **137**, 554–558 (1903).

[17] M. G. Mittag-Leffler, Sopra la funzione $\mathrm{E}_\alpha(x)$, Rend. Acc. Lincei **13**, 3–5 (1904).

[18] M. G. Mittag-Leffler, Sur la representation analytique d'une branche uniforme d'une fonction monogène, Acta Math. **29**, 101–181 (1905).
[19] A. Wiman, Über den fundamentalsatz in der teorie der funktion $E_\alpha(x)$, Acta Math. **29**, 191–201 (1905).
[20] A. Wiman, Über die nulstellen der funktion $E_\alpha(x)$, Acta Math. **29**, 217–234 (1905).
[21] R. K. Saxena, A. M. Mathai, and H. J. Haubold, On fractional kinetic equations, Astrophys. Space Sci. **282**, 281–287 (2002).
[22] A. Erdélyi, *Higher transcendental functions* (McGraw-Hill, New York, 1953).
[23] R. P. Agarwal, A propos d'une note de M. Pierre Humbert, Comptes Rendus Acad. Sci. Paris **236**, 2031–2032 (1953).
[24] P. Humbert, Quelques résultats relatifs à la fonction de Mittag-Leffler, Comptes Rendus Acad. Sci. Paris **236**, 1467–1468 (1953).
[25] P. Humbert and R. P. Agarwal, Sur la fonction de Mittag-Leffler et quelques-unes de ses généralisations, Bull. Sci. Math. **77**(2), 180–185 (1953).
[26] M. M. Dzherbashyan, *Integral transform representations of functions in the complex domain* (Nauka, Moscow, 1966).
[27] T. R. Prabhakar, A singular integral equation with a generalized Mittag-Leffler function in the kernel, Yokohama Math. J. **19**, 7–15 (1971).
[28] A. A. Kilbas, M. Saigo, and R. K. Saxena, Solution of Volterra integro-differential equations with generalized Mittag-Leffler functions in the kernels, J. Int. Equat. Appl. **14**, 377–396 (2002).
[29] H. J. Haubold, A. M. Mathai, and R. K. Saxena, Mittag-Leffler functions and their applications, J. Appl. Math. **2011**, 298628 (2011).
[30] A. M. Mathai, R. K. Saxena, and H. J. Haubold, *The* H*-function: Theory and applications* (Springer, New York, 2009).
[31] E. M. Wright, On the coefficients of power series having essential singularities, J. London Math. Soc. **8**, 71–80 (1933).
[32] E. M. Wright, The asymptotic expansion of the generalized Bessel function, Proc. London Math. Soc. **38**, 257–270 (1934).
[33] E. M. Wright, The generalized Bessel function of order greater than one, Quart. J. Math. Oxford Ser. **11**, 36–48 (1940).
[34] R. Gorenflo, Y. Luchko, and F. Mainardi, Analytical properties and applications of the Wright function, Fract. Calc. Appl. Anal. **2**, 383–414 (1999).
[35] G. Pagnini and E. Scalas, Historical notes on the M-Wright/Mainardi function, Comm. Appl. Ind. Math. **6**/1, e-495 (2015).
[36] F. Mainardi, *Fractional calculus and waves in linear viscoelesticity: An introduction to mathematical models* (Imperial College Press, London, 2010).
[37] R. Gorenflo, Y. Luchko, and F. Mainardi, Wright functions as scale-invariant solutions of the diffusion-wave equation, J. Comput. Appl. Math. **118**, 175–191 (2000).
[38] C. Fox, The *G* and *H* functions as symmetrical Fourier kernels, Trans. Amer. Math. Soc. **98**, 395–429 (1961).
[39] B. L. J. Braaksma, Asymptotic expansions and analytical continuations for a class of Barnes-integrals, Compositio Math. **15**, 239–341 (1962–1963).
[40] F. Mainardi, G. Pagnini, and R. K. Saxena, Fox *H* functions in fractional diffusion, J. Comp. Appl. Math. **178**, 321–331 (2005).
[41] R. Metzler and J. Klafter, The random walk's guide to anomalous diffusion: A fractional dynamics approach, Phys. Rep. **339**, 1–77 (2000).
[42] R. Metzler and T. F. Nonnenmacher, Space and time fractional diffusion and wave equations, fractional Fokker–Planck equations, and physical motivation, Chem. Phys. **284**, 67–90 (2002).

[43] A. A. Inayat-Hussain, New properties of hypergeometric series derivable from Feynman integrals. I. Transformation and reduction formulae, J. Phys. A **20**, 4109–4117 (1987).

[44] A. A. Inayat-Hussain, New properties of hypergeometric series derivable from Feynman integrals. II. A generalisation of the H-function, J. Phys. A **20**, 4119–4128 (1987).

[45] W. R. Schneider and W. Wyss, Fractional diffusion and wave equations, J. Math. Phys. **30**, 134–144 (1989).

[46] I. Podlubny, *Fractional differential equations: An introdution to fractional derivatives, fractional differential equations, to methods of their solutions and some of their applications* (Academic Press, San Diego, 1998).

[47] G. H. Pertz and C. J. Gerhardt, editors, *Leibnizens gesalmmelte Werke, Leibnizens mathematische Schriften*, Erste Abtheilung, Band II, pp. 301–302. Dritte Folge Mathematik (Erster Band). (A. Asher & Comp., Berlin, 1849).

[48] L. Euler, *De progressionibus transcendentibus seu quarum termini generales algebraice dari nequeunt.* The Euler archive at http://eulerarchive.maa.org/pages/E019.html. A translation by Stacy Langton, titled *On transcendental progressions that is, those whose general terms cannot be given algebraically*, is available at http://home.sandiego.edu/~langton/eg.pdf.

[49] J. L. Lagrange, *Sur une nouvelle espèce de calcul relatif a la diffentiation et a l'intégration des quantités vairables*, in *Ouvres de Lagrange, Tome Troisième*, edited by J.-A. Serret (Guthier-Villars, Paris, 1869), p. 448.

[50] P. S. Laplace, *Théorie analytique des probabilités*, 2nd edition (Courcier, Paris, 1814), 85–86.

[51] S. F. Lacroix, *Traité du calcul différentiel et du calcul intégral* (Courcier, Paris, 1819), Tome III, pp. 408–409.

[52] J. B. J. Fourier, *Théorie analytique de la chaleur* (Didot, Paris, 1822), pp. 561–562.

[53] M. Weilbeer, *Efficient numerical methods for fractional differential equations and their analytical background* (Ph.D. thesis, Technischen Universität Braunschweig, 2005).

[54] N. H. Abel, *Solution de quelques problèmes à l'aide d'integrales définies* in Oeuvres Complètes (Grondahl & Son, Christiania, 1881), pp. 15–18.

[55] S. G. Samko, A. A. Kilbas, and O. I. Marichev, *Fractional integrals and derivatives: Theory and applications* (Gordon and Breach, Yverdon, 1993).

[56] B. Ross, The development of fractional calculus 1695–1900, Historia Mathematica **4**, 75–89 (1977).

[57] J. Liouville, Mémoire sur quelches questions de géométrie et de mécanique, et sur un noveau genre de calcul pour résoudre ces questions, J. l' École Roy. Polytéchn., **13**, 1–69 (1832).

[58] H. T. Davis, The application of fractional operators to functional equations, Amer. J. Math. **49**, 123–142 (1927).

[59] J. Liouville, Mémoire sur le théorème des fonctions complémentaires, J. reine angew. Math., **11**, 1–19 (1834).

[60] G. F. B. Riemann, *Versuch einer Auffassung der integration und differentiation*, in *Gesammelte mathematische Werke und wissenschaftlicher Nachlass–Bernhard Riemann*, edited by H. Weber (Teubner, Leipzig, 1876), pp. 331–344.

[61] N. Y. Sonin, On differentiation with arbitrary index, Moscow Matem. Sbornik **6**, 1–38 (1869).

[62] H. Laurent, Sur le calcul des dérivées à indicies quelconques, Nouv. Ann. Math. **3**, 240–252 (1884).

[63] A. V. Letnikov, An explanation of the concepts of the theory of differentiation of arbitrary index (in Russian), Moscow Matem. Sbornik **6**, 413–445 (1872).

[64] A. K. Grünwald, Über "begrenzte" derivation und deren Anwendung, Z. angew. Math. Phys. **12**, 441–480 (1867)
[65] A. V. Letnikov, Theory of differentiation with an arbitrary index (in Russian), Moscow Matem. Sbornik **3**, 1–66 (1868).
[66] M. Caputo, Linear models of dissipation whose Q is almost frequency independent–II, Geophys. J. Royal Astr. Soc. **13**, 529–539 (1967).
[67] B. West, M. Bologna, and P. Grigolini, *Physics of fractal operators* (Springer, New York, 2003).
[68] T. Graham, A short account of experimental researches on the diffusion of gas through each other, and their separation by mechanical means, Quart. J. Sci. **2**, 74 (1829).
[69] T. Graham, LX. On the molecular mobility of gases, Phil. Mag. **26**/177, 409–434 (1863).
[70] J. C. Maxwell, *On the dynamical theory of gases* (1866), in *The scientific papers of James Clerk Maxwell*, edited by W. D. Niven (Dover, New York, 1965), p. 59.
[71] A. Fick, Über diffusion, Poggendorff's Annalen der Physik und Chemie **94**, 59–86 (1855). English version: On liquid diffusion, Phil. Mag. **10**, 30–39 (1855). A more recent English version may be found in J. Memb. Sci. **100**, 33–38 (1995).
[72] J. B. J. Fourier, *Analytical theory of heat* (Great Books of the Western World, vol. 45) (Encyclopedia Britannica, Chicago, 1952), p. 232.
[73] W. C. Roberts-Austen, Bakerian lecture – On the diffusion of metals, Philos. T. R. Soc. London **187**, 383–425 (1896).
[74] J. Philibert, One and a half century of diffusion: Fick, Einstein, before and beyond, Diffusion Fundamentals **2**, 1.1–1.10 (2005).
[75] Lucretius: *The nature of things* (Penguin Classics), A. E. Stallings, trans. (Penguin Books, London, 2007).
[76] A. Einstein, Über die von der molekularkinetischen theorie der wärme geforderte bewegung von in ruhenden flüssigkeiten suspendierten teilchen, Ann. Phys. **17**, 549–560 (1905). See also the English translation, *Investigations on the theory of the Brownian movement*, edited with notes by R. Fürth (Dover Publications, 1956).
[77] J. B. Perrin, Discontinuous structure of matter, Nobel Lectures, Physics 1922–1941 (Elsevier, Amsterdam, 1965).
[78] M. Smoluchowski, Sur le chemin moyen parcouru par les molécules d'un gaz et son rapport avec la théorie de la diffusion (On the mean path of molecules of gas and its relationship to the theory of diffusion), Bull. Int. Acad. Sci. Cracovie, 202–213 (1906); Essai d'une théorie cinétique du mouvement Brownien et des milieux troubles (Outline of the kinetic theory of Brownian motion of suspensions), ibid., pp. 577–602 (1906); Phys. Z. **17**, 557 (1916); ibid., 585 (1916).
[79] A. Fuliński, On Marian Smoluchowski's life and contribution to physics, Acta Phys. Pol. B **29**, 1523–1537 (1998).
[80] H. Risken, *The Fokker–Planck equation* (Springer-Verlag, Berlin, 1996).
[81] P. Langevin, Sur la théorie du mouvement brownien, C. R. Acad. Sci. (Paris) **146**, 530–533 (1908).
[82] D. S. Lemons and A. Gythiel, Paul Langevin's 1908 paper "On the theory of Brownian motion" [Sur la théorie du mouvement brownien, C. R. Acad. Sci. (Paris) **146**, 530–533 (1908)], Am. J. Phys. **65**, 1079–1081 (1997).
[83] L. Borland, Microscopic dynamics of the nonlinear Fokker–Planck equation: A phenomenological model, Phys. Rev. E **57**, 6634–6642 (1998).

[84] L. Borland, F. Pennini, A. R. Plastino, and A. Plastino, The nonlinear Fokker–Planck equation with state-dependent diffusion – a nonextensive maximum entropy approach, Eur. Phys. J. B **12**, 285–297 (1999).
[85] T. D. Frank, *Nonlinear Fokker–Planck equations: Fundamentals and applications*, (Springer-Verlag, Heidelberg, 2010).
[86] J. Masoliver, K. Lindenberg, and B. J. West, First-passage times for non-Markovian process, Phys. Rev. A **33**, 2177–2184 (1986).
[87] K. Pearson, The problem of the random walk, Nature **72**, 294 (1905).
[88] Lord Rayleigh, The problem of the random walk, Nature **72**, 318 (1905).
[89] K. Pearson, The problem of the random walk, Nature **72**, 294 (1905).
[90] L. E. Reichl, *A modern course in statistical physics* (Wiley, New York, 1998), p. 198.
[91] L. R. Richardson, Atmospheric diffusion shown on a distance neighbour graph, Proc. R. Soc. Lond. A **110**, 709–737 (1926).
[92] E. W. Montroll and H. Scher, Random walks on lattices. IV. Continuous-time walks and influence of absorbing boundaries, J. Stat. Phys. **9**, 101–135 (1973).
[93] H. Scher and E. W. Montroll, Anomalous transit-dispersion in amorphous solids, Phys. Rev. B **12**, 2455–2477 (1975).
[94] M. F. Schlesinger, Asymptotic solutions of continuous-time random walks, J. Stat. Phys. **10**, 421–434 (1974).
[95] B. O'Shaughnessy and I. Procaccia, Analytical solutions for diffusion on fractal objects, Phys. Rev. Lett. **54**, 455–458 (1985).
[96] T. Tomé and M. J. Oliveira, *Stochastic dynamics and irreversibility* (Springer, Heidelberg, 2015).
[97] A. T. Silva, E. K. Lenzi, L. R. Evangelista, M. K. Lenzi, H. V. Ribeiro, and A. A. Tateishi, Exact propagator for a Fokker–Planck equation, first passage time distribution, and anomalous diffusion, J. Math. Phys. **52**, 083301 (2011).
[98] J. Klafter, I. M. Sokolov, *First steps in random walks: From tools to applications* (Oxford University Press, Oxford, 2011).
[99] R. Metzler, E. Barkai, J. Klafter, Deriving fractional Fokker–Planck equations from a generalised master equation, Europhys. Lett. **46**, 431–436 (1999).
[100] M. A. Zahran, On the derivation of fractional diffusion equation with an absorbent term and a linear external force, Appl. Math. Mod. **33**, 3088–3092 (2009).
[101] A. V. Chechkin, R. Gorenflo, and I. M. Sokolov, Retarding subdiffusion and accelerating superdiffusion governed by distributed-order fractional diffusion equations, Phys. Rev. E **66**, 046129 (2002).
[102] E. K. Lenzi, R. S. Mendes, and C. Tsallis, Crossover in diffusion equation: Anomalous and normal behaviors, Phys. Rev. E **67**, 031104 (2003).
[103] C. W. Gardiner, *Handbook of stochastic methods for physics, chemistry, and the natural sciences*. Springer Series in Synergetics, H. Haken (editor), (Springer-Verlag, Heidelberg, 2004).
[104] F. Mota-Furtado and P. F. O'Mahony, Exact propagator for generalised Ornstein–Uhlenbeck processes, Phys. Rev. E **75**, 041102 (2007).
[105] F. Mota-Furtado and P. F. O'Mahony, Eigenfunctions and matrix elements for a class of eigenvalue problems with staggered ladder spectra, Phys. Rev. A **74**, 044102 (2006).
[106] V. Zaburdaev, S. Denisov, and J. Klafter, Lévy walks, Rev. Mod. Phys. **87**, 483–530 (2015).
[107] T. Srokowski, Fractional Fokker–Planck equation for Lévy flights in nonhomogeneous environments, Phys. Rev. E **79**, 040104(R) (2009).
[108] T. Srokowski, Anomalous diffusion in nonhomogeneous media: Time-subordinated Langevin equation approach, Phys. Rev. E **89**, 030102(R) (2014)

[109] T. Srokowski, Lévy flights in nonhomogeneous media: Distributed-order fractional equation approach, Phys. Rev. E **78**, 031135 (2008).

[110] T. Srokowski, Lévy flights and nonhomogenous memory effects: Relaxation to a stationary state, Phys. Rev. E **92**, 012125 (2015).

[111] L. C. Malacarne, R. S. Mendes, E. K. Lenzi, and M. K. Lenzi, General solution of the diffusion equation with a nonlocal diffusive term and a linear force term, Phys. Rev. E **74**, 042101 (2006)

[112] P. E. Strizhak, Macrokinetics of chemical processes on porous catalysts having regard to anomalous diffusion, Theor. Exp. Chem. **40**, 203–208 (2004).

[113] D. Avnir, *The fractal approach to heterogeneous chemistry* (Wiley-Interscience, New York, 1990).

[114] R. I. Masel, *Principles of adsorption and reaction on solid surfaces* (Wiley-Interscience, New York, 1996)

[115] G. Fibich, I. Gannot, A. Hammer, and S. Schochet, Chemical kinetics on surfaces: A singular limit of a reaction–diffusion system, SIAM J. Math. Anal. **38**, 1371–1388 (2006).

[116] E. K. Lenzi, M. A. F. dos Santos, D. S. Vieira, R. S. Zola, and H. V. Ribeiro, Solutions for a sorption process governed by a fractional diffusion equation, Physica A **443**, 32–41 (2016).

[117] S. A. Bradford, Y. Wang, H. Kim, S. Torkzaban, and J. Simunek, Modeling microorganism transport and survival in the subsurface, J. Environ. Qual. **43**, 421–440 (2014).

[118] J. Crank, *The mathematics of diffusion* (Oxford Science Publications, England, 1975).

[119] X. Li and M. Xu, A model for reversible reaction in a sub-diffusive regime, J. Math. Phys. **50**, 102708 (2009).

[120] B. I. Henry, T. A. M. Langlands, and S. L. Wearne, Anomalous diffusion with linear reaction dynamics: From continuous-time random walks to fractional reaction-diffusion equations, Phys. Rev. E **74**, 031116 (2006).

[121] I. L. Novak, P. Kraikivski, and B. M. Slepchenko, Diffusion in cytoplasm: Effects of excluded volume due to internal membranes and cytoskeletal structures, Biophys. J. **97**, 758–767 (2009).

[122] E. K. Lenzi, M. A. F. dos Santos, M. K. Lenzi, and R. Menechini Neto, Solutions for a mass transfer process governed by fractional diffusion equations with reaction terms, Commun. Nonlinear Sci. Numer. Simul. **48**, 307–317 (2017).

[123] I. M. Sokolov, A.V. Chechkin, and J. Klafter, Fractional diffusion equation for a power-law-truncated Lévy process, Physica A **336**, 245–251 (2004).

[124] T. A. M. Langlands, B. I. Henry, and S. L. Wearne, Anomalous subdiffusion with multispecies linear reaction dynamics, Phys. Rev. E **77**, 021111 (2008).

[125] V. Gafiychuk and B. Datsko, Mathematical modeling of different types of instabilities in time fractional reaction-diffusion systems, Comp. Math. Appl. **59**, 1101–1107 (2010).

[126] D. del-Castillo-Negrete, B. A. Carreras, and V. E. Lynch, Front dynamics in reaction-diffusion systems with Lévy flights: A fractional diffusion approach, Phys. Rev. Lett. **91**, 018302 (2003).

[127] V. A. Volpert, Y. Nec, and A. A. Nepomnyashchy, Fronts in anomalous diffusion–reaction systems, Philos. T. Roy. Soc. A **371**, 20120179 (2015).

[128] D. Campos, S. Fedotov, and V. Mendez, Anomalous reaction-transport processes: The dynamics beyond the law of mass action, Phys. Rev. E **77**, 061130 (2008).

[129] A. A. Nepomnyashchy, Mathematical modelling of subdiffusion–reaction systems, Math. Model. Nat. Phenom. **11**, 26–36 (2016).

[130] I. M. Sokolov, M. G. W. Schmidt, and F. Sagués, Reaction–subdiffusion equations, Phys. Rev. **73**, 031102 (2006).
[131] A. Yadav and W. Horsthemke, Kinetic equations for reaction–subdiffusion systems: Derivation and stability analysis, Phys. Rev. E **74**, 066118 (2006).
[132] M. G. W. Schmidt, F. Sagués, and I. M. Sokolov, Mesoscopic description of reactions for anomalous diffusion: A case study, J. Phys. Cond. Matt. **19**, 065118 (2007).
[133] K. Seki, M.Wojcik, and M. Tachiya, Fractional reaction–diffusion equation, J. Chem. Phys. **119**, 2165–2170 (2003).
[134] K. Seki, M.Wojcik, and M. Tachiya, Recombination kinetics in subdiffusive media, J. Chem. Phys. **119**, 7525–7533 (2003).
[135] V. Mendez, D. Campos, and S. Fedotov, Front propagation in reaction-dispersal models with finite jump speed, Phys. Rev. E **70**, 036121 (2004).
[136] E. K. Lenzi, H. V. Ribeiro, J. Martins, M. K. Lenzi, G. G. Lenzi, and S. Specchia, Non-Markovian diffusion equation and diffusion in a porous catalyst, Chem. Eng. J. **172**, 1083–1087 (2011).
[137] A. V. Weigel, B. Simon, M. M. Tamkun, and D. Krapf, Ergodic and nonergodic processes coexist in the plasma membrane as observed by single-molecule tracking, Procl. Natl. Acad. Sci. U.S.A. **108**, 6439–6443 (2011).
[138] M. Giona and M. Giustiniani, Adsorption kinetics on fractal surfaces, J. Phys. Chem. **100**, 16690–16699 (1996).
[139] E. K. Lenzi, M. K. Lenzi, R. S. Zola, H. V. Ribeiro, F. C. Zola, L. R. Evangelista, and G. Gonçalves, Reaction on a solid surface supplied by an anomalous mass transfer source, Physica A **410**, 399–406 (2014).
[140] P. C. Bressloff, *Stochastic processes in cell biology* (Springer, Heidelberg, 2014).
[141] E. K. Lenzi, P. R. G. Fernandes, T. Petrucci, H. Mukai, and H. V. Ribeiro, Anomalous-diffusion approach applied to the electrical response of water, Phys. Rev. E **84**, 041128 (2011).
[142] F. Ciuchi, A. Mazzulla, N. Scaramuzza, E. K. Lenzi, and L. R. Evangelista, Fractional diffusion equation and the electrical impedance: Experimental evidence in liquid-crystalline cells, J. Phys. Chem. C **116**, 8773–8777 (2012).
[143] R. Rossato, M. K. Lenzi, L. R. Evangelista, and E. K. Lenzi, Fractional diffusion equation in a confined region: Surface effects and exact solutions, Phys. Rev. E **76**, 032102 (2007).
[144] X. L. Wun and A. Libchaber, Particle diffusion in a quasi-two-dimensional bacterial bath, Phys. Rev. Lett. **84**, 3017–3020 (2000).
[145] G. Grégoire, H. Chaté, and Y. Tu, Comment on "Particle diffusion in a quasi-two-dimensional bacterial bath," Phys. Rev. Lett. **86**, 556–556 (2001).
[146] X. L. Wun and A. Libchaber, Wu and Libchaber reply, Phys. Rev. Lett. **86**, 557–557 (2001).
[147] A. Caspi, R. Granek, and M. Elbaum, Enhanced diffusion in active intracellular transport, Phys. Rev. Lett. **85**, 5655–5658 (2000).
[148] A. Caspi, R. Granek, and M. Elbaum, Diffusion and directed motion in cellular transport, Phys. Rev. E **66**, 011916 (2002).
[149] V. Latora, A. Rapisarda, and S. Ruffo, Superdiffusion and out-of-equilibrium chaotic dynamics with many degrees of freedoms, Phys. Rev. Lett. **83**, 2104–2107 (1999).
[150] V. Latora, A. Rapisarda, and C. Tsallis, Non-Gaussian equilibrium in a long-range Hamiltonian system, Phys. Rev. E **64**, 056134 (2001).
[151] R. S. Zola, E. K. Lenzi, L. R. Evangelista, and G. Barbero, Memory effect in the adsorption phenomena of neutral particles, Phys. Rev. E **75**, 042601 (2007).

[152] G. Barbero and L. R. Evangelista, *Adsorption phenomena and anchoring energy in nematic liquid crystals* (Taylor & Francis, London, 2006).

[153] O. M. P. Agrawal, Solution for a fractional diffusion-wave equation defined in a bounded domain, Nonlinear Dyn. **29**, 145–155 (2002).

[154] I. V. Kotov, M. V. Khazimullin, and A. P. Krekhov, Flexoelectric instability in nematic liquid crystals between coaxial cylinders, Mol. Cryst. Liq. Cryst. **366**, 885–892 (2001).

[155] H. Tsuru, Stability analysis of nematics between two concentric cylinders, J. Phys. Soc. Jpn. **59**, 1600–1616 (1990).

[156] D. R. M. Williams and A. Halperin, Nematic liquid crystal in a tube: The Fréedericksz transition, Phys. Rev. E **48**, R2366–R2369 (1993).

[157] D. R. M. Williams, Nematic liquid crystals between antagonistic cylinders: Spirals with bend-splay director undulations, Phys. Rev. E **50**, 1686–1687 (1994).

[158] I. Jánossy, A. D. Lloyd, and B. S. Wherrett, Anomalous optical Freedericksz transition in an absorbing liquid crystal, Mol. Cryst. Liq. Cryst. **179**, 1–12 (1990).

[159] I. Jánossy, L. Csillag, and A. D. Lloyd, Temperature dependence of the optical Fréedericksz transition in dyed nematic liquid crystals, Phys. Rev. A **44**, 8410–8413 (1991).

[160] I. Jánossy and T. Kósa, Influence of anthraquinone dyes on optical reorientation of nematic liquid crystals, Opt. Lett. **17**, 1183–1185 (1992).

[161] I. Jánossy, Molecular interpretation of the absorption-induced optical reorientation of nematic liquid crystals, Phys. Rev. E **49**, 2957–2963 (1994).

[162] R. Muenster, M. Jarasch, X. Zhuang, and Y. R. Shen, Dye-induced enhancement of optical nonlinearity in liquids and liquid crystals, Phys. Rev. Lett. **78**, 42–45 (1997).

[163] L. S. Lucena, L. R. Silva, L. R. Evangelista, M. K. Lenzi, R. Rossato, and E. K. Lenzi, Solutions for a fractional diffusion equation with spherical symmetry using Green function approach, Chem. Phys. **344**, 90–94 (2008).

[164] M. A. Lomholt, I. M. Zaid, and R. Metzler, Subdiffusion and weak ergodicity breaking in the presence of a reactive boundary, Phys. Rev. Lett. **98**, 200603 (2007).

[165] S. Condamin, O. Bénichou, and J. Klafter, First-passage time distributions for subdiffusion in confined geometry, Phys. Rev. Lett. **98**, 250602 (2007).

[166] M. O. Vlad, Fractional diffusion equation on fractals: Self-similar stationary solutions in a force field derived from a logarithmic potential, Chaos, Solitons & Fractals **4**, 191–199 (1994).

[167] Y. Z. Povstenko, Fractional radial diffusion in a cylinder, J. Mol. Liq. **137**, 46–50 (2008).

[168] B. N. Narahari Achar and J. W. Hanneken, Fractional radial diffusion in a cylinder, J. Mol. Liq. **114**, 147–151 (2004).

[169] T. Bickel, A note on confined diffusion, Physica A **377**, 24–32 (2007) 24; Erratum, Physica A **381**, 532 (2007).

[170] G. M. Barrow, *Physical chemistry* (McGraw-Hill, New York, 1961).

[171] E. M. Gomes, R. R. L. Araújo, M. K. Lenzi, F. R. G. B. Silva, and E. K. Lenzi, Parametric analysis of a heavy metal sorption isotherm based on fractional calculus, Math. Probl. Eng. **2013**, 642101 (2013).

[172] J. R. Fernández and M. C. Muniz, Numerical analysis of surfactant dynamics at air–water interface using the Henry isotherm, J. Math. Chem. **49**, 1624–1645 (2011).

[173] E. K. Lenzi, C. A. R. Yednak, and L. R. Evangelista, Non-Markovian diffusion and the adsorption–desorption process, Phys. Rev. E **81**, 011116 (2010).

[174] A. W. Adamson and A. P. Gast, *Physical chemistry of surfaces*, 6th edition (Wiley, New York, 1997).

[175] E. K. Lenzi, H. V. Ribeiro, A. A. Tateishi, R. S. Zola, and L. R. Evangelista, Anomalous diffusion and transport in heterogeneous systems separated by a membrane, Proc. R. Soc. A **472**, 20160502 (2016).

[176] V. G. Guimarães, H. V. Ribeiro, Q. Li, L. R. Evangelista, E. K. Lenzi, and R. S. Zola, Unusual diffusing regimes caused by different adsorbing surfaces, Soft Matter **11**, 1658–1666 (2015).

[177] R. Kopelman, Fractal reaction kinetics, Science **241**, 1620–1626 (1988).

[178] S. Schnell and T. E. Turner, Reaction kinetics in intracellular environments with macromolecular crowding: Simulations and rate laws, Prog. Biophys. Mol. Biol. **85**, 235–260 (2004).

[179] H. Berry, Monte Carlo simulations of enzyme reactions in two dimensions: Fractal kinetics and spatial segregation, Biophys. J. **83**, 1891–1901 (2002).

[180] R. Grima and S. Schnell, A systematic investigation of the rate laws valid in intracellular environments, Biophys. Chem. **124**, 1–10 (2006).

[181] M. T. Klann, A. Lapin, and M. Reuss, Agent-based simulation of reactions in the crowded and structured intracellular environment: Influence of mobility and location of the reactants, BMC Syst. Biol. **5**, 71 (2011).

[182] M. N. Berberan-Santos and J. M. G. Martinho, A linear response approach to kinetics with time-dependent rate coefficients, Chem. Phys. **164**, 259–269 (1992).

[183] E. O. Voit, H. A. Martens, and S. W. Omholt, 150 years of the mass action law, PLoS Comput. Biol. **11**, 1004012 (2015).

[184] V. A. Parsegian, *Van der Waals forces: A handbook for biologists, chemists, engineers, and physicists* (Cambridge University Press, Cambridge, 2006).

[185] I. M. Zaid, M. A. Lomholt, and R. Metzler, How subdiffusion changes the kinetics of binding to a surface, Biophys. J. **97**, 710–721 (2009).

[186] C. Tsallis and D. J. Bukman, Anomalous diffusion in the presence of external forces: Exact time-dependent solutions and their thermostatistical basis, Phys. Rev. E **54**, R2197–R2200 (1996).

[187] M. Bologna, C. Tsallis, and P. Grigolini, Anomalous diffusion associated with nonlinear fractional derivative Fokker–Planck-like equation: Exact time-dependent solutions, Phys. Rev. E **62**, 2213–2218 (2000).

[188] C. Tsallis, Non-extensive thermostatistics: Brief review and comments, Physica A **221**/1–3, 277–290 (1995).

[189] L. C. Malacarne, I. T. Pedron, R. S. Mendes, and E. K. Lenzi, Nonlinear equation for anomalous diffusion: Unified power-law and stretched exponential exact solution, Phys. Rev. E **63**, 30101R (2001).

[190] P. C. Assis, Jr, L. R. da Silva, E. K. Lenzi, L. C. Malacarne, and R. S. Mendes, Nonlinear diffusion equation, Tsallis formalism and exact solutions, J. Math. Phys. **46**, 123303 (2005).

[191] W. F. Ames, *Nonlinear partial differential equations in engineering* (Academic Press, New York, 1965).

[192] M. Muskat, *The flow of homogeneous fluids through porous media* (McGraw-Hill, New York, 1937).

[193] P. Y. Polubarinova-Kochina, *Theory of ground water movement* (Princeton University Press, Princeton, 1962).

[194] H. Spohn, Surface dynamics below the roughening transition, J. Phys. France I, **3**, 69–81 (1993).

[195] J. Buckmaster, Viscous sheets advancing over dry beds, J. Fluid Mech. **81**, 735–756 (1977).

[196] W. P. Johnson, Some applications of the q-exponential formula, Discrete Math. **157**, 207–225 (1996).

[197] C. Tsallis, *Introduction to nonextensive statistical mechanics: Approaching a complex world* (Springer-Verlag, New York, 2009).

[198] R. Metzler, W. G. Glockle, and T. F. Nonnenmacher, Fractional model equation for anomalous diffusion, Physica A **211**, 13–24 (1994).

[199] A. Caillard, P. Brault, J. Mathias, C. Charles, R. W. Boswell, and T. Sauvage, *Deposition and diffusion of platinum nanoparticles in porous carbon assisted by plasma sputtering*, Surf. Coat. Technol. **200**, 391–394 (2005).

[200] A. G. Cherstvy and R. Metzler, Ergodicity breaking, ageing, and confinement in generalized diffusion processes with position and time dependent diffusivity, J. Stat. Mech. Theory Exper., **2015**, P05010 (2015).

[201] E. K. Lenzi, L. R. Silva, M. K. Lenzi, M. A. F. Santos, H. V. Ribeiro, and L. R. Evangelista, Intermittent motion, nonlinear diffusion equation and Tsallis formalism, Entropy **19**, 42 (2017).

[202] V. Mendez, D. Campos, and F. Bartumeus, *Stochastic foundations in movement ecology* (Springer-Verlag, Heidelberg, 2014).

[203] C. Tsallis, S. V. F. Levy, A. M. C. Souza, and R. Maynard, Statistical-mechanical foundation of the ubiquity of Lévy distributions in nature, Phys. Rev. Lett. **75**, 3589–3593 (1995).

[204] E. K. Lenzi, N. R. Menechini, A. A. Tateishi, M. K. Lenzi, and H. V. Ribeiro, Fractional diffusion equations coupled by reaction terms, Physica A **458**, 9–16 (2016).

[205] W. H. Press, B. P. Flannery, and S. A. Teukolsky, *Numerical recipes in fortran: The art of scientific computing* (Cambridge University Press, New York, 1992).

[206] N. Shigesada, Spatial distribution of dispersing animals, J. Math. Biol. **9**, 85–96 (1980).

[207] C. Tsallis, Possible generalisation of Boltzmann–Gibbs statistics, J. Stat. Phys. **52**, 479–488 (1988).

[208] E. K. Lenzi, R. S. Mendes, G. Gonçalves, M. K. Lenzi, and L. R. da Silva, Fractional diffusion equation and Green function approach: Exact solutions, Physica A **360**, 215–266 (2006).

[209] E. K. Lenzi, R. S. Mendes, J. S. Andrade, Jr, L. R. da Silva, and L. S. Lucena, $\mathcal{N}$-dimensional fractional diffusion equation and Green function approach: Spatially dependent diffusion coefficient and external force, Phys. Rev. E **71**, 052101 (2005).

[210] E. K. Lenzi, L. R. Evangelista, G. Barbero, and F. Mantegazza, Anomalous diffusion and the adsorption–desorption process in anisotropic media, Europhys. Lett. **85**, 28004 (2009).

[211] A. Pekalski and K. Sznajd-Weron, eds., *Anomalous diffusion: From basics to applications*, Lecture Notes in Physics (Springer, Berlin, 1999).

[212] E. Hatta, Anomalous diffusion of fatty acid vesicles driven by adhesion gradients, J. Phys. Chem. B **112**, 8571–8577 (2008).

[213] H. Stark, Physics of colloidal dispersions in nematic liquid crystals, Phys. Rep. **351**, 387–474 (2001).

[214] H. E. Roman, A. Bunde, and W. Dieterich, Conductivity of dispersed ionic conductors: A percolation model with two critical points, Phys. Rev. B **34**, 3439–3445 (1986).

[215] F. Brochard and P. G. de Gennes, Theory of magnetic suspensions in liquid crystals, J. Phys. (Paris) **31**, 691–708 (1970).

[216] R. Eidenschink and W. H. de Jeu, Static scattering in filled nematic: New liquid crystal display technique, Electron. Lett. **27**, 1195–1196 (1991).

[217] P. Poulin, V. A. Raghunathan, P. Richetti, and D. Roux, On the dispersion of latex particles in a nematic solution. I. Experimental evidence and a simple model, J. Phys. II (France) **4**, 1557–1569 (1994).
[218] J. C. Loudet, P. Barois, and P. Poulin, Colloidal ordering from phase separation in a liquid-crystalline continuous phase, Nature **407**, 611–613 (2000).
[219] A. V. Chechkin, V. Yu. Gonchar, R. Gorenflo, N. Korabel, and I. M. Sokolov, Generalized fractional diffusion equations for accelerating subdiffusion and truncated Lévy flights, Phys. Rev. E **78**, 021111 (2008).
[220] G. Barbero and L. R. Evangelista, Adsorption phenomenon of neutral particles and a kinetic equation at the interface, Phys. Rev. E **70**, 031605 (2004).
[221] E. R. Weeks and D. A. Weitz, Subdiffusion and the cage effect studied near the colloidal glass transition, Chem. Phys. **284**, 361–367 (2002).
[222] G. Marty and O. Dauchot, Subdiffusion and cage effect in a sheared granular material, Phys. Rev. Lett. **94**, 015701 (2005).
[223] M. F. de Andrade, E. K. Lenzi, L. R. Evangelista, R. S. Mendes, and L. C. Malacarne, Anomalous diffusion and fractional diffusion equation: Anisotropic media and external forces, Phys. Lett. A **347**, 160–169 (2005).
[224] L. R. da Silva, A. A. Tateishi, M. K. Lenzi, E. K. Lenzi, and P. C. da Silva, Green's function for a non-Markovian Fokker–Planck equation: Comb-model and anomalous diffusion, Braz. J. Phys. **39**, 483–487 (2009).
[225] J. F. Douglas, *Polymer science applications of path-integration, integral equations, and fractional calculus*, in *Applications of fractional calculus in physics*, edited by R. Hilfer (World Scientific, Singapore, 2008).
[226] G. A. Mendes, E. K. Lenzi, R. S. Mendes, and L. R. da Silva, Anisotropic fractional diffusion equation, Physica A **346**, 271–283 (2005).
[227] L. R. da Silva, L. S. Lucena, P. C. da Silva, E. K. Lenzi, and R. S. Mendes, Multidimensional nonlinear diffusion equation: Spatial time dependent diffusion coefficient and external forces, Physica A **357**, 103–108 (2005).
[228] Y. Gefen, A. Aharony, and S. Alexander, Anomalous diffusion on percolating clusters, Phys. Rev. Lett. **50**, 77–80 (1983).
[229] H. E. Stanley, Cluster shapes at the percolation threshold: An effective cluster dimensionality and its connection with critical-point exponents, J. Phys. A **10**, L211 (1977).
[230] S. R. White and M. Barma, Field-induced drift and trapping in percolation networks, J. Phys. A **17**, 2995–3008 (1984).
[231] A. Bunde, S. Havlin, H. E. Stanley, B. Trus, and G. H. Weiss, Diffusion in random structures with a topological bias, Phys. Rev. B **34**, 8129–8132 (1986).
[232] V. E. Arkhincheev and É. M. Baskin, Anomalous diffusion and drift in a comb model of percolation clusters, Zh. Eksp. Teor. Fiz **100**, 292–300 (1991).
[233] A. Iomin, Fractional-time quantum dynamics, Phys. Rev. E **80**, 022103 (2009).
[234] A. Iomin, Fractional-time Schrödinger equation: Fractional dynamics on a comb, Chaos Solitons & Fractals **44**, 348–352 (2011).
[235] K. V. Chukbar, Quasidiffusion of a passive scalar, JETP **82**, 719–726 (1996).
[236] A. Iomin, Superdiffusion of cancer on a comb structure, J. Phys. Conf. Ser. **7**, 57–67 (2005).
[237] A. Iomin, Toy model of fractional transport of cancer cells due to self-entrapping, Phys. Rev. E **73**, 061918 (2006).
[238] O. A. Dvoretskaya and P. S. Kondratenko, Anomalous transport regimes and asymptotic concentration distributions in the presence of advection and diffusion on a comb structure, Phys. Rev. E **79**, 041128 (2009).

[239] E. K. Lenzi, L. R. da Silva, A. A. Tateishi, M. K. Lenzi, and H. V. Ribeiro, Diffusive process on a backbone structure with drift terms, Phys. Rev. E **87**, 012121 (2013).
[240] V. E. Arkhincheev, Random walk on hierarchical comb structures, JETP **88**, 710–715 (1999).
[241] A. M. Reynolds, On anomalous transport on comb structures, Physica A **334**, 39–45 (2004).
[242] V. E. Arkhincheev, Anomalous diffusion and charge relaxation on comb-model: Exact solutions, Physica A **280**, 304–314 (2000).
[243] V. E. Arkhincheev, Random walks on the comb model and its generalizations, Chaos **17**, 043102 (2007).
[244] S. C. Weber, J. A. Theriot, and A. J. Spakowitz, Subdiffusive motion of a polymer composed of subdiffusive monomers, Phys. Rev. E **82**, 011913 (2010).
[245] J.-H. Jeon and R. Metzler, Analysis of short subdiffusive time series: Scatter of the time-averaged mean-squared displacement, J. Phys. A **43**, 252001 (2010).
[246] W. F. Marshall, A. Straight, J. F. Marko, J. Swedlow, A. Dernburg, A. Belmont, A. W. Murray, D. A. Agard, and J. W. Sedat, Interphase chromosomes undergo constrained diffusional motion in living cells, Curr. Biol. **7**, 930–939 (1997).
[247] I. Golding and E. C. Cox, Physical nature of bacterial cytoplasm, Phys. Rev. Lett. **96**, 098102 (2006).
[248] N. Leijnse, J.-H. Jeon, S. Loft, R. Metzler, and L. B. Oddershede, Diffusion inside living human cells, Eur. Phys. J. – Special Topics **204**, 75–84 (2012).
[249] J.-H. Jeon, V. Tejedor, S. Burov, E. Barkai, C. Selhuber-Unkel, K. Berg-Sorensen, L. Oddershede, and R. Metzler, *In vivo* anomalous diffusion and weak ergodicity breaking of lipid granules, Phys. Rev. Lett. **106**, 048103 (2011).
[250] D. ben-Avraham and S. Havlin, *Diffusion and reactions in fractals and disordered systems* (Cambridge University Press, Cambridge, 2000).
[251] D. Villamaina, A. Sarracino, G. Gradenigo, A. Puglisi, and A. Vulpiani, On anomalous diffusion and the out of equilibrium response function in one-dimensional models, J. Stat. Mech. L01002 (2011).
[252] D. Panja, Anomalous polymer dynamics is non-Markovian: Memory effects and the generalized Langevin equation formulation, J. Stat. Mech. P06011 (2010).
[253] D. Panja, Generalized Langevin equation formulation for anomalous polymer dynamics, J. Stat. Mech. L02001 (2010).
[254] D. Panja, Probabilistic phase space trajectory description for anomalous polymer dynamics, J. Phys. Cond. Matt. **23**, 105103 (2011).
[255] B. N. Narahari Achar, B. T. Yale, and J. W. Hanneken, Time fractional Schrödinger equation revisited, Adv. Math. Phys. **2013**, 290216 (2013).
[256] B. Baeumer, M. M. Meerschaert, and M. Naber, Stochastic models for relativistic diffusion, Phys. Rev. E **82**, 011132 (2010).
[257] E. K. Lenzi, B. F. de Oliveira, N. G. C. Astrath, L. C. Malacarne, R. S. Mendes, M. L. Baesso, and L. R. Evangelista, Fractional approach, quantum statistics, and non-crystalline solids at very low temperatures, Eur. Phys. J. B **62**, 155–158 (2008).
[258] X. Jiang, H. Qi, and M. Xu, Exact solutions of fractional Schrödinger-like equation with a nonlocal term, J. Math. Phys. **52**, 042105 (2011).
[259] E. K. Lenzi, M. K. Lenzi, R. Rossato, and L. C. M. Filho, Solutions for diffusion equation with a nonlocal term, Acta Sci. Technol. **31**, 81–86 (2009).
[260] E. K. Lenzi, B. F. de Oliveira, L. R. da Silva, and L. R. Evangelista, Solutions for a Schrödinger equation with a nonlocal term, J. Math. Phys. **49**, 032108 (2008).
[261] N. Laskin, Lévy flights over quantum paths, Commun. Nonlinear Sci. Nonlinear Simul. **12**, 2–18 (2007).

[262] T. A. M. Langlands, Solution of a modified fractional diffusion equation, Physica A **367**, 136–144 (2006).
[263] N. Laskin, Fractional quantum mechanics and Lévy path integrals, Phys. Lett. A **268**, 298–305 (2000).
[264] N. Laskin, Fractional quantum mechanics, Phys. Rev. E **62**, 3135–3145 (2000).
[265] N. Laskin, Fractals and quantum mechanics, Chaos **10**, 780–790 (2000).
[266] N. Laskin, Fractional Schrödinger equation, Phys. Rev. E **66**, 056108 (2002).
[267] M. Naber, Time fractional Schrödinger equation, J. Math. Phys. **45**, 3339 (2004).
[268] J. Dong, Green's function for the time-dependent scattering problem in the fractional quantum mechanics, J. Math. Phys. **52**, 042103 (2011).
[269] H. Ertik, D. Demirhan, H. Sirin, and F. Büyükkiliç, Time fractional development of quantum systems, J. Math. Phys. **51**, 082102 (2010).
[270] S. S. Bayin, Time fractional Schrödinger equation: Fox's H-functions and the effective potential, J. Math. Phys. **54**, 012103 (2013).
[271] S. S. Bayin, On the consistency of the solutions of the space fractional Schrödinger equation, J. Math. Phys. **53**, 042105 (2012).
[272] E. Capelas de Oliveira, F. Silva Costa, and J. Vaz, The fractional Schrödinger equation for delta potentials, J. Math. Phys. **51**, 123517 (2010).
[273] E. K. Lenzi, H. V. Ribeiro, H. Mukai, and R. S. Mendes, Continuous-time random walk as a guide to fractional Schrödinger equation, J. Math. Phys. **51**, 092102 (2010).
[274] M. Jeng, S.-L.-Y. Xu, E. Hawkins, and J. M. Schwarz, On the nonlocality of the fractional Schrödinger equation, J. Math. Phys. **51**, 062102 (2010).
[275] Y. Luchko, Fractional Schrödinger equation for a particle moving in a potential well, J. Math. Phys. **54**, 012111 (2013).
[276] E. K. Lenzi, H. V. Ribeiro, M. A. F. dos Santos, R. Rossato, and R. S. Mendes, Time dependent solutions for a fractional Schrödinger equation with delta potentials, J. Math. Phys. **54**, 082107 (2013).
[277] E. C. de Oliveira and Jayme Vaz, Jr, Tunneling in fractional quantum mechanics, J. Phys. A Math. Theor. **44**, 185303 (2011).
[278] S. I. Muslih, Solutions of a particle with fractional δ-potential in a fractional dimensional space, Int. J. Theor. Phys. **49**, 2095–2104 (2010).
[279] M. Kleber, Exact solutions for time-dependent phenomena in quantum mechanics, Phys. Rep. **236**, 331–393 (1994).
[280] S. M. Blinder, Green's function and propagator for the one-dimensional δ-function potential, Phys. Rev. A **37**, 973–976 (1988).
[281] I. Cacciari and P. Moretti, Propagator for the double delta potential, Phys. Lett. A **359**, 396–401 (2006).
[282] J. Martins, H. V. Ribeiro, L. R. Evangelista, L. R. da Silva, and E. K. Lenzi, Fractional Schrödinger equation with noninteger dimensions, App. Math. Comp. **219**, 2313–2319 (2012).
[283] G. W. Weiss, *Aspects and applications of the random walk* (North Holland, Amsterdam, 1994).
[284] E. Barkai, Fractional Fokker–Planck equation, solution, and application, Phys. Rev. E, **63**, 046118 (2001).
[285] R. Gorenflo and F. Mainardi, Random walk models for space-fractional diffusion processes, Fractional Calculus Appl. Anal. **1**, 167–191 (1998).
[286] F. Mainardi, G. Pagnini, and R. Gorenflo, Some aspects of fractional diffusion equations of single and distributed order, App. Math. Comp. **187**, 295–305 (2007).
[287] P. Brockmann and I. M. Sokolov, Lévy flights in external force fields: From models to equations, Chem. Phys. **284**, 409–421 (2002).

[288] N. Krepysheva, L. Di Pietro, and M.-C. Néel, Space-fractional advection–diffusion and reflective boundary condition, Phys. Rev. E **73**, 021104 (2006).

[289] A. V. Chechkin, R. Metzler, V. Y. Gonchar, J. Klafter, and L. V. Tanatarov, First passage and arrival time densities for Lévy flights and the failure of the method of images, J. Phys. A **36**, L537 (2003).

[290] X. Guo and M. Xu, Some physical applications of fractional Schrödinger equation, J. Math. Phys. **47**, 082104 (2006).

[291] J. Dong and M. Xu, Some solutions to the space fractional Schrödinger equation using momentum representation method, J. Math. Phys. **48**, 072105 (2007).

[292] J. Dong and M. Xu, Applications of continuity and discontinuity of a fractional derivative of the wave functions to fractional quantum mechanics, J. Math. Phys. **49**, 052105 (2008).

[293] T. Sandev, I. Petreska, and E. K. Lenzi, Time-dependent Schrödinger-like equation with nonlocal term, J. Math. Phys. **55**, 092105 (2016).

[294] A. A. Tateishi, E. K. Lenzi, L. R. da Silva, H. V. Ribeiro, S. Picoli, Jr, and R. S. Mendes, Different diffusive regimes, generalized Langevin and diffusion equations, Phys. Rev. E **85**, 011147 (2012).

[295] I. M. Sokolov and J. Klafter, From diffusion to anomalous diffusion: A century after Einstein's Brownian motion, Chaos **15**, 026103–026109 (2005).

[296] W. Feller, *An introduction to probability theory and its applications*, vol. II (Wiley, New York, 1968).

[297] A. N. Kochubei, General fractional calculus, evolution equations, and renewal processes, Integr. Equ. Oper. Theory **71**, 583 (2011).

[298] F. Mainardi, A. Mura, R. Gorenflo, and M. Stojanovic, The two forms of fractional relaxation of distributed order, J. Vib. Control **13**, 1249–1268 (2007).

[299] A. V. Chechkin, J. Klafter, and I. M. Sokolov, Fractional Fokker–Planck equation for ultraslow kinetics, Europhys. Lett. **63**, 326–332 (2003).

[300] T. Sandev and Z. Tomovski, Langevin equation for a free particle driven by power-law behaviour type of noises, Phys. Lett. A **378**, 1–9 (2014).

[301] E. K. Lenzi, N. G. C. Astrath, R. Rossato, and L. R. Evangelista, Nonlocal effects on the thermal behavior of non-crystalline solids, Braz. J. Phys. **39**, 507–510 (2009).

[302] R. C. Zeller and R. O. Pohl, Thermal conductivity and specific heat of non-crystalline solids, Phys. Rev. B **4**, 2029–2041 (1971).

[303] S. Takeno and M. Goda, A theory of phonon-like excitations in non-crystalline solids and liquids, Prog. Theor. Phys. **47**, 790–806 (1972).

[304] S. Takeno and M. Goda, Frequency spectrum and low-temperature specific heat of non-crystalline solids, Prog. Theor. Phys. **48**, 1468–1473 (1972).

[305] W. H. Tanttila, Anomalous thermal properties of glasses, Phys. Rev. Lett. **39**, 554–557 (1977).

[306] W. H. Tanttila, J. Cooper, and D. J. Toms, Demonstration of possible excitations in liquids, Am. J. Phys. **45**, 395 (1977).

[307] L. P. Kadanoff and G. Baym, *Quantum statistical mechanics* (Benjamin, New York, 1962).

[308] C. M. Varma, R. C. Dynes, and J. R. Banavar, Thermally created tunnelling states in glasses, J. Phys. C **15**, L1221 (1982).

[309] A. Nittke, P. Esquinazi, H.-C. Semmelhak, A. L. Burin, and A. Z. Patashinski, Thermodynamic properties of small amorphous and crystalline silica particles at low temperatures, Eur. Phys. J. B **8**, 19–30 (1999).

[310] C. Talón, M. A. Ramos, and S. Vieira, Low-temperature specific heat of amorphous, orientational glass, and crystal phases of ethanol, Phys. Rev. B **66**, 012201 (2002).

[311] N. G. C. Astrath, A. C. Bento, M. L. Baesso, E. K. Lenzi, and L. R. Evangelista, Semiclassical approximation for the specific heat of non-crystalline solids at intermediate temperatures, Phil. Mag. **87**, 291–297 (2007).
[312] J. R. Macdonald and W. B. Johnson, *Fundamental of impedance spectroscopy*, in *Impedance spectroscopy, theory, experiment, and applications*, edited by E. Barsoukov and J. R. Macdonald (Wiley, New York, 2005), pp. 1–26.
[313] I. D. Raistrick, J. R. Macdonald, and D. R. Franceschetti, *Theory*, in *Impedance spectroscopy, theory, experiment, and applications*, edited by E. Barsoukov and J. R. Macdonald (Wiley, New York, 2005).
[314] E. K. Lenzi, R. S. Zola, H. V. Ribeiro, D. S. Vieira, F. Ciuchi, A. Mazzulla, N. Scaramuzza, and L. R. Evangelista, Ion motion in electrolytic cells: Anomalous diffusion evidences, J. Phys. Chem. B **121**/13, 2882–2886 (2017).
[315] E. K. Lenzi, R. S. Zola, R. Rossato, H. V. Ribeiro, D. S. Vieira, and L. R. Evangelista, Asymptotic behaviours of the Poisson–Nernst–Planck model, generalizations and best adjust of experimental data, Electrochim. Acta **226**, 40–45 (2017).
[316] R. S. Zola, F. C. M. Freire, E. K. Lenzi, L. R. Evangelista, and G. Barbero, Kinetic equation with memory effect for adsorption–desorption phenomena, Chem. Phys. Lett. **438**, 144–147 (2007).
[317] J. Bisquert and A. Compte, Theory of the electrochemical impedance of anomalous diffusion, J. Electroanal. Chem. **499**, 112–120 (2001).
[318] J. Bisquert, G. Garcia-Belmonte, and A. Pitarch, An explanation of anomalous diffusion patterns observed in electroactive materials by impedance methods, ChemPhysChem **4**, 287–292 (2003).
[319] J. Bisquert, Interpretation of a fractional diffusion equation with nonconserved probability density in terms of experimental systems with trapping or recombination, Phys. Rev. E **72**, 011109 (2005).
[320] J. Bisquert, Fractional diffusion in the multiple-trapping regime and revision of the equivalence with the continuous-time random walk, Phys. Rev. Lett. **91**, 010602 (2003).
[321] E. Warburg, Ueber die polarisationscapacität des platins, Ann. Phys. **311**, 125–135 (1901).
[322] J. N. Israelachvili, *Intermolecular and surface forces*, 3rd edition (Elsevier–Academic Press, Amsterdam, 2017).
[323] G. Barbero and M. Scalerandi, Similarities and differences among the models proposed for real electrodes in the Poisson–Nernst–Planck theory, J. Chem. Phys. **136**, 136, 084705 (2012).
[324] E. K. Lenzi, L. R. Evangelista, and G. Barbero, Fractional diffusion equation and impedance spectroscopy of electrolytic cells, J. Phys. Chem. B **113**, 11371–11374 (2009).
[325] I. M. Sokolov, A. V. Chechkin, and J. Klafter, Distributed-order fractional kinetics, Acta Phys. Pol. B **35**, 1323–1341 (2004).
[326] F. Mainardi and G. Pagnini, The role of the Fox–Wright functions in fractional sub-diffusion of distributed order, J. Comput. Appl. Math. **207**, 245–257 (2007).
[327] T. Pajkossy and L. Nyikos, Scaling-law analysis to describe the impedance behavior of fractal electrodes, Phys. Rev. B **42**, 709–719 (1990).
[328] H. Sanabria and J. H. Miller, Relaxation processes due to the electrode-electrolyte interface in ionic solutions, Phys. Rev. E **74**, 051505 (2006).
[329] T. Kosztolowicz and K. D. Lewandowska, Hyperbolic subdiffusive impedance, J. Phys. A Math. Theor. **42**, 055004 (2009).
[330] J. R. Macdonald, The impedance of a galvanic cell with two plane-parallel electrodes at a short distance, J. Electroanal. Chem. **32**, 317–328 (1971).

[331] A. Alexe-Ionescu and G. Barbero, Role of the diffuse layer of the ionic charge on the impedance spectroscopy of a cell of liquid, Liq. Cryst. **32**, 943–949 (2005).
[332] J. R. Macdonald, Utility of continuum diffusion models for analyzing mobile-ion immittance data: Electrode polarization, bulk, and generation–recombination effects, J. Phys. Cond. Matt. **22**, 495101 (2010).
[333] J. R. Macdonald, L. R. Evangelista, E. K. Lenzi, and G. Barbero, Comparison of impedance spectroscopy expressions and responses of alternate anomalous Poisson–Nernst–Planck diffusion equations for finite-length situations, J. Phys. Chem. C **115**, 7648–7655 (2011).
[334] L. R. Evangelista, E. K. Lenzi, and G. Barbero, The Kramers–Kronig relations for usual and anomalous Poisson–Nernst–Planck models, J. Phys. Cond. Matt. **46**, 465104 (2013).
[335] J. R. Macdonald and D. R. Franceschetti, Theory of small-signal ac response of solids and liquids with recombining mobile charge, J. Chem. Phys. **68**, 1614–1637 (1978).
[336] P. A. Santoro, J. L. de Paula, E. K. Lenzi, and L. R. Evangelista, Anomalous diffusion governed by a fractional diffusion equation and the electrical response of an electrolytic cell, J. Chem. Phys. **135**, 114704 (2011).
[337] D. L. Sidebottom, Colloquium: Understanding ion motion in disordered solids from impedance spectroscopy scaling, Rev. Mod. Phys. **81**, 999–1014 (2009).
[338] F. Batalioto, A. R. Duarte, G. Barbero, and A. M. F. Neto, Dielectric dispersion of water in the frequency range from 10 mHz to 30 MHz, J. Phys. Chem. B **114**, 3467–3471 (2010).
[339] G. Barbero, Influence of adsorption phenomenon on the impedance spectroscopy of a cell of liquid, Phys. Rev. E **71**, 062201 (2005).
[340] L. R. Evangelista, E. K. Lenzi, G. Barbero, and J. R. Macdonald, Anomalous diffusion and memory effects on the impedance spectroscopy for finite-length situations, J. Phys. Cond. Matt. **23**, 485005 (2011).
[341] P. A. Santoro, E. K. Lenzi, L. R. Evangelista, F. Ciuchi, A. Mazzulla, and N. Scaramuzza, Anomalous diffusion effects on the electrical impedance response of liquid-crystalline systems, Mol. Cryst. Liq. Crist. **576**, 23–31 (2013).
[342] J. R. Macdonald and L. D. Potter, A flexible procedure for analysing impedance spectroscopy results: Description and illustrations, Solid State Ionics **24**, 61–79 (1987).
[343] J. R. Macdonald, Equivalent circuits for the binary electrolyte in the Warburg region, Electroanal. Chem. Interfacial Electrochem. **47**, 182–189 (1973).
[344] J. R. Macdonald, Three to six ambiguities in immittance spectroscopy data fitting, J. Phys. Cond. Matt. **24**, 175004 (2012).
[345] J. Jorcin, M. E. Orazem, N. Pébére, and B. Tribollet, CPE analysis by local electrochemical impedance spectroscopy, Electrochim. Acta **51**, 1473–1479 (2006).
[346] P. Córdoba-Torres, T. J. Mesquita, O. Devos, B. Tribollet, V. Roche, and R. P. Nogueira, On the intrinsic coupling between constant-phase element parameters alpha and in electrochemical impedance spectroscopy, Electrochim. Acta **72**, 172–178 (2012).
[347] H. S. Liu, Fractal model for ac response of a rough interface, Phys. Rev. Lett. **55**, 529–532 (1985).
[348] U. Rammelt and G. Reinhard, On the applicability of a constant phase element (CPE) to the estimation of roughness of solid metal-electrodes, Electrochim. Acta, **35**, 1045–1049 (1990).

[349] B. Sapoval, *Transport across irregular interfaces: Fractal electrodes, membranes and catalysts*, in *Fractals and disordered systems*, edited by A. Bunde and S. Havlin (Springer-Verlag, Heidelberg, 1996), pp. 233–261.

[350] C. Hitz and A. Lasia, Experimental study and modeling of impedance of the her on porous Ni electrodes, J. Electroanal. Chem. **500**, 213–222 (2001).

[351] V. M. Huang, V. Vivier, M. E. Orazem, N. Pébère, and B. Tribollet, The apparent constant-phase-element behaviour of an ideally polarized blocking electrode – A global and local impedance analysis, J. Electrochem. Soc. **152**, C81–C88 (2007).

[352] J. Bisquert, Analysis of the kinetics of ion intercalation ion trapping approach to solid-state relaxation processes, Electrochim. Acta **47**, 2435–2449 (2002).

[353] J. Bisquert, Beyond the quasistatic approximation: Impedance and capacitance of an exponential distribution of traps, Phys. Rev. B **77**, 235203 (2008).

[354] E. K. Lenzi, J. L. de Paula, F. R. G. B. Silva, and L. R. Evangelista, A connection between anomalous Poisson–Nernst–Planck model and equivalent circuits with constant phase elements, J. Phys. Chem. C **117**, 23685–23690 (2013).

[355] B. Sapoval, J.-N. Chazalviel, and J. Peyriere, Electrical response of fractal and porous interfaces, Phys. Rev. A **38**, 5867–5887 (1988).

[356] L. Nyikos and T. Pajkossy, Fractal dimension and fractional power frequency-dependent impedance of blocking electrodes, Electrochim. Acta **30**, 1533–1540 (1985).

[357] B. Sapoval, Fractal electrodes and constant phase-angle response – Exact examples and counter examples, Solid State Ionics **23**, 253–259 (1987).

[358] T. Pajkossy, Electrochemistry at fractal surfaces, J. Electroanal. Chem. **300**, 1–11 (1991).

[359] B. Y. Park, R. Zaouk, C. Wang, and M. J. Madou, A case for fractal electrodes in electrochemical applications, J. Electrochem. Soc. **154**, P1–P5 (2007).

[360] T. C. Halsey and M. Leibig, The double-layer impedance at a rough-surface – Theoretical results, Ann. Phys. **219**, 109–147 (1992).

[361] B. Sapoval, General formulation of Laplacian transfer across irregular surfaces, Phys. Rev. Lett. **73**, 3314–3316 (1994).

[362] D. S. Grebenkov, M. Filoche, and B. Sapoval, Mathematical basis for a general theory of Laplacian transport towards irregular interfaces, Phys. Rev. E **73**, 021103 (2006).

[363] E. K. Lenzi, P. R. G. Fernandes, T. Petrucci, H. Mukai, H. V. Ribeiro, M. K. Lenzi, and G. Gonçalves, Anomalous diffusion and electrical response of ionic solutions, Int. J. Electrochem. Sci. **8**, 2849–2862 (2013).

[364] E. K. Lenzi, M. K. Lenzi, F. R. G. B. Silva, G. Gonçalves, R. Rossato, R. S. Zola, and L. R. Evangelista, A framework to investigate the immittance responses for finite-length-situations: Fractional diffusion equation, reaction term, and boundary conditions, J. Electroanal. Chem. **712**, 82–88 (2014).

[365] J. Klafter, A. Blumen, and M. F. Shlesinger, Stochastic pathway to anomalous diffusion, Phys. Rev. A **35**, 3081–3085 (1987).

[366] G. Derfel, E. K. Lenzi, C. R. Yednak, and G. Barbero, Electrical impedance of an electrolytic cell in the presence of generation and recombination of ions, J. Chem. Phys. **132**, 224901 (2010).

[367] J. L. de Paula, P. A. Santoro, R. S. Zola, E. K. Lenzi, L. R. Evangelista, F. Ciuchi, A. Mazzulla, and N. Scaramuzza, Non-Debye relaxation in the dielectric response of nematic liquid crystals: Surface and memory effects in the adsorption–desorption process of ionic impurities, Phys. Rev. E **86**, 051705 (2012).

[368] J. D. Murray, *Mathematical biology I: An introduction* (Springer-Verlag, Heidelberg, 2002).

[369] J. D. Murray, *Mathematical biology II: Spatial models and biomedical applications* (Springer-Verlag, Heidelberg, 2003).

[370] G. Barbero and J. R. Macdonald, Transport process of ions in insulating media in the hyperbolic diffusion regime, Phys. Rev. E **81**, 051503 (2010).

[371] C. Criado, P. Galan Montenegro, P. Velasquez, and J. R. Ramos Barrado, Diffusion with general boundary conditions in electrochemical systems, J. Electroanal. Chem. **488**, 59–63 (2000).

[372] J. R. Ramos Barrado, P. Galan Montenegro, and C. Criado Gambom, A generalized Warburg impedance for a nonvanishing relaxation process, J. Chem. Phys. **105**, 2813–2815 (1996).

[373] K. D. Lewandoska and T. Kosztolowicz, Application of generalized Cattaneo equation to model subdiffusion impedance, Acta Phys. Pol. B **39**, 1211–1220 (2008).

[374] F. Batalioto, O. G. Martins, A. R. Duarte, and A. M. Figueiredo Neto, The influence of adsorption phenomena on the impedance spectroscopy of an electrolytic cell, Eur. Phys. J. E **34**, 10 (2011).

[375] T. Basu, M. M. Goswami, T. R. Middya, and S. Tarafdar, Morphology and ion-conductivity of gelatin-$LiClO_4$ films: Fractional diffusion analysis, J. Phys. Chem. B **116**, 11362–11369 (2012).

[376] D. R. Franceschetti and J. R. Macdonald, Numerical analysis of electrical response: Statics and dynamics of space-charge regions at blocking electrodes, J. Appl. Phys. **50**, 291–302 (1979).

[377] H. Chang and G. Jaffé, Polarization in electrolytic solutions. Part I. Theory, J. Chem. Phys. **20**, 1071–1077 (1952).

[378] R. J. Friauf, Polarization effects in the ionic conductivity of silver bromide, J. Chem. Phys. **22**, 1329–1338 (1954).

[379] J. R. Macdonald, Theory of space-charge polarization and electrode-discharge effects, J. Chem. Phys. **58**, 4982–5001 (1973); Erratum: Theory of space-charged polarization and electrode-discharge effects, **60**, 343 (1974).

[380] J. R. Macdonald, Effects of various boundary conditions on the response of Poisson–Nernst–Planck impedance spectroscopy analysis models and comparison with a continuous time random walk model, J. Phys. Chem. A **115**, 13370–13380 (2011).

Index